Handbook of Industrial Engineering

Handbook of Industrial Engineering

Edited by **Michelle Vine**

CLANRYE
INTERNATIONAL

New Jersey

Published by Clanrye International,
55 Van Reypen Street,
Jersey City, NJ 07306, USA
www.clanryeinternational.com

Handbook of Industrial Engineering
Edited by Michelle Vine

International Standard Book Number: 978-1-63240-274-5 (Hardback)

Contents

Preface

Industrial engineering is an engineering discipline that gives you an opportunity to work on a variety of applications. Some of the industrial engineering applications are manufacturing supreme automobiles, creating mega machines, or simply streaming an operating room. All these applications have only one goal that is to save money and increase the efficiency. Today, industrial engineering is a key to maintaining continuous productivity and quality improvement to survive the competitive world market. The main reason behind this is; only industrial engineers are trained as specialists in productivity and quality improvement. The motive behind the progress of industrial engineering is to figure out ways to do things better by engineering the system and processes to improve quality and productivity. The aim is to save time, money, materials, energy, and other inputs. Some of the important approaches in the field of industrial engineering are Fuzzy Quality Function Deployment, Configuration of Knowledge-Based Networks Using QFD, ANP, and Mixed-Integer Programming Model, Collaborative Decision-Making in Product Design.

This book illustrates some significant researches which examine the advanced industrial engineering approaches. I hope that this book can provide industrial engineers with a solid grounding in this important field so that they can utilize the best approach to make an active and involved engineering raise, a reality for them and to the world that requires more to be added to it.

I wish to thank all the contributing authors who shared their knowledge with us. I also wish to thank the publisher and the publishing team for their consistent efforts and constant support.

Editor

Collaborative Decision-Making in Product Design: An Interactive Multiobjective Approach

Chantal Baril,[1] Soumaya Yacout,[2] and Bernard Clément[2]

[1] Department of Industrial Engineering, Université du Québec à Trois-Rivières, 3351 Boulevard des Forges, Trois-Rivières, QC, Canada G9A 5H7
[2] Department of Mathematical and Industrial Engineering, École Polytechnique de Montréal, 2900 Boulevard Édouard-Montpetit, Campus de l'Université de Montréal, Montréal, QC, Canada H3T 1J4

Correspondence should be addressed to Chantal Baril; chantal.baril@uqtr.ca

Academic Editor: C. K. Kwong

This paper presents a new procedure to solve multiobjective problems, where the objectives are distributed to various working groups and the decision process is centralized. The approach is interactive and considers the preferences of the working groups. It is based on two techniques: an interactive technique that solves multi-objective problems based on goal programming, and a technique called "linear physical programming" which considers the preferences of the working groups. The approach generates Pareto-optimal solutions. It guides the director in the determination of target values for the objective functions. The approach was tested on two problems that present its capacity to generate Pareto-optimal solutions and to show the convergence to compromise solutions for all the working groups.

1. Introduction

The process of product design is often organized in a hierarchical structure where the specialists are separated by discipline in several working groups. As shown in Figure 1, the working groups are supervised by a director who coordinates the design activities. The role of the director is to collect information provided by the groups and to use computational method to finding an optimal design. The working groups are considered as experts that have the technical knowledge in their proper discipline.

According to their competencies, each working group is responsible of achieving specific design objectives expressing the customer's requirements. Often these objectives are functions of the same set of design variables and in certain cases, they may be conflicting. For that reason, it is necessary to find an optimization procedure that takes into consideration that knowledge and includes it in the solution.

In this paper, we develop a new Interactive Multiobjective approach taking into account the working group's Preferences (IMOP). The original contributions of the IMOP algorithm are the as follows.

(i) It has the ability to define a reduced set of target values that can be divided into degrees of desirability to capture the working groups' preferences. This is an important contribution because it is a challenging issue in multi-objective optimization.

(ii) It generates Pareto-optimal solutions corresponding to the working groups' preferences.

(iii) It subtracts the stability set from the reduced set of target values at each iteration, thus ensuring a different Pareto-optimal solution each time.

The proposed approach is particularly interesting when the decision process is centralized and involves many working groups who are collaborating in order to find a best compromise solution.

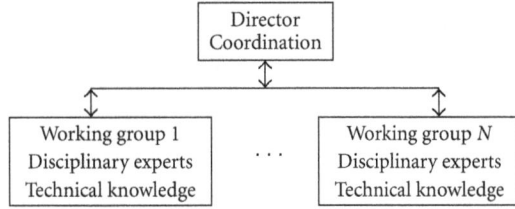

FIGURE 1: Organizational structure in product design.

2. Multiobjective Problem and Pareto-Optimal Concept

The purpose of a general multi-objective optimization problem is to find the design variables that optimize a vector objective function $F(X) = \{f_1(X), f_2(X), \ldots, f_k(X)\}$ over the feasible design space. The minimization problem formulation in standard form is as follows [1]:

$$\text{Minimize} \quad F(X) = \{f_1(X), f_2(X), \ldots, f_k(X)\}$$

$$\text{subject to} \quad h_s(X) = 0, \quad s = 1, \ldots, t,$$

$$g_r(X) \geq 0, \quad r = 1, \ldots, m, \tag{1}$$

$$X^l \leq X \leq X^u.$$

The aim of solving a multi-objective problem is to get a Pareto-optimal solution or a set of Pareto-optimal solutions. Conceptually, a Pareto-optimal solution is one which is not dominated by any other feasible solution. Mathematically, for a minimization problem with k objective functions f_i, $i = 1, \ldots, k$, a vector X^* is Pareto-optimal if there is no other feasible X such that $f(X) \leq f(X^*)$, meaning that $f_i(X) \leq f_i(X^*)$ for all $i = 1, \ldots, k$ with strict inequality for at least one i [2]. In general, the optimal solutions obtained by the individual optimization of the objectives are not the same. It is then necessary to find solutions to the multi-objective problem which are Pareto-optimal.

There are several techniques for solving multi-objective optimization problems. Some methods have been developed to find an exact Pareto set, or an approximation of it, inside of which one of the generated Pareto optimal solutions is chosen for implementation. These methods include compromise programming [3], the weight method, and the constraints method [4]. Several metaheuristics approaches have also been used to solve multi-objective problems like simulated annealing [5, 6] particles swarm optimization [7] and evolutionary algorithms [8–10].

However, as the number of competing objectives increases, the problem of finding the best compromise solution becomes increasingly complex. Hence, it can become overwhelming to analyze the entire Pareto-optimal solution set to select one solution for implementation. It becomes attractive to reduce the size of the solution set, and to assist the decision maker in selecting a final solution [11]. Some methods attempt to quantify the decision maker's preferences, and with this information, the solution that best satisfies the decision maker's preferences is then identified. These methods include among others goal programming,

and linear physical programming [12]. Linear physical programming is a method for generating a preferred Pareto solution during multi-objective optimization. It is an extension of goal programming. The initial development of the physical programming methodology is presented in Messac et al. [12]. Physical programming captures the decision maker's preferences, a priori, in a mathematically consistent manner using a preference function. The decision maker (DM) classifies each objective function into the four soft and the four hard classes as shown in Table 1.

The DM specifies the degrees of desirability $(t^-_{i5}, t^-_{i4}, t^-_{i3}, t^-_{i2}, t^-_{i1}$ and/or $t^+_{i1}, t^+_{i2}, t^+_{i3}, t^+_{i4}, t^+_{i5})$ for each objective function f_i in the soft category. For classes 1S through 4S, there are, respectively, five, nine, and ten such values as shown in Tables 2, 3, 4, and 5.

For classes 1H through 4H, these values are, respectively, $t_{l.\max}$, $t_{l,\min}$, $t_{l,\text{val}}$, and $t_{l,\min}$ and $t_{l.\max}$. The physical programming method involves converting a multi-objective problem into a single objective problem by using preference functions that capture the DM's preferences. Given the DM's input in the form of range boundaries (or targets) for each objective, Messac et al. [12] suggest an algorithm to generate the weights $\widetilde{w}^-_{is}$ and $\widetilde{w}^+_{is}$. The following problem is then solved (d^-_{is} and d^+_{is} are the deviational variables):

$$\underset{d^-_{is}, d^+_{is}, X}{\text{Min}} \quad J = \sum_{i=1}^{k} \sum_{s=2}^{5} \left(\widetilde{w}^-_{is} d^-_{is} + \widetilde{w}^+_{is} d^+_{is} \right)$$

$$\text{subject to} \quad f(X)_i - d^+_{is} \leq t^+_{i(s-1)}$$

$\forall i \in$ classes 1S, 3S, 4S; $i = 1, 2, \ldots, k$;

$$s = 2, \ldots, 5,$$

$d^+_{is} \geq 0 \quad \forall i \in$ classes 1S, 3S, 4S;

$$i = 1, 2, \ldots, k; \ s = 2, \ldots, 5$$

$f_i(X) \leq t^+_{i5} \quad \forall i \in$ classes 1S, 3S, 4S;

$$i = 1, 2, \ldots, k,$$

$f_i(X) + d^-_{is} \geq t^-_{i(s-1)}, \quad \forall i \in$ classes 2S, 3S, 4S;

$$i = 1, 2, \ldots, k; \ s = 2, \ldots, 5,$$

$d^-_{is} \geq 0, \quad \forall i \in$ classes 2S, 3S, 4S;

$$i = 1, 2, \ldots, k; \ s = 2, \ldots, 5,$$

$f_i(X) \geq t^-_{i5}, \quad \forall i \in$ classes 2S, 3S, 4S;

$$i = 1, 2, \ldots, k,$$

$f_l(X) \leq t_{l,\max}, \quad \forall l \in$ class 1H;

$$l = 1, 2, \ldots, L,$$

$f_l(X) \geq t_{l,\min}, \quad \forall l \in$ class 2H;

$$l = 1, 2, \ldots, L,$$

$f_l(X) = t_{l,\text{val}}, \quad \forall l \in$ class 3H;

$$l = 1, 2, \ldots, L,$$

$$t_{l,\min} \leq f_l(X) \leq t_{l,\max}, \quad \forall l \in \text{class } 4H;$$

$$l = 1, 2, \ldots, L,$$

$$X_{\min} \leq X \leq X_{\max}.$$

(2)

The limitation of the physical programming is that it requires a priori selection of range parameters for each of the objective functions and provides information for only one design scenario (i.e., a single Pareto solution). Tappeta et al. [13] have twinned linear physical programming with an interactive algorithm [1]. Their algorithm finds a Pareto solution and can generate other Pareto designs in the neighbourhood of the current Pareto solution. No means are provided to help the DM to specify his/her initial preferences in the form of region limits defined in physical programming.

Some authors have suggested several interactive multi-objective optimization methods [1, 14–16]. These methods allow the decision maker to express his/her preferences by using a reference point or by classifying the objectives, functions. The disadvantages of using traditional multi-objective methods are as follows [1]: (1) require a priori selection of weights or targets for each of the objective functions, (2) provide only a single Pareto-optimal solution, and (3) are unable to generate proper Pareto-optimal points for non convex problems (the weights method). Abdel Haleem [17] developed an interactive nonlinear goal programming algorithm (INLGP) that helps the decision maker to determine reference points for the goals. The decision maker does not need to do any ranking of classification of these goals. The advantages of this INLGP algorithm are as follows: (1) it reduces the parametric space of the target values by limiting each parameter with minimum and maximum values rather than by choosing any random values from the whole parametric space, and (2) the algorithm is guaranteed to generate Pareto-optimal solutions at each iteration. The INLGP algorithm was used for the design of a low-pass electrical circuit [18]. However, no means are provided to divide the reduced parametric space. Realizing these limitations, an interactive multi-objective approach is proposed which attempts to address the issues mentioned above.

3. An Interactive Multiobjective Approach Taking into Account the Working Groups' Preferences (IMOP Approach)

The IMOP approach is based on the interactive nonlinear goal programming algorithm (INLGP) of Abdel Haleem [17] combined to the linear physical programming introduced by Messac et al. [12]. The IMOP approach has the following advantages: (1) it provides means to capture the working group's preferences, (2) it offers the possibility of interaction between the director and his/her working groups, (3) it generates several Pareto-optimal solutions (several design scenarios), and (4) it fits with the industries organizational structure.

Before using the IMOP approach, it is necessary to distribute the objective functions among the working groups

TABLE 1: Objective function classification.

Soft	Class 1S	Small is better (minimization)
	Class 2S	Larger is better (maximization)
	Class 3S	Value is better
	Class 4S	Range is better
Hard	Class 1H	Must be smaller ($f_l \leq t_{l,\max}$)
	Class 2H	Must be larger ($f_l \geq t_{l,\min}$)
	Class 3H	Must be equal ($f_l = t_{l,\text{val}}$)
	Class 4H	Must be in range ($t_{l,\min} \leq f_l \leq t_{l,\max}$)

according to their respective disciplinary competencies. More than one objective can be assigned to the same working group. The multi-objective optimization process is centralized at the director level. The director coordinates the activities between the working groups. The working groups collaborate to the resolution process by defining their preferences and providing the target values for their objective functions. The following are the steps involved in the application of the new interactive multi-objective approach.

Step 1. Each working group classifies his objective functions into four classes (Table 6).

Step 2. For the k objective functions, each working group solves its objective optimization problem individually according to a category chosen in 1. The optimal solutions are X^{*i}, $i = 1, \ldots, k$. The optimal values of the objective functions are f_i^*, $i = 1, \ldots, k$. The working groups know the best possible values of each objective function under their control. These values are returned to the director.

Step 3. The director evaluates the value of the other $k-1$ objective functions at the k optimal solutions X^{*i}, $i = 1, \ldots, k$, and constructs $k \times k$ table of the objectives values as shown in Table 7. From this table, the director will know the best and the worst possible values of each objective function f_i^*, $i = 1, \ldots, k$ that corresponding to $b_{i\min}$, $b_{i\max}$, $i = 1, \ldots, k$ of each objective function (for a minimization problem). The approach proceeds by determining the reduced solvability set denoted by D' where $D' = \{D \mid b_{i\min} \leq b_i \leq b_{i\max}, i = 1, \ldots, k\}$ and D is the set of parameter values for which the problem is solvable. The reduced solvability set will be used by the working groups to define their preference according to their competencies.

Step 4. The director presents the reduced solvability set D' to the working groups to seek preferences for each objective function.

(i) For class 1S (minimization), the preferences are highly desirable (t_{i1}^+), desirable (t_{i2}^+), tolerable (t_{i3}^+), undesirable, and (t_{i4}^+) and highly undesirable (t_{i5}^+).

(ii) For class 2S (maximization), the preferences are highly desirable (t_{i1}^-), desirable (t_{i2}^-), tolerable (t_{i3}^-), undesirable (t_{i4}^-), and highly undesirable (t_{i5}^-).

(iii) For class 3S (value is better), the preferences are highly desirable (t_{i1}), desirable (t_{i2}^- and t_{i2}^+), tolerable (t_{i3}^- and t_{i3}^+), undesirable (t_{i4}^- and t_{i4}^+) and highly undesirable (t_{i5}^- and t_{i5}^+).

(iv) For class 4S (range is better), the preferences are highly desirable (t_{i1}^- and t_{i1}^+), desirable (t_{i2}^- and t_{i2}^+), tolerable (t_{i3}^- and t_{i3}^+), undesirable (t_{i4}^- and t_{i4}^+), and highly undesirable (t_{i5}^- and t_{i5}^+).

For multi-objective problem, the director has not all the necessary competencies to choose these values. It is why the collaboration of the working groups is important. For example, the following scenario can be used to define the degrees of desirability for a pure mathematical minimization problem. Supposing that $t_{i1}^+ = b_{i\,Min}$, $t_{i5}^+ = b_{i\,Max}$ and $(b_{i\,Max} - b_{i\,Min})/4 = v_i$, the reduced solvability set D' can be divided as follows: $t_{i1}^+ = b_{i\,Min}$, $t_{i2}^+ = t_{i1}^+ + v_i$, $t_{i3}^+ = t_{i2}^+ + v_i$, $t_{i4}^+ = t_{i3}^+ + v_i$, and $t_{i5}^+ = b_{i\,Max}$. For design problem, these values are set according to the working groups' competencies and customer's requirements.

Step 5. Set solution $j = 1$. Each working group selects the target value b_i for each of their objective functions and transfers these values to the director.

Step 6. The director uses the algorithm proposed by Dauer and Krueger [19] to solve the following multiobjective goal programming problem and to obtain the Pareto-optimal solution $\overline{X}_j$. This algorithm is detailed in Appendix A. The last attainment problem for goal k twinned with the linear physical programming is (P_k)

Minimize d_k

subject to $M(b) = \{X \in R^n \mid g_r(X) \le b_r, r = 1, \dots, m, X \ge 0\}$
$$\tag{3}$$

and for classes 1S, 3S, and 4S

$$g_{m+i}(X) \equiv f_i(X) - d_i \le b_i, \quad 1 \le i \le k,$$
$$d_i = d_i^*, \quad 1 \le i \le k-1,$$
$$f_i(X) \le t_{i5}^+, \quad 1 \le i \le k, \tag{4}$$
$$d_k \ge 0$$

and for classes 2S, 3S, and 4S

$$g_{m+i}(X) \equiv f_i(X) + d_i \ge b_i, \quad 1 \le i \le k,$$
$$d_i = d_i^*, \quad 1 \le i \le k-1,$$
$$f_i(X) \ge t_{i5}^-, \quad 1 \le i \le k, \tag{5}$$
$$d_k \ge 0.$$

For all classes

$$X_{min} \le X \le X_{max}. \tag{6}$$

Note. The constraints $g_{m+i}(X)$ are called the goals constraints.

This step permits to find a solution that meets as much as possible the working group's preferences.

Step 7. If the working groups are satisfied with this solution, stop and go to Step 13, if not, go to Step 8. It is suggested to generate a certain number of optimal solutions, which are Pareto optimal before stopping.

Step 8. The director formulates the KKT conditions of the problem (P_k) and determines the values of the Kuhn Tucker multipliers associated with the goals constraints: $\bar{u}_r$, $r = 1, \dots, k + m$.

Step 9. According to the values $\bar{u}_r$ and by using the algorithm presented in Appendix B, the director determines the stability set of the first kind $G(\overline{X}_j)$ which is the set of parameter values for which the optimal solution remains optimal.

Step 10. The director uses the sets subtraction algorithm presented in Appendix C to obtain the new reduced solvability set $\{D' - \bigcup_{p=1}^{j} G(\overline{X}_p)\}$ which excludes the stability set. Steps 8, 9, and 10 are necessary to ensure that the work groups will choose target values leading to other Pareto-optimal solution.

Step 11. If no values can be chosen in $\{D' - \bigcup_{p=1}^{j} G(\overline{X}_p)\}$, stop and go to Step 13, otherwise go to Step 12.

Step 12. Set $j = j + 1$. The working groups select other target values $b_i \in \{D' - \bigcup_{p=1}^{j-1} G(\overline{X}_p)\}$ and go to Step 6. One can use these rules to select the values and to obtain other Pareto optimal solutions.

(i) Rule no. 1: It is always necessary to improve the objective function having the worst value by choosing its target value in a better zone and by sacrificing the other objectives by choosing their target values in a less desirable zone. The aim of these choices is to obtain, if possible, all the objective's values in the tolerable zone (or better).

(ii) Rule no. 2: Once in the tolerable zone, try other values in this zone in order to obtain other Pareto-optimal solutions. The selected values should cover all the zone. For example, choose a value at one end of the tolerable zone and the other values at the other end. One can also try to choose one of the target values in the desirable zone while leaving the other target values in the tolerable zone.

(iii) Rule no. 3: if it is impossible to follow the first rule due to the reduced solvability set, try all the possibilities to find the best choice.

Step 13. The director presents all the Pareto-optimal solutions to the working groups and tries to get consensus for the best compromise. If other solutions are necessary, go to Step 12.

TABLE 2: Degrees of desirability for class 1S.

Class 1S—smaller is better (i.e., minimization)				
$f_i \leq t_{i1}^+$	$t_{i1}^+ < f_i \leq t_{i2}^+$	$t_{i2}^+ < f_i \leq t_{i3}^+$	$t_{i3}^+ < f_i \leq t_{i4}^+$	$t_{i4}^+ < f_i \leq t_{i5}^+$
Highly desirable	Desirable	Tolerable	Undesirable	Highly undesirable

TABLE 3: Degrees of desirability for class 2S.

Class 2S—larger is better (i.e., maximization)				
$t_{i5}^- \leq f_i < t_{i4}^-$	$t_{i4}^- \leq f_i < t_{i3}^-$	$t_{i3}^- \leq f_i < t_{i2}^-$	$t_{i2}^- \leq f_i < t_{i1}^-$	$t_{i1}^- \leq f_i$
Highly undesirable	Undesirable	Tolerable	Desirable	Highly desirable

TABLE 4: Degrees of desirability for class 3S.

Class 3S—value is better (i.e., seek value)								
$t_{i5}^- \leq f_i < t_{i4}^- \leq f_i < t_{i3}^- \leq f_i < t_{i2}^- \leq f_i < t_{i1}^- < f_i \leq t_{i2}^+ < f_i \leq t_{i3}^+ < f_i \leq t_{i4}^+ < f_i \leq t_{i5}^+$								
Highly undesirable	Undesirable	Tolerable	Desirable	Highly desirable	Desirable	Tolerable	Undesirable	Highly undesirable

TABLE 5: Degrees of desirability for class 4S.

Class 4S—range is better (i.e., seek range)								
$t_{i5}^- \leq f_i < t_{i4}^- \leq f_i < t_{i3}^- \leq f_i < t_{i2}^- \leq f_i < t_{i1}^- \leq f_i \leq t_{i1}^+ < f_i \leq t_{i2}^+ < f_i \leq t_{i3}^+ < f_i \leq t_{i4}^+ < f_i \leq t_{i5}^+$								
Highly undesirable	Undesirable	Tolerable	Desirable	Highly desirable	Desirable	Tolérable	Undesirable	Highly undesirable

TABLE 6: Classification of the objective functions.

Class 1S	Small is better (minimization)
Class 2S	Larger is better (maximization)
Class 3S	Value is better (seek value)
Class 4S	Range is better (seek range)

TABLE 7: Table of objectives values.

Optimal solutions	Objective functions			
X^{*1}	$f_1^*(X^{*1})$	$f_2^*(X^{*1})$	$\cdots$	$f_i^*(X^{*1})$
X^{*2}	$f_1^*(X^{*2})$	$f_2^*(X^{*2})$	$\cdots$	$f_i^*(X^{*2})$
$\cdots$	$\cdots$	$\cdots$	$\cdots$	$\cdots$
X^{*i}	$f_1^*(X^{*i})$	$f_2^*(X^{*i})$	$\cdots$	$f_i^*(X^{*i})$

4. Numerical Examples

In this section, the interactive multi-objective procedure is applied to two design problems. The first problem consists of a set of simple analytical expressions for its objective and constraint functions and was presented by Tappeta et al. [13]. This problem is chosen to illustrate the key features of the approach and to compare with the results obtained by those authors. The second problem is the design of a two-bar structure that is subjected to a force, F, at a point that vertically deflects by an amount d. In both cases, the IMOP approach is implemented in Matlab 7.0.4.365 (R14) and the optimization process was conducted on Pentium D duo core 3.4 GHz and 2 GB RAM. The computational time is less than 1 minute.

4.1. Test Problem 1. This problem was introduced by Tappeta et al. [13] and has three design variables, three objective functions, and a constraint. The problem definition in standard form and the application of the IMOP approach are as follows:

$$\text{Minimize} \quad F(X) = \{f_1(X), f_2(X), f_3(X)\}$$

$$\text{subject to} \quad g_1(X) = 12 - x_1^2 - x_2^2, \quad (7)$$

$$X \geq 0,$$

where

$$f_1(X) = 10 - \frac{\left(x_1^3 + x_1^2(1 + x_2 + x_3) + x_2^3 + x_3^3\right)}{10},$$

$$f_2(X) = 15 - \frac{\left(x_1^3 + 2x_2^3 + x_2^2(2 + x_1 + x_3) + x_3^3\right)}{10}, \quad (8)$$

$$f_3(X) = 20 - \frac{\left(x_1^3 + x_2^3 + 3x_3^3 + x_3^2(3 + x_1 + x_2)\right)}{10}.$$

For this example, we suppose that $f_1(X)$ and $f_2(X)$ needed specific competencies so they are assigned to a working group and $f_3(X)$ need other competencies so it is assigned to another group. Therefore, the procedure proceeds with a director and two working groups.

Step 1. Each working group classifies its objective functions:

Working group 1 classifies $f_1(X)$ in class 1S,

Working group 1 classifies $f_2(X)$ in class 1S,

Working group 2 classifies $f_3(X)$ in class 1S.

Step 2. For the k objective functions, each working group solves its single optimization problem individually according

TABLE 8: Optimal values for the objective functions of test Problem 1.

X^{*i}	x_1	x_2	x_3	f_i^*
Working group 1	3.2539	0.8402	0.8402	3.5980
Working group 1	0.4651	3.4011	0.4651	3.7221
Working group 2	0.3169	0.3169	3.4350	3.5471

TABLE 9: Objective function values table for test Problem 1.

	x_1	x_2	x_3	f_1	f_2	f_3
X^{*1}	3.2539	0.8402	0.8402	3.5980	10.9465	15.8166
X^{*2}	0.4651	3.4011	0.4651	5.9405	3.7221	15.8771
X^{*3}	0.3169	0.3169	3.4350	5.8929	10.8797	3.5471

TABLE 10: The degrees of desirability specified by Tappeta et al. [13].

Criteria	Class	HD t_{i1}^+	D t_{i2}^+	T t_{i3}^+	U t_{i4}^+	HU t_{i5}^+
f_1	1S	3.0	4.25	6.0	7.5	9.0
f_2	1S	3.7	7.0	9.25	11.8	12.5
f_3	1S	6.0	12.0	15.0	18.0	20.0

HD: (highly desirable $\leq t_{i1}^+$), D: ($t_{i1}^+ <$ desirable $\leq t_{i2}^+$), T: ($t_{i2}^+ <$ tolerable $\leq t_{i3}^+$), ID: ($t_{i3}^+ <$ undesirable $\leq t_{i4}^+$), IA: ($t_{i4}^+ <$ highly undesirable $\leq t_{i5}^+$).

TABLE 11: The working group's preferences for test Problem 1.

Criteria	Class	I t_{i1}^+	D t_{i2}^+	T t_{i3}^+	ID t_{i4}^+	IA t_{i5}^+
f_1	1S	4.1836	4.7693	5.3549	5.9405	6.5261
f_2	1S	5.5282	7.3343	9.1404	10.9465	12.7526
f_3	1S	6.6296	9.7121	12.7946	15.8771	18.9596

HD: (highly desirable $\leq t_{i1}^+$), D: ($t_{i1}^+ <$ desirable $\leq t_{i2}^+$), T: ($t_{i2}^+ <$ tolerable $\leq t_{i3}^+$), ID: ($t_{i3}^+ <$ undesirable $\leq t_{i4}^+$), IA: ($t_{i4}^+ <$ highly undesirable $\leq t_{i5}^+$).

to the category chosen in 1. The optimal solutions are X^{*i}, $i = 1,\ldots,k$. The optimal values of the objective functions are noted to be f_i^*, $i = 1,\ldots,k$ and are presented in Table 8.

Step 3. The director evaluates the value of the other $k - 1$ objective function at the k optimal solutions and constructs the $k \times k$ table of the objective values. From Table 9, the director knows the best and the worst values for each objective function. These values are noted to be $b_{i\min}$, $b_{i\max}$, $i = 1,\ldots,k$. The approach proceeds by determining the reduced solvability set $D' = \{\{D\} \mid b_{i\min} \leq b_i \leq b_{i\max}, i = 1,\ldots,k\}$ where D is the set of parameters for which the problem is solvable.

The reduced solvability set is

$$3.5980 \leq b_1 \leq 5.9405,$$
$$3.7221 \leq b_2 \leq 10.9465,$$
$$3.5471 \leq b_3 \leq 15.8771.$$

Step 4. The director presents the reduced solvability set D' to the working groups to seek their preferences for each objective function. These values are set according to the working groups' knowledge and experience. For class 1S, each working group determines the degrees of desirability $t_{i1}^+, t_{i2}^+, t_{i3}^+, t_{i4}^+, t_{i5}^+$. Table 10 shows the degrees of desirability fixed by Tappeta et al. [13]. These degrees of desirability are used to be able to compare the results.

It is obvious that the degrees of desirability t_{i2}^+ for the objective functions f_1 and f_2 could never be reached, since they are not included in the reduced solvability set: the minimal value for the objective function f_1 is 3.5980 and for f_2 is 3.7221. This example shows that the degrees of desirability should not be given blindly to prevent the choice of scenarios which are not feasible. Table 11 shows more realistic degrees of desirability. These degrees of desirability are obtained by dividing the solvability set $3.5980 \leq b_1 \leq 5.9405$, $3.7221 \leq b_2 \leq 10.9465$ and $3.5471 \leq b_3 \leq 15.8771$ according to this scenario: we suppose that the worst value is undesirable ($t_{i4}^+ = b_{i\,\text{Max}}$) and we calculate $(b_{i\,\text{Max}} - b_{i\,\text{Min}})/4 = v_i$ to find the following degrees of desirability: $t_{i1}^+ = t_{i2}^+ - v_i$, $t_{i2}^+ = t_{i3}^+ - v_i$, $t_{i3}^+ = t_{i4}^+ - v_i$, $t_{i4}^+ = b_{i\,\text{Max}}$, and $t_{i5}^+ = t_{i4}^+ + v_i$.

We assume that preferences are uniformly distributed across the solvability set but it is not necessarily always the case.

Step 5. Set solution $j = 1$. The working groups select the target values b_i for each objective function. It is obvious that

each working group wants to obtain the better value for their objective functions. So they will choose target values in the highly desirable zone. We assume that the approach starts with the target values corresponding to t_{i1}^+:

Working group 1 sets the target value of b_1 at 4.1836 (highly desirable),

Working group 2 sets the target value of b_2 at 5.5282 (highly desirable),

Working group 3 sets the target value of b_3 at 6.6296 (highly desirable).

Step 6. With the target values supplied by the working groups, the director uses the algorithm proposed by Dauer and Krueger [19] given in Appendix A to solve the multi-objective goal programming problem and to obtain a first Pareto optimal solution $\overline{X}_1$:

$\overline{X}_1 = (2.8568, 1.8775, 0.5598)$,

$f_1 = 4.1836$ (the value of f_1 is in the highly desirable zone),

$f_2 = 9.4178$ (the value of f_2 is in the undesirable zone),

$f_3 = 16.7115$ (the value of f_3 is in the highly undesirable zone).

Step 7. If the working groups are satisfied with this solution, stop and go to Step 13, if not, go to Step 8. For this case, we assume that the working groups 1 and 2 are not satisfied since the values of their objective functions f_2 and f_3 are in the undesirable and highly undesirable zones, respectively, and want to generate another solution. Go to Step 8. Steps 8, 9, and 10 are necessary to ensure that the work groups will choose target values leading to other Pareto-optimal solution.

Step 8. The director formulates the KKT conditions of the problem and determines the values of the Kuhn Tucker multipliers associated with the goal constraints $\overline{u}_r$, $r = 1, \ldots, k + m$,

$$\overline{u}_2 = 1.2771 \times 10^5, \quad \overline{u}_3 = 5.6776 \times 10^4, \quad \overline{u}_4 = 1. \quad (9)$$

Step 9. According to the values $\overline{u}_r$, and by using the algorithm of Osman [20] given in Appendix B, the director determines the stability set $G(\overline{X}_1)$

Given $\overline{u}_2 > 0$ and $g_2 = 4.1836$ then $b_1 = 4.1836$,

Given $\overline{u}_3 > 0$ and $g_3 = 5.5282$ then $b_2 = 5.5282$,

Given $\overline{u}_4 > 0$ and $g_4 = 6.6296$ then $b_3 = 6.6296$.

Step 10. The director uses the sets subtraction algorithm proposed by Abdel Haleem [17] given in Appendix C to obtain the new reduced solvability set $\{D^{'} - \bigcup_{p=1}^{j} G(\overline{X}_p)\}$ given in Table 12.

Step 11. If no values can be selected in $\{D^{'} - \bigcup_{p=1}^{j} G(\overline{X}_p)\}$ stop and go to Step 13, otherwise go to Step 12. In this case, other values can be chosen in Table 10 so go to Step 12.

Step 12. Set $j = j + 1$. The working groups select other target values for their objective function in $b^j \in \{D^{'} - \bigcup_{p=1}^{j-1} G(\overline{X}_p)\}$ and go to Step 6. The solutions obtained are presented in Table 13.

The third Pareto-optimal solution seems to be the best one because all the objective values match the target values according to Table 10. This solution can be considered satisfactory and a good compromise for all the working groups.

Step 13. The director presents the Pareto-optimal solutions obtained to the working groups to select the best one for everyone (stop). If other solutions are necessary go to Step 12. Although the solutions obtained in the six iterations are Pareto optimal, the best Pareto-optimal solutions according to the working group's preferences (desirability) are solutions 3 and 4. These solutions can be retained for implementation.

Finally, it is also interesting to know if the solutions (Pareto points) obtained by this approach are close to certain targeted aspiration points. To do this, we compare the obtained results with the Pareto-optimal results obtained by

TABLE 12: The reduced solvability set for test Problem 1.

Set number	b_1 min	b_1 max	b_2 min	b_2 max	b_3 min	b_3 max
1	3.5980	4.1836	3.7221	10.9465	3.5471	15.8771
2	4.1836	5.9405	3.7221	10.9465	3.5471	15.8771
3	4.1836	4.1836	3.7221	5.5282	3.5471	15.8771
4	4.1836	4.1836	5.5282	10.9465	3.5471	15.8771
5	4.1836	4.1836	5.5282	5.5282	3.5471	6.6296
6	4.1836	4.1836	5.5282	5.5282	6.6296	15.8771

TABLE 13: Pareto-optimal solutions generated by the approach.

Solution j	Target values b_i	Objective function values f_i
1	$b_1 = 4.1836$ (highly desirable) $b_2 = 5.5282$ (highly desirable) $b_3 = 6.6296$ (highly desirable)	$f_1 = 4.1836$ (highly desirable) $f_2 = 9.4178$ (undesirable) $f_3 = 16.7115$ (highly undesirable)
2	$b_1 = 4.5$ (desirable) $b_2 = 8$ (tolerable) $b_3 = 13$ (undesirable)	$f_1 = 4.5$ (desirable) $f_2 = 8.6799$ (tolerable) $f_3 = 16.8579$ (highly undesirable)
3	$b_1 = 5.2$ (tolerable) $b_2 = 9$ (tolerable) $b_3 = 14$ (undesirable)	$f_1 = 5.2$ (tolerable) $f_2 = 9$ (tolerable) $f_3 = 14.8087$ (undesirable)
4	$b_1 = 5.3$ (tolerable) $b_2 = 9.1$ (tolerable) $b_3 = 12.5$ (tolerable)	$f_1 = 5.3$ (tolerable) $f_2 = 9.1$ (tolerable) $f_3 = 14.4381$ (undesirable)
5	$b_1 = 5.9$ (undesirable) $b_2 = 8.9$ (tolerable) $b_3 = 12.6$ (tolerable)	$f_1 = 5.9$ (undesirable) $f_2 = 8.9$ (tolerable) $f_3 = 13.0147$ (undesirable)
6	$b_1 = 5$ (tolerable) $b_2 = 10$ (undesirable) $b_3 = 12$ (tolerable)	$f_1 = 5$ (tolerable) $f_2 = 10$ (undesirable) $f_3 = 13.9899$ (undesirable)

Tappeta et al. [13] at specific aspiration points. The comparison is presented in Table 14.

For a minimization problem, we want to find a better (smaller) solution than or equal the aspiration values. The symbol (+) indicates that the solution obtained by our approach is worse (bigger) than the aspiration values, the symbol (=) indicates that the solution obtained is the same (equal) as the aspiration values, and the symbol (−) indicates that the solution obtained is better (smaller) than the aspiration values. The solutions obtained with our algorithm are considered better than or equal to those found by Tappeta et al. [13] if the number of symbols (−) and (=) exceeds the number of symbols (+). For the first solution, the number of symbols (−) and (=) is 2 for the IMOP algorithm and the number of symbols (−) and (=) is 1 for Tappet et al. [13]. These results are very encouraging because they demonstrate that our approach can find solutions closer to the working

TABLE 14: Aspiration points and Pareto data from Tappeta et al. [13].

Aspiration values (target values)			Pareto points Tappeta et al. [13]			Pareto points IMOP approach		
1	2	3	1	2	3	1	2	3
7.483	6.788	11.285	6.956 (−)	7.437 (+)	11.496 (+)	6.9722 (−)	6.7880 (=)	12.3239 (+)
5.400	6.788	16.927	5.413 (+)	6.916 (+)	16.218 (−)	4.4679 (−)	5.1878 (−)	16.3795 (−)
6.016	10.183	11.285	3.994 (−)	10.095 (−)	16.130 (+)	6.0002 (−)	9.9646 (−)	10.7528 (−)
3.933	10.183	16.927	4.708 (+)	8.689 (−)	16.259 (−)	3.9235 (−)	10.0191 (−)	16.4693 (−)

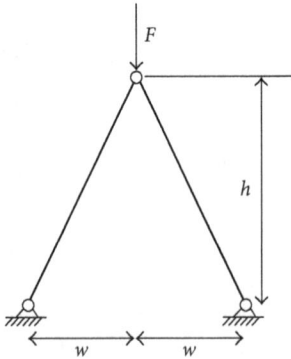

FIGURE 2: Two-bar truss example.

TABLE 15: Optimal values for the objective function of test Problem 2.

X^{*i}	x_1	x_2	f_i^*
Working group 1	39.2944	335.6810	3956
Working group 2	100	1000	119.3662
Working group 3	100	1000	0.8881

TABLE 16: Objective function values table for test Problem 2.

	x_1	x_2	f_1	f_2	f_3
X^{*1}	39.2944	335.6810	3956	595	6
X^{*2}	100	1000	15315	119.3662	0.8881
X^{*3}	100	1000	15315	119.3662	0.8881

group's requirements (aspiration values) than algorithms available in the literature.

4.2. Test Problem 2.

The second problem is the design of a two-bar structure that is subjected to a force, F, at a point that vertically deflects by an amount, d. This optimization problem involves the minimization of the mass, m, the normal stress, s, and the vertical deflection, d, of a two-bar truss. The design variables are the diameter of the member, $x_1 = a$, and the height, $x_2 = h$. Normal stress must be less than the buckling stress, as a constraint. A graphical representation of the truss is shown in Figure 2 [21]. The specific parameter values are as follows: $F = 150$ kN, $t = 2.5$ mm, structure width $w = 750$ mm, mass density $\rho = 7.8 \times 10^{-3}$ g/mm^3, and elastic modulus $E = 210000$ N·mm^2.

The problem's formulation is as follows:

$$\text{Minimize} \quad f_1(X) = m = 2\pi\rho t x_1 \sqrt{w^2 + x_2^2},$$

$$f_2(X) = s = \frac{F}{2\pi t x_1 x_2} \sqrt{w^2 + x_2^2},$$

$$f_3(X) = d = \frac{F(w^2 + x_2^2)^{3/2}}{2\pi t E x_1 x_2^2}$$

$$\text{subject to} \quad g_1(X) = \frac{F}{2\pi t x_1 x_2} \sqrt{w^2 + x_2^2} - \frac{1}{8}\pi^2 E \frac{t^2 + x_1^2}{w^2 + x_2^2} \le 0,$$

$$1 \le x_1 \le 100,$$

$$10 \le x_2 \le 1000.$$

$$(10)$$

For this example, we assume that each objective function needs specific competencies so one objective function is assigned to a working group. The procedure proceeds with a director and three working groups.

Step 1. Each working group classifies its objective function:

Working group 1 classifies $f_1(X)$ in class 1S,

Working group 2 classifies $f_2(X)$ in class 1S,

Working group 3 classifies $f_3(X)$ in class 1S.

Step 2. Each working group solves its single optimization problem according to the category chosen in 1. The optimal solutions are X^{*i}, $i = 1, \ldots, 3$. The optimal values of the objective functions are noted to be f_i^*, $i = 1, \ldots, 3$ and are presented in Table 15.

Step 3. The director evaluates the two other objective functions at the three optimal solutions and constructs the 3×3 table of the objective functions' values. From Table 16, the director knows the best and the worst values for each objective function. These values are noted to be $b_{i\min}, b_{i\max}$, $i = 1, \ldots, 3$. The approach proceeds by determining the reduced solvability set denoted by D' where $D' = \{D \mid b_{i\min} \le b_i \le b_{i\max}, i = 1, \ldots, 3\}$ and D is the set of parameters for which the problem is solvable.

The reduced solvability set is

$$3956 \le b_1 \le 15315,$$

$$119.3662 \le b_2 \le 595,$$

$$0.8881 \le b_3 \le 6.$$

TABLE 17: Working group's preferences for test Problem 2.

Function Class		HD t_{i1}^+	D t_{i2}^+	T t_{i3}^+	U t_{i4}^+	HU t_{i5}^+
f_1	1S	4450	4550	4650	4750	4850
f_2	1S	370	390	400	450	500
f_3	1S	2	2.5	3	3.5	4

HD: (highly desirable $\leq t_{i1}^+$), D: ($t_{i1}^+ <$ desirable $\leq t_{i2}^+$), T: ($t_{i2}^+ <$ tolerable $\leq t_{i3}^+$), ID: ($t_{i3}^+ <$ undesirable $\leq t_{i4}^+$), IA: ($t_{i4}^+ <$ highly undesirable $\leq t_{i5}^+$).

Step 4. The director presents the reduced solvability set D' to the working groups to seek their preferences for each objective function. These values are set according to the working groups' knowledge and experience. For class 1S, each working group determines $t_{i1}^+, t_{i2}^+, t_{i3}^+, t_{i4}^+, t_{i5}^+$. Table 17 shows the degrees of desirability determined by Messac and Ismail-Yahaya [21]. These degrees of desirability are used to be able to compare the results. These degrees of desirability are realistic because they are inside the reduced solvability set determined in Step 3.

Step 5. Set solution $j = 1$. Each working group selects the target values b_i for its objective function. We assume that the working groups will not make a compromise, and they will choose target values in the highly desirable zone. The approach starts with the target value corresponding to t_{i1}^+:

Working group 1 sets target value b_1 at 4450 (highly desirable),

Working group 2 sets target value b_2 at 370 (highly desirable),

Working group 3 sets target value b_3 at 2 (highly desirable).

Step 6. With the target values supplied by the working groups, the director uses the algorithm proposed by Dauer and Krueger [19] given in Appendix A to solve the multi-objective goal programming problem and to obtain a first Pareto optimal solution $\overline{X}_1$:

$\overline{X}_1 = (37.8392\ 599.0083)$,

$f_1 = 4450$ (the value of f_1 is in the highly desirable zone),

$f_2 = 404.3889$ (the value of f_2 is in the undesirable zone),

$f_3 = 2.9618$ (the value of f_3 is in the tolerable zone).

Step 7. If the working groups are satisfied with this solution, stop and go to Step 13, if not, go to Step 8. For this problem, we assume that working group 2 is not satisfied since the value of its objective function is in the undesirable zone and wants to generate another solution. Go to Step 8. Steps 8, 9, and 10 are necessary to ensure that the working groups will choose target values leading to other Pareto-optimal solution.

TABLE 18: The reduced solvability set for test Problem 2.

Set no:	b_1 min	b_1 max	b_2 min	b_2 max	b_3 min	b_3 max
1	3956	4450	119	595	1	6
2	4450	15315	119	370	1	6
3	4450	15315	370	595	1	2

TABLE 19: Pareto-optimal solutions generated for Problem test 2.

Solution	Target values b_i	Objective function values f_i
1	$b_1 = 4450$ (highly desirable)	$f_1 = 4450$ (highly desirable)
	$b_2 = 370$ (highly desirable)	$f_2 = 404.3889$ (undesirable)
	$b_3 = 2$ (highly desirable)	$f_3 = 2.9618$ (tolerable)
2	$b_1 = 4600$ (tolerable)	$f_1 = 4600$ (tolerable)
	$b_2 = 395$ (tolerable)	$f_2 = 386.1490$ (desirable)
	$b_3 = 1.8$ (highly desirable)	$f_3 = 2.7917$ (tolerable)
3	$b_1 = 4565$ (tolerable)	$f_1 = 4565$ (tolerable)
	$b_2 = 369$ (highly desirable)	$f_2 = 390.0621$ (tolerable)
	$b_3 = 2.8$ (tolerable)	$f_3 = 2.8269$ (tolerable)

Step 8. The director formulates the KKT conditions for the problem and determines the values of the Kuhn Tucker multipliers associated with the goal constraints $\overline{u}_r$, $r = 1, \ldots, k + m$

$$\overline{u}_2 = 0, \qquad \overline{u}_3 = 0, \qquad \overline{u}_4 = 0. \tag{11}$$

Step 9. According to the values $\overline{u}_r$, and by using the algorithm of Osman [20] given in Appendix B, the director determines the stability set $G(\overline{X}_j)$:

Given $\overline{u}_2 = 0$ and $g_2 = 4450$ then $b_1 \geq 4450$,

Given $\overline{u}_3 = 0$ and $g_3 = 370$ then $b_2 \geq 370$,

Given $\overline{u}_4 = 0$ and $g_4 = 2$ then $b_3 \geq 2$.

Step 10. The director uses the sets subtraction algorithm proposed by Abdel Haleem [17] given in Appendix C to obtain the reduced solvability set $\{D' - \bigcup_{p=1}^{j} G(\overline{X}_p)\}$ given in Table 18.

TABLE 20: Results' comparison.

Two-bar structure characteristics	Results IMOP approach	Results of Messac and Ismail-Yahaya [21]
Diameter (x_1)	3.80 cm	3.80 cm
Height (x_2)	64.2 cm	63.26 cm
Mass (f_1)	4.600 kg (Tolerable)	4.565 kg (Tolerable)
Normal stress (f_2)	386 N (desirable)	390 N (tolerable)
Vertical deflection (f_3)	2.7917 (tolerable)	2.826 (tolerable)

Step 11. If no target values can be chosen in $\{D' - \bigcup_{p=1}^{j} G(\overline{X}_p)\}$ stop and go to Step 13, otherwise go to Step 12. In this case, other values can be chosen in Table 18 so go to Step 12.

Step 12. Set $j = j + 1$. The working groups select other target values for the goal vector $b^j \in \{D' - \bigcup_{p=1}^{j-1} G(\overline{X}_p)\}$ and go to Step 6. Some solution results are presented in Table 19.

The second Pareto-optimal solution seems to be the best one because all the objective values are in the tolerable or desirable zone according to Table 17. This solution can be considered satisfactory and a good compromise for all the working groups.

Step 13. The director presents the Pareto-optimal solutions to the working groups to select the best solution for everyone (stop). If other solutions are necessary go to Step 12.

Table 20 shows the results obtained for the characteristics of the two-bar structure.

For the normal stress function, the result obtained with the IMOP approach is in the desirable zone while the solution obtained by Messac et Ismail-Yahaya [21] is in the tolerable zone according Table 17. This means that working group 2 is better satisfied with our solution. For the other functions, both results are in the same zone according Table 17. The difference between the results is that our solution is obtained by an interactive and collaborative process between the DM and the working groups and it is possible to generate several design scenarios (Pareto-optimal solutions) without changing the degrees of desirability. Messac and Ismail-Yahaya [21] provide information for only one design scenario (i.e., a single Pareto solution). If we want another solution we have to change the degrees of desirability. This IMOP algorithm has permitted to convergence to a solution that is acceptable for all the working groups. As shown, this procedure offers more flexibility for the director and his/her working groups.

5. Conclusion

The IMOP approach developed in this paper is an extension of the interactive nonlinear goal programming algorithm of Abdel Haleem [17]. The first contribution of the IMOP

algorithm is the ability to define a reduced set of target values that can be divided into degrees of desirability to capture the working groups' preferences. This is an important contribution because it is a challenging issue in multi-objective optimization. It also subtracts the stability set from the reduced set of target values at each iteration, thus ensuring a different Pareto-optimal solution each time. Also, the distribution of the objective functions among working groups is beneficial to consider disciplinary knowledge and experience in determining the degrees of desirability. The IMOP approach generates as many new Pareto optimal solutions (design alternatives) as needed. These solutions meet as much as possible the requirements of the working groups. Also, the application of the decision that rules for choosing the target values permits the convergence to Pareto-optimal solutions in the same desirability zone (or better) for all the objectives. The approach has been successfully applied to two problems. It is true that these problems are simple but they make the application of the IMOP approach clear. In this paper, the multi-objective optimization process is centralized. Future work is also planned to use the IMOP algorithm in the case where the multi-objective optimization process is not under the control of the director but distributed to the working groups. We will be interesting by multidisciplinary optimization. Multidisciplinary optimization is a methodology used for designing complex systems that must satisfy many constraints and that must be carried out in a decentralized environment. Multidisciplinary optimization assumes a form of collaboration between the working groups because the decision variables are under the control of several working groups. The multidisciplinary optimization approaches are Concurrent subspace optimization [22–24] Bilevel integrated system synthesis [24, 25] Collaborative optimization [22, 24, 26] and Analytical Target Cascading method [27]. We are working to combine the IMOP algorithm with one of these optimization approaches.

Appendices

A. Algorithm of Dauer and Krueger [19]

We consider the classical nonlinear goal programming problem with k goals (objective functions), subject o a set of constraints $\{M\}$:

$$(\text{NLGP}): \quad f_1(X) \le b_1$$
$$f_2(X) \le b_2$$
$$\vdots$$
$$f_k(X) \le b_k$$
$$\text{subject to} \quad M = \{X \in R^n \mid g_r(X) \le 0, \ r = 1, \ldots, m, \ X \ge 0\},$$
$$\text{(A.1)}$$

where X is the vector of decision variables $\{x_1, x_2, \ldots, x_n\}$ and b_i, $i = 1, \ldots, k$ represent aspiration levels for objectives $f_i(X)$, $i = 1, \ldots, k$. The goals are arranged according to their priority levels, that is, if $i \le j$ then goal i, $f_i(X) \le b_i$ has a higher priority level than goal $j, f_j(X) \le b_j$. It is well known that the fundamental premise of goal programming is that goal i is sought to attain without regard to the attainability of the goals with lower priority level j. This idea has been used by Dauer and Krueger to develop an algorithm for solving linear, nonlinear, and integer goal programming problems. The algorithm solves k singles objective function problems successively. The first and last problems are as follows:

Solving the attainment problem for goal 1, P_1 is:

P_1:   Minimize d_1

subject to $f_1(X) - d_1 \leq b_1$

$g_r(X) \leq 0, r = 1, \ldots, m$ (A.2)

$d_1 \geq 0, X \geq 0,$

where d_1 is the positive deviation for objective $f_1(X)$ from its goal b_1. The solution of this problem is d_1^*, which is the over attainment of goal 1.

The last attainment problem for goal k, P_k is

P_k: Minimize d_k

subject to $f_i(X) - d_i \leq b_i,   1 \leq i \leq k$

$d_i = d_i^*, 1 \leq i \leq k - 1$ (A.3)

$g_r(X) \leq 0,   r = 1, \ldots, m,$

$d_k \geq 0, X \geq 0.$

By letting $d_i = x_{n+k}, i = 1, \ldots, k$ the last attainment problem can be written in the form

$P'(G)$:   Minimize x_{n+k}

subject to $g_r(X) \leq b_r,   r = 1, \ldots, m$

$x_a = x_a^*,   n + 1 \leq a \leq n + k - 1$

$x_{n+k} \geq 0, x_i \geq 0, i = 1, \ldots, n,$

(A.4)

where $X \in R^{n+k}$ and $d_i^*, 1 \leq i \leq k - 1$ is replaced by $x_a^*, n + 1 \leq a \leq n + k - 1$. The solution of this problem denoted by $\overline{X} = (\overline{x}_1, \overline{x}_2, \ldots, \overline{x}_{n+k})$ is the optimal solution for the NLGP problem under consideration. Problem $P'(G)$ can be considered as a parametric programming problem having parameters b_r in the RHS of the constraints and can be written in the form

$P(G)$:   Minimize $f(X) \equiv x_{n+k}$

subject to $M(b) = \left\{ X \in R^{n+k} \mid g_r(X) \leq b_r, \right.$

$\left. r = 1, \ldots, k + m, X \geq 0 \right\},$

(A.5)

where b_r is any arbitrary real number, and $x_a^*, n + 1 \leq a \leq n + k - 1$ have been directly substituted in the inequality constraints of $P'(G)$. The solution of problem $P(G)$ is thus the same as the solution of the $P'(G)$ and NLGP, and the stability sets of problem $P(G)$ can be calculated.

B. The Determination of the Stability Set

Osman (1977) presented the following algorithm for the determination of the stability set of the first kind:

(1) Select an arbitrary $\overline{b} \in D$ and solve $P(G)$ to obtain $\overline{X}$ and formulate the K.K.T. conditions.

(2) Determine the values of $\overline{u}_r$ using any available algorithm.

(3) According to the values of $\overline{u}_r$, the stability set of the first kind $G(\overline{X})$ can be determined as follows:

(a) For $\overline{u}_r = 0, r = 1, \ldots, k + m, G_1(\overline{X}) = \{b \mid b_r \geq g_r(\overline{X})\}$

(b) For $\overline{u}_r > 0, r = 1, \ldots, k + m, G_2(\overline{X}) = \{b \mid b_r = g_r(\overline{X})\}$

(c) For $\overline{u}_r = 0, r \in J \subseteq \{1, \ldots, k + m\}, \overline{u}_r > 0, r \notin J,$ $G_J(\overline{X}) = \{b \mid b_r \geq g_r(\overline{X}), r \in J, b_r = g_r(\overline{X}), r \in J\}, G_3(\overline{X}) = \bigcup_{\text{possible}J} G_J(\overline{X}).$

C. The Sets Subtraction Algorithm [17]

Let $b_i, i = 1, \ldots, k$ be the elements of the universal set V in the k dimensional space. V is considered a universal set from which some other sets $S_i, i = 1, \ldots, I$ are subtracted. Let $V_s = \{\bigcup_{i=1}^{I} S_i, i = 1, \ldots, I\}$ be the subtracted set. The elements contained in the universal set and the subtracted sets are used to determine the lower and the upper bounds for each set in each dimension. These values represent the input to the sets subtraction algorithm. Each set is represented as a record containing the lower and the upper bounds for each dimension as shown in Table 2.

The function of the algorithm is to get the difference between the universal set and the subtracted set $\{V - V_s\}$. This difference is defined as those elements that are contained in the set V and not contained in V_s. The subtraction is done in steps. First, the algorithm gets the difference between V and V_s, where $V_s = S_1$, thus getting $\{V - S_1\}$. Then the set S_2 is subtracted from $\{V - S_1\}$, thus getting $\{\{V - S_1\} - S_2\}$ and $V_s = \bigcup_{i=1}^{2} S_i$, and so on.

References

[1] R. V. Tappeta and J. E. Renaud, "Interactive multiobjective optimization procedure," *AIAA Journal*, vol. 37, no. 7, pp. 881–889, 1999.

[2] V. Chankong and Y. Y. Haimes, *Multiobjective Decision Making: Theory and Methodology*, vol. 8 of *North-Holland Series in System Science and Engineering*, Elsevier, New York, NY, USA, 1983.

[3] F. Mistree, O. F. Hughes, and B. Bras, *Compromise Decision Support Problem and the Adaptive Linear Programming Algorithm, Structural Optimization: Status and Promise*, vol. 50 of *Progress in Astronautics and Aeronautics*, American Institute, Washington, DC, USA, 1993.

[4] U. Diwekar, *Introduction to Applied Optimization*, vol. 80, Kluwer Academic, Boston, Mass, USA, 2003.

[5] B. Suman and P. Kuman, "A survey of simulated annealing as a tool for single and multiobjective optimization," *Journal of the Operational Society*, vol. 57, pp. 1143–1160, 2006.

[6] A. Suppapitnarm, K. A. Seffer, and G. T. Parks, "A simulated annealing algorithm for multiobjective optimization," *Engineering Optimization*, vol. 33, no. 1, pp. 59–85, 2000.

[7] M. Reyes-Sierra and C. A. Coello Coello, "Multiobjective particle swarm optimizers: a survey of state-of-the-art," *International Journal of Computational Intelligence Research*, vol. 2, no. 3, pp. 287–308, 2006.

[8] C. A. C. Coello, D. A. VanVeldhuizen, and G. Lamonr, *Evolutionary Algorithms for Solving Multi-Objective Problems*, Kluwer Academic, Boston, Mass, USA, 2002.

[9] D. E. Salazar and C. M. Rocco, "Solving advanced multiobjective robust designs by means of multiple objective evolutionary algorithms (MOEA): a reliability application," *Reliability Engineering and System Safety*, vol. 92, no. 6, pp. 697–706, 2007.

[10] W. Gong and Z. Cai, "An improved multiobjective differential evolution based on Pareto-adaptive ε-dominance and orthogonal design," *European Journal of Operational Research*, vol. 198, no. 2, pp. 576–601, 2009.

[11] H. A. Taboada, F. Baheranwala, D. W. Coit, and N. Wattanapongsakorn, "Practical solutions for multi-objective optimization: an application to system reliability design problems," *Reliability Engineering and System Safety*, vol. 92, no. 3, pp. 314–322, 2007.

[12] A. Messac, S. M. Gupta, and B. Akbulut, "Linear physical programming: a new approach to multiple objective optimization," *Transactions on Operational Research*, vol. 8, pp. 39–59, 1996.

[13] R. V. Tappeta, J. E. Renaud, A. Messac, and G. J. Sundararaj, "Interactive physical programming: tradeoff analysis and decision making in multicriteria optimization," *AIAA Journal*, vol. 38, no. 5, pp. 917–926, 2000.

[14] V. Vassilev, S. C. Narula, and V. G. Gouljashki, "An interactive reference direction algorithm for solving multi-objective convex nonlinear integer programming problems," *International Transactions in Operational Research*, vol. 8, no. 4, pp. 367–380, 2001.

[15] R. V. Tappeta and J. E. Renaud, "Interactive multiobjective optimization design strategy for decision based design," *Journal of Mechanical Design*, vol. 123, no. 2, pp. 205–215, 2001.

[16] K. Miettinen and M. M. Mäkelä, "Synchronous approach in interactive multiobjective optimization," *European Journal of Operational Research*, vol. 170, no. 3, pp. 909–922, 2006.

[17] B. Abdel Haleem, *A study on interactive multiple criteria decision making problems [Ph.D. thesis]*, Mechanical Design and Production Departement, Faculty of Engineering, Cairo University, 1991.

[18] A. Lamghabbar, S. Yacout, and M. S. Ouali, "Concurrent optimization of the design and manufacturing stages of product development," *International Journal of Production Research*, vol. 42, no. 21, pp. 4495–4512, 2004.

[19] J. P. Dauer and R. J. Krueger, "An Iterative approach to goal programming," *Operational Research Quarterly*, vol. 28, no. 3, pp. 671–681, 1977.

[20] M. S. A. Osman, "Characterization of the stability set of the first kind with parameters in the objective function," in *Proceedings of the 10th International Conference on Mathematical Programming*, Montreal, Canada, 1979.

[21] A. Messac and A. Ismail-Yahaya, "Multiobjective robust design using physical programming," *Structural and Multidisciplinary Optimization*, vol. 23, no. 5, pp. 357–371, 2002.

[22] J. Sobieszczanski-Sobieski and R. T. Haftka, "Multidisciplinary aerospace design optimization: survey of recent developments," *Structural Optimization*, vol. 14, no. 1, pp. 1–23, 1997.

[23] R. V. Tappeta, S. Nagendra, and J. E. Renaud, "Concurrent Sub-Space optimization (CSSO) MDO Algorithm in iSIGHT: validation and testing," GE research & Development Center, 1998.

[24] I. Kroo, *Distributed Multidisciplinary Design and Collaborative Optimization*, VKI lecture series on Optimization Methods & Tools for multicriteria/multidisciplinary Design, Stanford University, 2004.

[25] J. Sobieszczanski-Sobieski, D. T. Altus, M. Philips, and R. Sandusky, "Bi-level System Synthesis (BLISS) for concurrent and distributed processing," AIAA-2002-5409, American Institute of Aeronautics and Astronautics, 2002.

[26] R. D. Braun, *Collaborative optimization: an architecture for large-scale decentralized design [Ph.D. thesis]*, Stanford University, Stanford, Calif, USA, 1996.

[27] H. M. Min, N. F. Michelena, P. Y. Papalambros, and T. Jiang, "Target cascading in optimal system design," *Journal of Mechanical Design*, vol. 125, no. 3, pp. 474–480, 2003.

Material Handling Equipment Selection Using Weighted Utility Additive Theory

Prasad Karande[1] and Shankar Chakraborty[2]

[1] Mechanical Engineering Department, Government Polytechnic Mumbai, Maharashtra 400 051, India
[2] Department of Production Engineering, Jadavpur University, Kolkata, West Bengal 700 032, India

Correspondence should be addressed to Shankar Chakraborty; s_chakraborty00@yahoo.co.in

Academic Editor: Xueqing Zhang

Better utilization of manpower, providing product flexibility, increasing productivity, decreasing lead time, reduction in handling cost, increased efficiency of material flow, and enhancement of production process are some of the most important issues influencing material handling (MH) equipment selection decision. As a wide variety of MH equipment is available today, selection of the proper equipment for a designed manufacturing system is a complicated task. Selection of suitable MH equipment for a typical handling environment is found to be a multicriteria decision-making (MCDM) problem. As the selection process is found to be unstructured, characterized by domain dependent knowledge, there is a need to apply an efficient MCDM tool to select the most suitable MH equipment for the given application. This paper applies weighted utility additive (WUTA) method to solve an MH equipment selection problem. The ranking obtained using the WUTA method is compared with that derived by the past researchers which proves its potentiality, applicability, and accuracy to solve complex decision-making problems.

1. Introduction

Material handling (MH) is an activity that uses the right method to provide the right amount of the right material at the right place, at the right time, in the right sequence, in the right position, and at the right cost [1]. An MH system is responsible for transporting materials between workstations with minimum obstruction and joins all the workstations and workshops in a manufacturing system by acting as a basic integrator. The MH task accounts for 30–75% of the total cost of a product, and efficient MH can be responsible for reducing the manufacturing system operations cost by 15–30% [2]. These figures justify the importance of MH cost as an element in improving the cost structure of a manufacturing organization. An efficient MH system greatly improves the competitiveness of a product through the reduction of handling cost, enhances the production process, increases production and system flexibility, increases efficiency of material flow, improves facility utilization, provides effective utilization of manpower, and decreases lead time [3].

The functions performed by MH equipment can be classified into four broad categories, that is, (a) transport, (b) positioning, (c) unit formation, and (d) storage. Usually, all the MH functions are composed of one or more combinations of these four primary functions. Equipment in transport category simply moves materials from one point to another, which includes conveyors, industrial trucks, cranes, and so forth. Unlike transport equipment, positioning equipment is usually employed at workstations to aid machining operations. Robots, index tables, rotary tables, and so forth are the examples of this type of equipment. Unit formation equipment is used for holding or carrying materials in standardized unit load forms for transport and storage and generally includes bins, pallets, skids, and containers. Storage equipment is used for holding or buffering materials over a period of time. Typical examples that perform this function are AS/RS, pallet racks, and shelves.

The MH equipment selection is an important function in the design of an MH system and, thus, a crucial step for facility planning. The determination of an MH system involves both the selection of suitable MH equipment and the assignment of MH operations to each individual piece of equipment. As a wide variety of equipment is available today, each having distinct characteristics and cost that distinguish from others, selection of the proper equipment for a designed manufacturing system is a very complicated task and is often influenced by the ongoing development of new technology, practices, and equipment. While choosing the best MH equipment, the successful solution would likely involve matching the best solution with the existing or contemplated physical facilities and environment. The major factors contributing to the complexity of MH selection process are constraints imposed by the facility and materials, multiple conflicting design criteria, uncertainty in the operational environment, and the wide variety of equipment types and models available.

When implementing a new MH equipment, the decision makers are faced with the following issues, that is, (a) selection of an MH equipment that would give the desired benefits to the manufacturing organization with due consideration to its objectives and operating characteristics, (b) financial justification of the investment, and (c) development of a plan to ensure that the set objectives are met when the selected MH equipment is implemented and evaluated. For these reasons, the decision makers have to consider various quantitative (load capacity, energy consumption, reliability, cost, etc.) and qualitative (flexibility, performance, environmental hazard, safety, load shape, load type, etc.) criteria. On the other hand, some of the selection criteria are beneficial (higher values are preferred) and some are nonbeneficial (lower values are desired). Therefore, MH equipment selection can be viewed as a multicriteria decision-making (MCDM) problem in the presence of many conflicting criteria.

As the MH equipment selection is a difficult and knowledge intensive process, various mathematical tools can be effectively applied to solve this problem. However, it is always observed that the evaluation criteria involved in MH equipment selection problems have contradictory effects on the performance of the alternatives, are versatile in nature, and often expressed in different units with varying ranges. Therefore, a strong and unprejudiced mathematical model is essential for selection of the most appropriate MH equipment for a given industrial application. The weighted utility additive (WUTA) method having a sound mathematical background, ability to incorporate preferences for the selection criteria, and competency to handle mixed (cardinal and ordinal) data is a perfect choice to rank and select the best suited MH equipment. In this method, the reference ranking of the alternatives is formulated, and the indifference as well as preference relations between the alternatives are utilized for ranking purpose, deriving almost accurate results. It enhances the strengths of the conventional utility additive (UTA) method by incorporating criteria weights, which are usually observed as essential for solving the decision-making problems. Thus, the aim of this paper is set to show the viability of the WUTA method to solve decision-making problems with any number of selection criteria and candidate alternatives, with special emphasis on MH equipment selection. It is a variant of UTA family of models. The effectiveness

and solution accuracy of any MCDM method can only be validated by comparing the derived rankings with those obtained by the earlier researchers. Here, the rank orderings of the alternatives derived by the past researchers act like some benchmarks. The cited example, already solved using different MCDM methods for ranking of MH equipment alternatives, thus provides sufficient ground for comparison of the performance of the proposed WUTA method.

2. Literature Review

Since 1990s, research concentrating on the selection and assignment of MH equipment has been carried out, and significant achievements have been attained. Chakraborty and Banik [4] applied analytic hierarchy process (AHP) for selecting the best MH equipment under a specific handling environment. The relative importance of each criteria and subcriteria was measured using pair-wise comparison matrices, and the overall rankings of all the alternative equipment were then determined. To identify the most critical and robust criteria in the MH equipment selection process, sensitivity analysis was also performed. Sujono and Lashkari [5] proposed a method for simultaneously determining the operation allocation and MH system selection in a flexible manufacturing environment with multiple performance objectives. A 0-1 integer programming model was developed to select machines, assign operations of part types to the selected machines, allocate MH equipment to transport the parts from machine to machine, and as to handle the part at a given machine. Onut et al. [6] proposed an integrated fuzzy analytic network process (F-ANP) and fuzzy technique for order performance by similarity to ideal solution (F-TOPSIS) methodology for evaluating and selecting the most suitable MH equipment types for a manufacturing organization. Komljenovic and Kecojevic [7] applied coefficient of technical level and AHP methods for selection of rail-mounted boom type bucket wheel reclaimers and stacker-reclaimers as used for material handling at the stockyards. Tuzkaya et al. [8] suggested an integrated F-ANP and fuzzy preference ranking organization method for enrichment evaluation (F-PROMETHEE) approach for solving the MH equipment selection problems. Sawant et al. [9] applied preference selection index (PSI) method to choose automated guided vehicle (AGV) in a given manufacturing environment. An AGV selection index was proposed to evaluate and rank the considered alternatives. Maniya and Bhatt [10] used AHP to assign the relative importance between different AGV selection criteria and then applied modified grey relational analysis (M-GRA) method to determine the corresponding index values for AGV selection.

On the other hand, some researchers have attempted to develop knowledge-based systems for proper selection of equipment used for varying handling tasks. Welgama and Gibson [11] proposed a methodology for automating the selection of an MH system while combining the knowledge-base and optimization approaches. Chu et al. [12] developed a microcomputer-based system called "ADVISOR" to help user to design, select, and evaluate the proper MH equipment for a production shop. Chan et al. [13] proposed an intelligent MH equipment selection advisor (MHESA), composed of a database to store equipment types with their specifications, knowledge-based expert system for assisting MH equipment selection, and an AHP model to choose the most appropriate MH equipment. Yaman [14] described a knowledge-based approach for MH equipment selection and re-design of equipment in a given facility layout. Fonseca et al. [15] developed a prototype expert system for industrial conveyor selection which would provide the user with a list of conveyor solutions for their MH needs along with a list of suppliers for the suggested conveyors. Conveyor types were selected on the basis of a suitability score, which was a measure of the fulfillment of MH requirements by the characteristics of the conveyor. Kulak [2] developed a fuzzy multiattribute MH equipment selection system consisting of a database, a rule-based system, and multiattribute decision-making modules. A fuzzy information axiom approach was also introduced and used in the selection of MH equipment in a real case. Cho and Edbelu [16] developed a web-based system, called as "DESIGNER" for the design of integrated MH systems in a manufacturing environment, which could model and automate the MH system design process, including the selection of MH equipment. Mirhosseyni and Webb [17] presented a hybrid method for selection and assignment of the most appropriate MH equipment. At first, the system would select the most appropriate MH equipment type for every MH operation in a given application using a fuzzy knowledge-based expert system, and in the second phase, a genetic algorithm would search throughout the feasible solution space, constituting of all possible combinations of the feasible equipment specified in the previous phase, in order to discover the optimal solution. The main disadvantage of the knowledge-based expert systems is that, in these approaches, as the set rules are static in nature and domain-specific, it is very difficult for the decision makers to know how the decision for the best MH equipment has been arrived.

3. The WUTA Method

The WUTA method which is an extension of the UTA approach, proposed by Jacquet-Lagreze and Siskos [18], aims at inferring one or more additive value functions from a given ranking on reference set, A_R. In MCDM problems, the decision makers usually consider a set of alternatives, called A, which is valued by a family of criteria, $g = (g_1, g_2, \ldots, g_n)$. A classical operational attitude of assessing a model of overall preference of the decision makers leads to the aggregation of all the criteria into a unique criterion, called a utility function $(U(g))$.

$$U(g) = U(g_1, g_2, \ldots, g_n). \qquad (1)$$

Let P be the strict preference relation, and let I be the indifference relation, and if $g(a) = [g_1(a), g_2(a), \ldots, g_n(a)]$ and $g(b) = [g_1(b), g_2(b), \ldots, g_n(b)]$ are the multicriteria evaluations of the alternatives "a" and "b", respectively, then

the following properties generally hold for the utility function $(U(g))$.

$$U[g(a)] > U[g(b)] \iff aPb,$$
$$U[g(a)] = U[g(b)] \iff aIb. \qquad (2)$$

The relation $R = P \cup I$ is a weak order.

The utility function is additive if it is of the following form.

$$X(g) = \sum_{i=1}^{n} x_i(g_i), \qquad (3)$$

where $x_i(g_i)$ is the marginal utility of the attribute, g_i for the given alternative. When different weight values (relative importance) are assigned to the attributes, the weighted utility function can be expressed as follows:

$$U(g) = \sum_{i=1}^{n} u_i(g_i), \qquad (4)$$

where $u_i(g_i) = w_i x_i(g_i)$, and w_i is the weight for ith attribute.

Again, for alternative "a," (4) can be written as

$$U[g(a)] = \sum_{i=1}^{n} u_i[g_i(a)]. \qquad (5)$$

Now, g_i^+ and g_i^-, respectively, denote the most and the least preferred value of ith attribute. The most common normalization constraints using the additive form of (4) are as follows:

$$\sum_{i=1}^{n} u_i(g_i^+) = 1, \qquad (6)$$
$$u_i(g_i^-) = 0 \quad \forall i.$$

On the basis of the additive model, as shown in (5) and taking into account the preference conditions, the value of each alternative, $a \in A_R$ can be written as follows:

$$U'[g(a)] = \sum_{i=1}^{n} u_i[g_i(a)] + \sigma(a) \quad \forall a \in A_R, \qquad (7)$$

where $\sigma(a) \geq 0$ is a potential error relative to the utility.

To estimate the marginal value functions in a piecewise linear approach, a linear interpolation method is proposed [18–20]. For each attribute, the interval $[g_i^-, g_i^+]$ is divided into $(\alpha_i - 1)$ equal segments. The end points, (g_i^j) are given as follows:

$$g_i^j = g_i^- + \frac{j-1}{\alpha_i - 1}(g_i^+ - g_i^-) \quad \forall j = 1, 2, \ldots, \alpha_i. \qquad (8)$$

Now, the variable to estimate is $u_i(g_i^j)$. The marginal utility of an alternative is approximated by a linear interpolation method, and thus, for $g_i(a) \in [g_i^j, g_i^{j+1}]$,

$$u_i[g_i(a)] = u_i\left(g_i^j\right)$$
$$+ \frac{g_i(a) - g_i^j}{g_i^{j+1} - g_i^j}\left[u_i\left(g_i^{j+1}\right) - u_i\left(g_i^j\right)\right]. \qquad (9)$$

The set of preference alternatives, $A_R = \{a_1, a_2, \ldots, a_m\}$, is also rearranged in such a way that a_1 is the head of the ranking (best) and a_m is its tail (worst). Since, the ranking has the form of a weak order, R for each pair of alternatives (a_k, a_{k+1}), it holds either $a_k \succ a_{k+1}$ or $a_k \approx a_{k+1}$. Thus if

$$\Delta\left(a_k, a_{k+1}\right) = U'\left[g\left(a_k\right)\right] - U'\left[g\left(a_{k+1}\right)\right] \quad (10)$$

then one of the following relationships holds:

$$\begin{aligned} \Delta\left(a_k, a_{k+1}\right) \geq \delta &\quad \text{if } a_k \succ a_{k+1} \text{ (preference)}, \\ \Delta\left(a_k, a_{k+1}\right) = 0 &\quad \text{if } a_k \approx a_{k+1} \text{ (indifference)}, \end{aligned} \quad (11)$$

where δ is a small positive number to discriminate significantly two successive equivalence classes of R.

The marginal value functions are finally estimated using the following linear program (LP), in which the objective function depends on $\sigma(a)$, indicating the amount of total deviation.

$$[\min] \quad F = \sum_{a \in A_R} \sigma(a)$$

subject to

$$\Delta\left(a_k, a_{k+1}\right) \geq \delta \quad \text{if } a_k \succ a_{k+1} \; \forall k,$$

$$\Delta\left(a_k, a_{k+1}\right) = 0 \quad \text{if } a_k \approx a_{k+1} \; \forall k,$$

$$u_i\left(g_i^{j+1}\right) - u_i\left(g_i^j\right) \geq 0 \quad \text{for } i = 1, 2, \ldots, n, \; j = 1, 2, \ldots, \alpha_i,$$

$$\sum_{i=1}^n u_i\left(g_i^+\right) = 1,$$

$$u_i\left(g_i^-\right) = 0; \quad u_i\left(g_i^j\right) \geq 0; \quad \sigma(a) \geq 0; \quad \forall a \in A_R; \; \forall i, j. \quad (12)$$

This LP model is solved to obtain the marginal utility values. Then, the utility value $(U[g(a)])$ for each alternative is calculated. The higher the $U[g(a)]$ value, the better the alternative.

It is observed that the WUTA method generally copes well with noisy or inconsistent data [21], it is least sensitive to changes in preferences for the considered criteria, and it is not as time consuming, redundant, and boring as the other MCDM methods where the decision makers have to define certain preference functions to evaluate the superiority of one alternative over the other. This method is based on two fundamental concessions; that is, (a) it does not allow any situation of incomparability between two alternatives, and (b) it addresses the evaluation (assessment) problem in a synthesizing, exhaustive, and definite way. It has also several interesting features, like it makes possible estimation of a nonlinear additive function which is obtained by the use of a linear program that provides convenient piecewise linear approximation of the function, and the only information required from the decision makers is the global stated preferences between different alternatives of the reference set. It also perfectly fits in those situations where there are

difficulties in directly obtaining from the users the values of the preference model.

Generally, the data available for various criteria in a decision-making problem are expressed in different dimensional units with varying ranges. In order to eradicate these effects, it is required to normalize the criteria values within a range of 0 to 1. On the contrary, if criteria values are not normalized, those criteria with higher weights will be more prone to affect the final ranking of the alternatives. In case of nonnormalized data, the change in unit for a particular criterion will directly affect the values of the weighted decision matrix, which will ultimately have an effect on the ranking of the alternatives. In order to consider the effects of higher and lower preferences of beneficial and non-beneficial criteria in a decision-making problem, they have to be treated separately, which is taken into care in the normalization procedure. In the WUTA method, if the criteria values are not normalized, all the criteria will be treated as beneficial because in this method, the final ranking of the alternatives is based on the reference ranking, which is obtained by adding the weighted normalized criteria values.

4. Illustrative Example

In order to illustrate and validate the applicability of the WUTA method for solving MH equipment selection problems, a real time example considering the selection of a conveyor [2] is cited here. The final ranking of the alternative MH equipment as obtained using the WUTA method is also compared with that derived by the past researchers.

This MH equipment selection problem is aimed to determine the most appropriate conveyor among the alternatives of the same type. The related objective and subjective data of the attributes are given in Table 1 [2]. The attributes considered are fixed cost per hour (FC), variable cost per hour (VC), speed of conveyor (SC), item width (IW), item weight (W), and flexibility (F). Among those six attributes, flexibility was defined subjectively. Rao [22] converted the linguistic terms for flexibility criterion into corresponding fuzzy scores, and appropriate objective values were subsequently assigned. The conveyor should have low fixed and variable costs, higher speed, ability to handle large item widths and weights, and have higher flexibility. FC and VC are nonbeneficial attributes (where lower values are desired), and the remaining four attributes are considered as beneficial (where higher values are preferred).

Rao [22] considered equal weights for all the six criteria and obtained the best and the worst choices as conveyor 3 and conveyor 1, respectively, while solving this problem using simple additive weighting (SAW), weighted product method (WPM), AHP, graph theory and matrix approach (GTMA), TOPSIS and modified TOPSIS methods. Giving equal weights to the considered criteria may sometimes lead to wrong and biased decisions. Hence, the criteria weights are recalculated here using AHP method, as shown in Table 2, and are used for subsequent WUTA method-based analysis.

TABLE 1: Quantitative data for the conveyor selection problem [2].

Conveyor	Fixed cost per hour (FC)	Variable cost per hour (VC)	Speed of conveyor (m/min) (SC)	Item width (cm) (IW)	Item weight (kg) (W)	Flexibility (F)
A_1	2	0.45	12	15	10	Very good (0.745)
A_2	2.3	0.44	13	20	10	Excellent (0.955)
A_3	2.25	0.45	11	30	20	Excellent (0.955)
A_4	2.4	0.46	10	25	15	Very good (0.745)

TABLE 2: Criteria weights for conveyor selection problem.

Attributes	FC	VC	SC	IW	W	F
Weights	0.1049	0.1260	0.1260	0.2402	0.2245	0.1782

TABLE 3: Normalized decision matrix for conveyor selection problem.

Conveyor	FC	VC	SC	IW	W	F
A_1	1.0000	0.9778	0.9231	0.5000	0.5000	0.7801
A_2	0.8696	1.0000	1.0000	0.6667	0.5000	1.0000
A_3	0.8889	0.9778	0.8462	1.0000	1.0000	1.0000
A_4	0.8333	0.9565	0.7692	0.8333	0.7500	0.7801

For solving this problem using the WUTA method, the criteria values of Table 1 are first normalized using the following equations.

For beneficial attribute:

$$y_{ij} = \frac{x_{ij}}{\max\left(x_{ij}\right)} \quad \text{for } i = 1, 2, \ldots, m; \; j = 1, 2, \ldots, n. \quad (13)$$

For nonbeneficial attribute:

$$y_{ij} = \frac{\min\left(x_{ij}\right)}{x_{ij}}, \quad (14)$$

where x_{ij} is the performance of ith alternative with respect to jth criterion, and y_{ij} is the normalized value of x_{ij}.

Then, the weighted normalized criteria values (r_{ij}) are obtained using the following expression:

$$r_{ij} = w_i y_{ij}. \quad (15)$$

The normalized and weighted normalized decision matrices are given in Tables 3 and 4, respectively.

From the $\sum r_{ij}$ values, the reference sequence (A_R) for the alternative conveyors is observed as $A_3 - A_2 - A_1 - A_4$. Now, the range $[g_i^-, g_i^+]$ for each conveyor selection criterion is divided into equal intervals. The number of intervals and the interval difference for each criterion, as calculated using (8), with their corresponding g_i^- and g_i^+ values are given in Table 5. The number of intervals (α_i) is selected in such a way that the interval for each criterion is almost equal. However, a large number of intervals may cause an increase in the computational complexity as well as time. In order to minimize the computation time, a minimum possible number of intervals is selected here, so that the final results

are not affected. For example, in case of "VC" criterion, the value for the number of intervals is chosen as 2 because it has the lowest range between $[g_i^-, g_i^+]$. As the criteria "FC", "SC", and "F" have ranges close to that for "VC" criterion, the same number of intervals is also selected for those criteria. On the other hand, for "IW" and "W" criteria, the number of intervals is selected as 3 for their wider ranges.

Now, we have the following set of equations.

For attribute FC:

$$u_1 (0.0874) = u_{11} = 0,$$

$$u_1 (0.0874 + 0.0087) = u_1 (0.0962) = u_{12}, \quad (16)$$

$$u_1 (0.0962 + 0.0087) = u_1 (0.1049) = u_{13}.$$

For attribute VC:

$$u_2 (0.1205) = u_{21} = 0,$$

$$u_2 (0.1205 + 0.0027) = u_2 (0.1233) = u_{22}, \quad (17)$$

$$u_2 (0.1233 + 0.0027) = u_2 (0.1260) = u_{23}.$$

For attribute SC:

$$u_3 (0.0969) = u_{31} = 0,$$

$$u_3 (0.0969 + 0.0145) = u_3 (0.1115) = u_{32}, \quad (18)$$

$$u_3 (0.1115 + 0.0145) = u_3 (0.1260) = u_{33}.$$

For attribute IW:

$$u_4 (0.1201) = u_{41} = 0,$$

$$u_4 (0.1201 + 0.0400) = u_4 (0.1602) = u_{42},$$

$$u_4 (0.1602 + 0.0400) = u_4 (0.2002) = u_{43}, \quad (19)$$

$$u_4 (0.2002 + 0.0400) = u_4 (0.2402) = u_{44}.$$

For attribute W:

$$u_5 (0.1123) = u_{51} = 0,$$

$$u_5 (0.1123 + 0.0374) = u_5 (0.1497) = u_{52},$$

$$u_5 (0.1497 + 0.0374) = u_5 (0.1871) = u_{53}, \quad (20)$$

$$u_5 (0.1871 + 0.0374) = u_5 (0.2245) = u_{54}.$$

For attribute F:

$$u_6 (0.1390) = u_{61} = 0,$$

$$u_6 (0.1390 + 0.0196) = u_6 (0.1586) = u_{62}, \quad (21)$$

$$u_6 (0.1586 + 0.0196) = u_6 (0.1782) = u_{63}.$$

TABLE 4: Weighted normalized decision matrix.

Conveyor	FC	VC	SC	IW	W	F	$\sum r_{ij}$	Rank
A_1	0.1049	0.1232	0.1163	0.1201	0.1123	0.1390	0.7159	3
A_2	0.0912	0.1260	0.1260	0.1602	0.1123	0.1782	0.7940	2
A_3	0.0933	0.1232	0.1066	0.2402	0.2245	0.1782	0.9662	1
A_4	0.0874	0.1205	0.0969	0.2002	0.1684	0.1390	0.6735	4

TABLE 5: Most and least preferred values with interval difference for each criterion.

Attribute	FC	VC	SC	IW	W	F
g_i^+	0.1049	0.1260	0.1260	0.2402	0.2245	0.1782
g_i^-	0.0874	0.1205	0.0969	0.1201	0.1123	0.1390
$(g_i^+ - g_i^-)$	0.0175	0.0055	0.0291	0.1201	0.1123	0.0392
Intervals (α_i)	2	2	2	3	3	2
$[(g_i^+ - g_i^-)/\alpha_i]$	0.0087	0.0027	0.0145	0.0400	0.0374	0.0196

The utility values for the alternative conveyors are now calculated as below:

$$U\left[g\left(A_1\right)\right] = u_1\left(0.1049\right) + u_2\left(0.1232\right) + u_3\left(0.1163\right)$$
$$+ u_4\left(0.1201\right) + u_5\left(0.1123\right) + u_6\left(0.1390\right),$$

$$U\left[g\left(A_2\right)\right] = u_1\left(0.0912\right) + u_2\left(0.1260\right) + u_3\left(0.1260\right)$$
$$+ u_4\left(0.1602\right) + u_5\left(0.1123\right) + u_6\left(0.1782\right),$$

$$U\left[g\left(A_3\right)\right] = u_1\left(0.0933\right) + u_2\left(0.1232\right) + u_3\left(0.1066\right)$$
$$+ u_4\left(0.2402\right) + u_5\left(0.2245\right) + u_6\left(0.1782\right),$$

$$U\left[g\left(A_4\right)\right] = u_1\left(0.0874\right) + u_2\left(0.1205\right) + u_3\left(0.0969\right)$$
$$+ u_4\left(0.2002\right) + u_5\left(0.1684\right) + u_6\left(0.1390\right). \tag{22}$$

Now, after solving the above set of equations using (9), the following results are derived.

For alternative A_1:

$$u_1\left(0.1049\right) = u_{13}, \qquad u_2\left(0.1232\right) = u_{22},$$

$$u_3\left(0.1163\right) = u_{32} + \left(\frac{0.1163 - 0.1115}{0.0145}\right)\left(u_{33} - u_{32}\right),$$

$$u_4\left(0.1201\right) = u_{41} = 0, \qquad u_5\left(0.1123\right) = u_{51} = 0,$$

$$u_6\left(0.1390\right) = u_{61} = 0. \tag{23}$$

For alternative A_2:

$$u_1\left(0.0912\right) = u_{11} + \left(\frac{0.0912 - 0.0874}{0.0087}\right)\left(u_{12} - u_{11}\right),$$

$$u_2\left(0.1260\right) = u_{23}, \qquad u_3\left(0.1260\right) = u_{33},$$

$$u_4\left(0.1602\right) = u_{42},$$

$$u_5\left(0.1123\right) = u_{51} = 0,$$

$$u_6\left(0.1782\right) = u_{63}. \tag{24}$$

For alternative A_3:

$$u_1\left(0.0933\right) = u_{11} + \left(\frac{0.0933 - 0.0874}{0.0087}\right)\left(u_{12} - u_{11}\right),$$

$$u_2\left(0.1232\right) = u_{22},$$

$$u_3\left(0.1066\right) = u_{31} + \left(\frac{0.1066 - 0.0969}{0.0145}\right)\left(u_{32} - u_{31}\right),$$

$$u_4\left(0.2402\right) = u_{44}, \quad u_5\left(0.2245\right) = u_{54}, \quad u_6\left(0.1782\right) = u_{63}. \tag{25}$$

For alternative A_4:

$$u_1\left(0.0874\right) = u_{11} = 0, \quad u_2\left(0.1205\right) = u_{21} = 0,$$

$$u_3\left(0.0969\right) = u_{31} = 0, \quad u_4\left(0.2002\right) = u_{43},$$

$$u_5\left(0.1684\right) = u_{52} + \left(\frac{0.1684 - 0.1497}{0.0374}\right)\left(u_{53} - u_{52}\right),$$

$$u_6\left(0.1390\right) = u_{61} = 0. \tag{26}$$

Now, the utility values for all the four alternatives are calculated using (7) and are shown below:

$$U'\left[g\left(A_1\right)\right] = u_{13} + u_{22} + 0.6687u_{32} + 0.3313u_{33} + \sigma_1,$$

$$U'\left[g\left(A_2\right)\right] = 0.4291u_{12} + u_{23} + u_{33} + u_{42} + u_{63} + \sigma_2,$$

$$U'\left[g\left(A_3\right)\right] = 0.6692u_{12} + u_{22} + 0.6643u_{32} + u_{44} + u_{54} + u_{63} + \sigma_3,$$

$$U'\left[g\left(A_4\right)\right] = u_{43} + 0.5003u_{52} + 0.4997u_{53} + \sigma_4. \tag{27}$$

The mathematical model for the problem is formulated as below:

$$\text{Min} \qquad (F) = \sigma_1 + \sigma_2 + \sigma_3 + \sigma_4$$

$$\text{subject to} \quad \Delta\left(3,2\right) \geq \delta, \ \Delta\left(2,1\right) \geq \delta, \ \Delta\left(1,4\right) \geq \delta,$$

$$u_{13} - u_{12} \geq 0, \qquad u_{23} - u_{22} \geq 0,$$

$$u_{33} - u_{32} \geq 0, \qquad u_{44} - u_{43} \geq 0,$$

$$u_{43} - u_{42} \geq 0, \qquad u_{54} - u_{53} \geq 0,$$

$$u_{53} - u_{52} \geq 0, \qquad u_{63} - u_{62} \geq 0,$$

$$u_{13} + u_{23} + u_{33} + u_{44} + u_{54} + u_{63} = 1,$$

$$u_{12}, u_{13}, u_{22}, u_{23}, u_{32}, u_{33}, u_{42}, u_{43},$$

$$u_{44}, u_{52}, u_{53}, u_{54}, u_{62}, u_{63}, \sigma_1, \sigma_2, \sigma_3, \sigma_4 \geq 0. \tag{28}$$

TABLE 6: Sensitivity analysis for conveyor selection problem.

Conveyor	Decreased weights of IW and W by		Basic solution	Increased weights of IW and W by	
	25%	10%		10%	25%
A_1	3	3	3	3	4
A_2	2	2	2	2	2
A_3	1	1	1	1	1
A_4	4	4	4	4	3

TABLE 7: Rankings of conveyor alternatives using different MCDM methods.

Conveyor	FUMAHES	GTMA	VIKOR	PROMETHEE	ELECTRE	WUTA
A_1	2	4	4	2	4	3
A_2	1	2	2	3	3	2
A_3	3	1	1	1	1	1
A_4	4	3	3	4	2	4

Now considering the value of δ = 0.0001, the final mathematical formulation for the given conveyor selection problem is written as follows:

Minimize $(F) = \sigma_1 + \sigma_2 + \sigma_3 + \sigma_4$

subject to

$0.2401u_{12} + u_{22} - u_{23} + 0.6643u_{32} - u_{33}$

$\quad - u_{42} + u_{44} + u_{54} + \sigma_3 - \sigma_2 \geq 0.0001,$

$0.4291u_{12} - u_{13} - u_{22} + u_{23} - 0.6687u_{32}$

$\quad + 0.6687u_{33} + u_{42} + u_{63} + \sigma_2 - \sigma_1 \geq 0.0001,$

$u_{13} + u_{22} + 0.6687u_{32} + 0.3313u_{33} - u_{43} - 0.5003u_{52}$

$\quad - 0.4997u_{53} + \sigma_1 - \sigma_4 \geq 0.0001,$ (29)

$u_{13} - u_{12} \geq 0, \quad u_{23} - u_{22} \geq 0, \quad u_{33} - u_{32} \geq 0,$

$u_{44} - u_{43} \geq 0, \quad u_{43} - u_{42} \geq 0,$

$u_{54} - u_{53} \geq 0, \quad u_{53} - u_{52} \geq 0, \quad u_{63} - u_{62} \geq 0,$

$u_{13} + u_{23} + u_{33} + u_{44} + u_{54} + u_{63} = 1,$

$u_{12}, u_{13}, u_{22}, u_{23}, u_{32}, u_{33}, u_{42}, u_{43}, u_{44}, u_{52},$

$\quad u_{53}, u_{54}, u_{62}, u_{63}, \sigma_1, \sigma_2, \sigma_3, \sigma_4 \geq 0.$

This LP problem is solved using LINDO software which gives the results as $F = 0, u_{12} = 0, u_{13} = 0, u_{22} = 0, u_{23} = 0, u_{32} = 0,$ $u_{33} = 0.0003, u_{42} = 0, u_{43} = 0, u_{44} = 0.9997, u_{52} = 0, u_{53} = 0,$ $u_{54} = 0, u_{62} = 0,$ and $u_{63} = 0.$

Now, applying (4), the utility values of the alternative conveyors are calculated as follows:

$$U[g(A_1)] = 0.0001, \quad U[g(A_2)] = 0.0003,$$
$$U[g(A_3)] = 0.9997, \quad U[g(A_4)] = 0.0000.$$ (30)

As the optimal solution of the objective function in the LP problem results in a zero value, the utility functions

are perfectly compatible with the reference sequence. After arranging these utility values in descending order, the final ranking of the four conveyors is $A_3 - A_2 - A_1 - A_4$, suggesting that A_3 is the best conveyor among the considered alternatives, followed by A_2. A_4 is the worst choice. Rao [22] also obtained A_3 as the best choice and a total ranking for the conveyors as $A_3 - A_2 - A_4 - A_1$. The Spearman's rank correlation coefficient (r_s) between these two rank orderings is calculated as 0.8, which represents the capability of the WUTA method for solving this conveyor selection problem.

Often the criteria weights in MCDM problems are challenged because of assortment and uncertainty involved in their calculations. Therefore, in order to deal with this issue, a sensitivity analysis is performed to study the impact of different criteria weights on the final ranking of the alternative conveyors. In this example, the weights for "IW" and "W" criteria are maximum, and hence, they are selected for increasing and decreasing their values in steps on either side. The weights of these two criteria are subsequently changed by −25%, −10%, +10%, and +25% in steps, and the weights of the remaining criteria are equally adjusted, so that the sum of all the criteria weights must add up to one. The results of this sensitivity analysis are exhibited in Table 6. It is observed from this table that changes in weights of the two most important criteria by +10%, +25%, and −10% do not show any variation in the final rankings of the alternative conveyors, but when the weights of the two selected criteria are changed by −25%, the positions of the last two alternative conveyors are just reversed. In all the cases, the best chosen conveyor remains unaffected. This result proves the robustness of the WUTA method for solving such types of MCDM problems.

Table 7 compares the rankings of the alternative conveyors as obtained by WUTA and other popular MCDM methods, like VIKOR (VIse Kriterijumska Optimizacija kompromisno Resenje), PROMETHEE, and ELECTRE (ELimination and Et Choice Translating Reality). These derived rankings are also compared with those obtained by Kulak [2] and Rao [22]. Kulak [2] developed a decision support system (FUMAHES: fuzzy multiattribute material handling equipment selection), and Rao [22] mainly applied GTMA method for solving this problem. It is observed that in most of the MCDM methods, the best and the least preferred alternative conveyors remain unchanged. Even though, the results obtained using different MCDM methods are quite similar, the WUTA method requires less computational time, as the LP-based mathematical formulations can be quickly solved employing LINDO software tool. A sound, systematic and logical base for this method provides almost robust rankings for the candidate alternatives as compared to other MCDM methods, which can be judged through the results of sensitivity analysis. In this method, the decision makers need not to perform tedious and repetitive pair-wise comparisons between the performance of different alternatives with respect to each criterion, thus saving computational time. In addition, the results obtained from this method are completely free from inconsistent and biased judgments of the decision makers. Thus, it may always be expected that this robust method would provide accurate ranking preorders for the alternatives, having minimally affected by the change in criteria weights and decision makers' perceptions.

5. Conclusions

The problem of selecting the most appropriate MH equipment for a specific task is a strategic issue, greatly influencing the performance and profitability of the manufacturing organizations. This paper presents the use of WUTA method for solving an MH equipment selection problem. It is observed that the WUTA method is a viable tool in solving the MH equipment selection problems. It allows the decision makers to rank the candidate alternatives more efficiently and accurately. As this method has a strong and sound mathematical foundation, it is capable of deriving more accurate ranking of the considered alternatives. It can not only help in just selecting the best MH equipment, but it can also be applied for any decision-making problem with any number of selection criteria and feasible alternatives while offering a more objective and straightforward approach. It is also observed that this method is quite robust against changes in the criteria weights.

References

[1] J. A. Tompkins, *Facilities Planning*, John Wiley and Sons, New York, NY, USA, 2010.

[2] O. Kulak, "A decision support system for fuzzy multi-attribute selection of material handling equipments," *Expert Systems with Applications*, vol. 29, no. 2, pp. 310–319, 2005.

[3] B. M. Beamon, "Performance, reliability, and performability of material handling systems," *International Journal of Production Research*, vol. 36, no. 2, pp. 377–393, 1998.

[4] S. Chakraborty and D. Banik, "Design of a material handling equipment selection model using analytic hierarchy process," *International Journal of Advanced Manufacturing Technology*, vol. 28, no. 11-12, pp. 1237–1245, 2006.

[5] S. Sujono and R. S. Lashkari, "A multi-objective model of operation allocation and material handling system selection in FMS design," *International Journal of Production Economics*, vol. 105, no. 1, pp. 116–133, 2007.

[6] S. Onut, S. S. Kara, and S. Mert, "Selecting the suitable material handling equipment in the presence of vagueness," *International Journal of Advanced Manufacturing Technology*, vol. 44, no. 7-8, pp. 818–828, 2009.

[7] D. Komljenovic and V. Kecojevic, "Multi-attribute selection method for materials handling equipment," *International Journal of Industrial and Systems Engineering*, vol. 4, no. 2, pp. 151–173, 2009.

[8] G. Tuzkaya, B. Gülsün, C. Kahraman, and D. Özgen, "An integrated fuzzy multi-criteria decision making methodology for material handling equipment selection problem and an application," *Expert Systems with Applications*, vol. 37, no. 4, pp. 2853–2863, 2010.

[9] V. B. Sawant, S. S. Mohite, and R. Patil, "A decision-making methodology for automated guided vehicle selection problem using a preference selection index method," *Communications in Computer and Information Science*, vol. 145, pp. 176–181, 2011.

[10] K. D. Maniya and M. G. Bhatt, "A multi-attribute selection of automated guided vehicle using the AHP/M-GRA technique," *International Journal of Production Research*, vol. 49, pp. 6107–6124, 2011.

[11] P. S. Welgama and P. R. Gibson, "A hybrid knowledge based/optimization system for automated selection of materials handling system," *Computers and Industrial Engineering*, vol. 28, no. 2, pp. 205–217, 1995.

[12] H. K. Chu, P. J. Egbelu, and C. T. Wu, "ADVISOR: a computer-aided material handling equipment selection system," *International Journal of Production Research*, vol. 33, no. 12, pp. 3311–3329, 1995.

[13] F. T. S. Chan, R. W. L. Ip, and H. Lau, "Integration of expert system with analytic hierarchy process for the design of material handling equipment selection system," *Journal of Materials Processing Technology*, vol. 116, no. 2-3, pp. 137–145, 2001.

[14] R. Yaman, "A knowledge-based approach for selection of material handling equipment and material handling system pre-design," *Turkish Journal of Engineering and Environmental Sciences*, vol. 25, no. 4, pp. 267–278, 2001.

[15] D. J. Fonseca, G. Uppal, and T. J. Greene, "A knowledge-based system for conveyor equipment selection," *Expert Systems with Applications*, vol. 26, no. 4, pp. 615–623, 2004.

[16] C. Cho and P. J. Edbelu, "Design of a web-based integrated material handling system for manufacturing applications," *International Journal of Production Research*, vol. 43, pp. 375–403, 2005.

[17] S. H. L. Mirhosseyni and P. Webb, "A hybrid fuzzy knowledge-based expert system and genetic algorithm for efficient selection and assignment of material handling equipment," *Expert Systems with Applications*, vol. 36, no. 9, pp. 11875–11887, 2009.

[18] E. Jacquet-Lagreze and J. Siskos, "Assessing a set of additive utility functions for multicriteria decision-making, the UTA method," *European Journal of Operational Research*, vol. 10, no. 2, pp. 151–164, 1982.

[19] Z. Hatush and M. Skitmore, "Contractor selection using multicriteria utility theory: an additive model," *Building and Environment*, vol. 33, no. 2-3, pp. 105–115, 1998.

[20] M. Beuthe and G. Scannella, "Comparative analysis of UTA multicriteria methods," *European Journal of Operational Research*, vol. 130, no. 2, pp. 246–262, 2001.

[21] N. Manouselis and D. Sampson, "Multi-criteria decision making for broker agents in e-learning environments," *Operational Research*, vol. 2, pp. 347–361, 2002.

[22] R. V. Rao, *Decision Making in the Manufacturing Environment Using Graph Theory and Fuzzy Multiple Attribute Decision Making Methods*, Springer, London, UK, 2007.

3

Analysis and Scheduling of Maintenance Operations for a Chain of Gas Stations

Mehmet Savsar

Kuwait University, College of Engineering and Petroleum, P.O. Box 5969, 13060 Safat, Kuwait

Correspondence should be addressed to Mehmet Savsar; msavsar@gmail.com

Academic Editor: Anis Chelbi

Maintenance is one of the central issues in operational activities, which involve any type of equipment. In this paper we have considered analysis, modeling, and scheduling of preventive maintenance operations for fuel dispensers in a chain of gas stations. A gas station company with more than 570 dispensers in more than 40 stations is considered and the maintenance problem is studied in detail. Operations research tools, including maintenance models and linear programming, were used to establish optimum schedules for preventive maintenance operations. Detailed cost analyses were carried out to determine feasibility of the proposed preventive maintenance schedules. Models and procedures presented in this paper could guide operation engineers and maintenance managers in solving similar problems for operational improvements.

1. Introduction

Complex equipment and devices used in any system constitute majority of the capital invested in industry. Equipment is subject to deterioration with usage and time and deterioration is often reflected in higher operation costs and lower service quality. In order to keep operational costs down while maintaining good service quality, preventive maintenance (PM) is often performed on a scheduled basis. The cost of maintenance-related activities in industrial facilities has been estimated by Mobley [1] as 15–40% of total operation costs and the trend toward increased automation has forced managers to pay even more attention to maintain complex equipment and keep them in available state. If the equipment is maintained only when it fails, it is called corrective maintenance (CM), while preplanned maintenance is called preventive maintenance (PM). Traditionally it is known that the probability of failure would increase as equipment is aged, and that it would sharply decrease after a planned preventive maintenance (PM) is implemented. However, as indicated by Savsar [2], the amount of reduction in failure rate due to introduction of a preventive maintenance has not been fully studied. In particular, it would be desirable to know the performance of a system before and after the introduction of PM. It is also desired to know the type and the rate at

which a preventive maintenance should be scheduled or the maintenance policy to be implemented.

A gas station includes several facilities and equipment that need to be maintained. In particular, dispensers (gasoline pumps), storage tanks, car wash equipment, and other ancillary equipment need to be kept in operational condition for effective performance and profitable service. The first gasoline pumps were developed in 1885 in Indiana by S. F. Bowser to be used for kerosene lamps and stoves. Later, when the automobiles were invented, these pumps were improved and used by adding a hose and several other safety measures. A modern fuel dispenser is divided into two main parts, including an electronic head and a mechanical section. Electronic head contains an embedded computer, which controls the action of the pump, drives the pump's displays, and communicates with an indoor sales system. The mechanical section, which is in a self-contained unit, has an electric motor, pumping unit, meters, pulsers, and valves to physically pump and control the fuel flow. There are many different variations of fuel dispensers in use today. The term "gas pump" is usually used as an informal way to refer to a fuel dispenser.

In order to maintain functionality of a gas station, all equipment has to be maintained by preventive maintenance at scheduled times or repairs when failures occur. Every fuel dispenser is a combination of various small components and it is essential to ensure the appropriate performance of each component for profitable and high quality service in a gas station. Typical maintenance activities in a gas station include

    (i) mechanical pumps, meters, and lubrication equipment repair,

    (ii) electronic dispenser and electronic control console repair,

    (iii) automated system repair including automatic tank gauging, release detection systems, and Point-of-Sale systems,

    (iv) (POS), price scanners, card readers, and communication links,

    (v) tank system repair including: tanks, pumps, leak detectors, piping, hoses, and nozzles,

    (vi) car wash system and ancillary equipment maintenance and repairs.

Preventive maintenance (PM) programs are implemented to reduce annual repair costs and costs associated with equipment downtime. While preventive maintenance and repair instructions of major equipments at gas stations are specified by the manufacturers, it is difficult to find specific studies related to maintenance analysis of such systems. However, extensive studies have been carried out in the areas of reliability and maintenance management in general. The existing body of theory on general system reliability and maintenance is scattered over a large number of scholarly journals belonging to a diverse variety of disciplines. In particular, mathematical sophistication of preventive maintenance models has increased in parallel to the growth in the complexity of modern manufacturing systems. Research work has been published in the areas of maintenance modeling, optimization, and management. Cho and Parlar [3] presented surveys of maintenance models for multiunit systems. Dekker [4] also presented an excellent review of maintenance optimization models. Sheu and Krajewski [5] presented a decision model based on simulation and economic analysis for corrective maintenance policy evaluation. Simeu-Abazi et al. [6] evaluated dependability of manufacturing systems. Waeyenbergh et al. [7] presented a case study for maintenance concept development. Also, Waeyenbergh and Pintelon [8] discussed procedures, knowledge-based concepts, and frameworks in maintenance policy development and implementation in industry. Komonen [9] presented a cost model of industrial maintenance for profitability analysis. Abdulmalek et al. [10] presented a simulation model and procedures for analyzing tool change policies in a manufacturing system. Chan et al. [11] also presented a maintenance implementation for total productive maintenance in the context of a case study in electronics industry. Kyriakidis and Dimitrakos [12] presented a maintenance model for a production system with intermediate buffers. Gómez de León Hijes and Cartagena [13] presented a maintenance strategy based on equipment classification using a multicriterion objective. Carnero [14] presented a procedure for setting up a predictive maintenance program using detailed system evaluation. Savsar [15] and

Savsar and Aldaihani [16] presented reliability models for flexible manufacturing cells (FMC) and obtained FMC availability assuming only corrective maintenance is performed. Savsar [17, 18] presented models for analysis of the effects of maintenance policies on FMC performance measures. Savsar [19] developed a discrete mathematical model to describe a serial production flow line with buffers and machine failures and incorporated it into a simulation model to study the effects of equipment failures and corrective maintenance operations on production line performance under different operational conditions.

While most of the studies related to reliability and maintenance analysis are theoretical, limited numbers of practical applications are published in the literature. In this study, maintenance modeling and analysis is carried out in the context of a chain of gas stations system. In particular, models are presented for ·determination of various maintenance-related activities. It is well known that equipment failures occur due to wearouts and random causes. Therefore, in the most general way, maintenance operations are classified as corrective maintenance (CM) and preventive maintenance (PM). There are some other variations of these maintenances. However, CM and PM are the most general procedures in industry. Causes of random failures, which result in CM, are not known and cannot be predicted. These types of failures occur even in new systems. However, wearout failures occur by the usage of the equipment and as the time passes. These failures can be predicted and mean time to failure can be estimated. PM operations are carried out before the expected failures. For example, time to failure of a gear or a belt due to usage can be estimated and the time to change the component can be specified. While random failures cannot be eliminated totally, wearout failures can be eliminated by PM operations and thus a reduction in CM can be achieved. The exact effects of PM operations in reducing CM frequency are modeled and applied to various components in the selected system. System down time and productivity are estimated before and after the introduction of PM. Finally, cost analysis of the system is carried out with respect to the effectiveness of PM operations. Models and analysis developed and presented in this paper will be useful for maintenance engineers and operation managers in practical applications.

2. Analysis of Preventive Maintenance Operations for Fuel Dispensers

The gas station company considered in this study had a chain of 41 gas stations and a total of 577 dispensers. In order to perform the maintenance analysis, data related to dispenser failures at each gas station were collected and gas stations were grouped into three categories as high-failure stations, medium-failure stations, and low-failure stations. Grouping was necessary since it was not practical to analyze each of 41 stations separately. Each group was analyzed based on average failure and maintenance rates in order to have applicable results.

It is well known that performing preventive maintenance at scheduled points in time before an asset loses optimum performance can help in providing acceptable levels of operability in efficient and cost-effective manners. As the preventive maintenance is increased, the need for corrective maintenance is reduced and subsequently, the down time of the equipments will be reduced too. A study of rescheduling the preventive maintenance was performed by analyzing the effect of increasing the preventive maintenance on the mean time between failures.

The study was applied on the dispenser area for the three categories of high-, medium-, and low-failure stations. The compound mean time between maintenance activities, $MTBM_{mt}$, must be obtained by combining the rates for corrective maintenance (CM) and preventive maintenance (PM) activities. Assuming that the CM and PM activities are independent, the compound maintenance rate for CM and PM activities is obtained by the adding the related rates. The following notations are used for the calculations:

λ: corrective maintenance rate,

π: preventive maintenance rate,

$MTBM_{ct} = 1/\lambda$: mean time between corrective maintenance (this is also referred to as MTBF),

$MTBM_{pt} = 1/\pi$: Mean time between preventive maintenance.

Combined maintenance rate is obtained from $\lambda + \pi$ and compound mean time between all maintenances is calculated from $1/(\lambda + \pi)$, which can also be expressed as

$$MTBM_{mt} = \frac{1}{1/MTBM_{ct} + 1/MTBM_{pt}} = \frac{1}{\lambda + \pi}. \quad (1)$$

This equation shows the relation between mean time between maintenance activities and the maintenance rates. It does not depend on the distribution of failures. Time between failures can follow gamma distribution and its special form, which is exponential distribution.

Mean rate of compound maintenance is defined as $\psi = 1/MTBM_{mt}$.

Failure rate can be expressed as a function of combined maintenance rate and the preventive maintenance rate by manipulating (1) as follows:

$$\lambda = \frac{1}{MTBM_{mt}} - \pi, \quad (2)$$

where $\pi = 1/MTBM_{pt}$. Since increasing preventive maintenance decreases the need for corrective maintenance, it is possible to alter the PM rate π in order o recalculate the estimated corrective maintenance. In this case, combined maintenance rate is kept constant and CM rate is estimated based on a given PM rate. Maintenance analysis is carried out for dispensers in each group of stations. Value of λ and π can be easily estimated in a real system. Time between failures of a component can be recorded over time and the mean time between failures can be estimated from the data. Failure rate can be estimated for the component or the system under consideration by $\lambda = 1/MTBF$. Maintenance rate, π can also be easily estimated based on mean time between preventive maintenances of the component or the system considered based on the expected MTBMs specified by the manufacturer.

TABLE 1: Increasing $MTBM_{ct}$ as a result of decreasing $MTBM_{pt}$ for high failure stations.

$MTBM_{pt}$	$MTBM_{ct}$ (hours)
4 months	459
3 months	485
2 months	546
1 month	877

2.1. High Failure Stations. High failure stations category consists of 7 stations with a total of 133 dispensers. The average failure rate was found to be $\lambda = 0.002179$ failure/hour/ dispenser which equals 2539 failures per year. Currently, $MTBM_{pt}$ equals 4 months (30 days per month and 24 hours per day) and the preventive maintenance rate was calculated as follows:

$$\pi = \frac{1}{4 * 30 * 24} = 0.000347. \quad (3)$$

By using the failure rate and the preventive maintenance rate, combined maintenance rate, $MTBM_{mt}$, was calculated using (1) and found to be 395.88 hours. By substituting the $MTBM_{mt}$ into (2) and changing the PM rate, several possible failure rates were estimated using the equation below:

$$\lambda = \frac{1}{395.88} - \pi. \quad (4)$$

Table 1 summarizes the $MTBM_{ct}$ for different preventive maintenance schedules. From the results in the table, it can be shown that for the high failure stations, $MTBF_{ct}$ increases slightly when performing the preventive maintenance once every two months, but it increases highly when performing preventive maintenance once every month. Effectively, failure rate $\lambda = 1/MTBM_{ct}$ decreases significantly as PM is performed more frequently, such as every month.

2.2. Medium Failure Stations. Medium failure stations category consists of 17 gas stations with a total of 248 dispensers. The average failure rate was found to be $\lambda = 0.001935$ failure/hour/dispenser which equals 4204 failures per year. Currently, $MTBM_{pt}$ equals 4 months and the preventive maintenance rate was calculated as follows:

$$\pi = \frac{1}{4 * 30 * 24} = 0.000347. \quad (5)$$

By utilizing the failure rate λ and the preventive maintenance rate π, combined maintenance rate, $MTBM_{mt}$, was calculated using (1) and found to be 438.212 hours. Finally, by utilizing this value and different values for PM schedules, a set of possible corrective maintenance rates were estimated using the equation below and tabulated in Table 2. From Table 2, it can be shown that for the medium failure stations, the $MTBM_{ct}$ increases slightly when performing the preventive maintenance once every two months, but it again increases sharply when performing the preventive maintenance once every month:

$$\lambda = \frac{1}{438.212} - \pi. \quad (6)$$

TABLE 2: Increasing MTBM_{ct} as a result of decreasing MTBM_{pt} for medium failure stations.

MTBM_{pt}	MTBM_{ct} (hours)
4 months	517
3 months	549
2 months	629
1 month	1121

TABLE 3: MTBM_{ct} based on MTBM_{pt} for low failure group of stations.

MTBM_{pt}	MTBM_{ct} (hours)
4 months	820
3 months	909
2 months	1145
1 month	5650

2.3. Low Failure Stations.

Low failure category of stations consists of 17 gas stations with a total of 196 dispensers. The average failure rate was found to be 0.00122 failure/hour/dispenser which equals 2095 failures per year. Currently, MTBM_{pt} equals 4 months and the preventive maintenance rate was determined as follows:

$$\pi = \frac{1}{4 * 30 * 24} = 0.000347. \tag{7}$$

By utilizing failure rate and preventive maintenance rate, combined maintenance rate, MTBM_{mt}, was calculated using (1) again and found to be 638.162 hours. Finally, failure rates were recalculated based on various values of preventive maintenances using the equation below:

$$\lambda = \frac{1}{638.162} - \pi, \tag{8}$$

Table 3 summarizes the MTBF_{ct} for different PMs. From Table 3, it can be seen that for the low failure stations, MTBF_{ct} increases slightly when performing the PM up to once every two months, but it increases again sharply when performing PM once every month. These results will be used in the next section to optimize PM schedules for the dispensers in three categories of stations.

3. Preventive Maintenance Scheduling for Fuel Dispensers Using LP Model

From the previous analysis, we found that the MTBF_{ct} for CM increases as the mean time between PM is decreased. But, we have some constraints that prevent us from performing the PM once every month for the 41 gas stations. In particular, numbers of technicians are limited and the PM cannot be performed on more than one station per day; also it is not preferred to perform the PM monthly for the dispenser area. As a result, by comparing different situations for the three categories, it was found that the high failure stations have the lowest mean time between failures and the highest number of failures. Consequently, it was suggested to give them higher priority than the other stations.

A linear programming model was constructed to determine optimum PM schedule for the dispenser area. First, a pair wise comparison method was applied to assign weight to each failure category that will be used to formulate the objective function. Pairwise comparison is a kind of divide-and-conquer problem-solving method. It allows one to determine the relative ranking of a group of items. The first step is to identify the criteria. In our case we have three criteria, such as high failure, medium failure, and low failure stations. The second step is to arrange the criteria in an $N \times N$ matrix and compare the item in the row with respect to each item in the rest of the row by putting the category that we consider the most important in each pair wise comparison. The matrix would be as shown in Table 4.

TABLE 4: Pair wise comparison matrix.

	High failure stations	Medium failure stations	Low failure stations
High failure stations	—	High	High
Medium failure stations	—	—	Medium
Low failure stations	—	—	—

The next step is to create the ranking of items by creating an ordered list of the items, ranked by the number of cells containing their names. This leads to

(i) high failure stations (2),

(ii) medium failure stations (1),

(iii) low failure stations (0).

In order to get the weights, we have assumed a linear proportion between all the weights and solved the equation below, which states that the total of all the weights must be 100%. The weight assignments between classes or elements have been based on linear proportion in the literature unless there is evidence and data to assume some other relationship. In our case, we did not have any information or data that contradicts to the assumption of linear proportion between the weights. Therefore, we have used a linear proportion which is commonly used in the literature. The equation for the weights is as follows

$$100 = 2x + 1x + 0x. \tag{9}$$

Therefore, $x = 33.33$.

This leads to the following approximate weights:

(i) high failure stations: 60%,

(ii) medium failure stations: 30%,

(iii) low failure stations: 10%.

Note that final percentages are approximated by rounding them down and the "10%" for low failure stations resulted from summing the round-off values from other calculations. These final assignments after rounding down and up are arbitrary and do not necessarily carry a specific meaning or significance. The only importance is that we have to give

TABLE 5: Various parameters and maintenance cost comparisons for 3 categories of dispensers.

Category	Number of failures/year	Number of PM/station/year	Number of stations	CMC	PMC	TMC	Percent reduction in cost
High failure stations—under current PM	3281	3	7	40161	1498	41695	
High failure stations—under proposed PM	2747	6	7	33696	2992	36688	14%
Medium failure stations—under current PM	2101	3	15	25718	900	26688	
Medium failure stations—under proposed PM	1395	4	15	17075	1199	18274	31%
Low failure stations	Current and proposed PM schedules are the same and thus the costs are not affected						

some small weight to the low failure stations to represent them in the objective function, which will be shown below. After getting the weights, a linear programming model was formulated for maintenance scheduling. The following notations are used in the LP model:

h: high failure stations; $h = 1, 2, 3, 4, 5, 6, 7$,

d: medium failure stations; $d = 8, 9, 10, 11, 12, 13, 14, 15, 16, 17, 18, 19, 20, 21, 22, 23, 24$,

l: low failure stations; $l = 25, 26, 27, 28, 29, 30, 31, 32, 33, 34, 35, 36, 37, 38, 39, 40, 41$,

m: months; $m = 1, 2, 3, 4, 5, 6, 7, 8, 9, 10, 11, 12$.

Objective Function. Consider:

$$\text{Max.} \quad z = 0.6 \sum_{h=1}^{7} \sum_{m=1}^{12} x_{hm} + 0.3 \sum_{d=8}^{24} \sum_{m=1}^{12} x_{dm}$$
$$+ 0.1 \sum_{l=25}^{41} \sum_{m=1}^{12} x_{lm}. \tag{10}$$

Decision Variables. Consider:

$$x_{hm} = \begin{cases} 1 & \text{if station } h \text{ is maintained in month } m \\ 0 & \text{otherwise}, \end{cases}$$

$$x_{dm} = \begin{cases} 1 & \text{if station } d \text{ is maintained in month } m \\ 0 & \text{otherwise}, \end{cases} \tag{11}$$

$$x_{lm} = \begin{cases} 1 & \text{if station } l \text{ is maintained in month } m \\ 0 & \text{otherwise}. \end{cases}$$

Constraints. Consider:

$$\sum_{m=1}^{12} x_{hm} \geq 3 \quad \forall h,$$

$$\sum_{m=1}^{12} x_{dm} \geq 3 \quad \forall d,$$

$$\sum_{m=1}^{12} x_{lm} \geq 3 \quad \forall l,$$

$$\sum_{m=1}^{12} x_{hm} \leq 6 \quad \forall h,$$

$$\sum_{m=1}^{12} x_{dm} \leq 6 \quad \forall d,$$

$$\sum_{m=1}^{12} \leq 6 \quad \forall l,$$

$$\sum_{h=1}^{7} x_{hm} + \sum_{d=8}^{24} x_{dm} + \sum_{l=25}^{41} x_{lm} \leq 13$$
$$\forall m = 3, 4, 6, 8, 9, 11,$$

$$\sum_{h=1}^{7} x_{hm} + \sum_{d=8}^{24} x_{dm} + \sum_{l=25}^{41} x_{lm} \leq 14$$
$$\forall m = 1, 5, 7, 12,$$

$$\sum_{h=1}^{7} x_{hm} + \sum_{d=8}^{24} x_{dm} + \sum_{l=25}^{41} x_{lm} \leq 12 \quad \forall m = 2,$$

$$x_{hm}, x_{dm}, x_{lm} = \{0, 1\}, \tag{12}$$

where the objective function maximizes the total number of preventive maintenances to be performed so that total corrective maintenances are minimized.

The first, second, and third constraints insure that the preventive maintenance is performed at least 3 times per year for the high failure stations, medium failure stations, and low failure stations, respectively (as the current situation).

The fourth, fifth, and sixth constraints insure that the preventive maintenance will not be performed every month, but at maximum of 6 times per year. This means that minimum time between PM will be 2 months.

The seventh, eight, and ninth constraints insure that the labor capacity in each month will not be exceeded, provided that it takes 2 days to perform the preventive maintenance in each station with the new schedule.

The model was solved using GAMS software and optimal preventive maintenance schedule was found to be as follows.

(i) For the high failure stations, 6 times per year.

(ii) For the medium failure stations, it was divided into three categories as

   (a) 6 times per year for 3 stations,
   (b) 4 times per year for 6 stations,
   (c) 3 times per year for 8 stations.

(iii) For low failure stations, 3 times per year.

As a result, 3 medium failure stations have the same schedule as the high failure stations and 8 medium failure stations have the same schedule as low failure stations. Therefore, we will assume that the 3 highest failure stations from the medium failure stations category will be added to the high failure stations, the 8 lowest failure stations from the medium failure stations category will be added to the low failure stations, and the remaining 6 stations will be kept as medium failure stations. In Section 5, a cost analysis will be performed to compare the optimum PM schedules with the current PM schedules.

4. Maintenance Cost Analysis

Preventive maintenance involves a basic tradeoff between the costs of conducting maintenance activities and the savings achieved by reducing the overall rate of occurrence of system failures. Although the gas company is paying a fixed amount of money per month regardless of the number of failures occurring per month, a study of the effects of changing the PM schedule on the total maintenance cost was performed to make sure that the new PM schedule will not cause them to pay more money. The total maintenance cost was found using:

$$TMC = CMC + PMC + DTC, \qquad (13)$$

where TMC is total maintenance cost per year

CMC is corrective maintenance cost per year calculated by

$$CMC = (\text{Number of Failures per Year})$$
$$* (\text{Corrective Maintenance Cost per Failure}), \qquad (14)$$

PMC is preventive maintenance cost per year calculated by

$$PMC = (\text{Number of Preventive Maintenance per Year})$$
$$* (\text{Cost per Preventive Maintenance}) \qquad (15)$$

DTC is down time cost per year calculated by:

$$DTC = \left[\left(\frac{\text{No. of failures}}{\text{year}} \right) \right.$$
$$* (\text{Average Repair Time} + \text{Average Waiting Time}) \Big]$$
$$* [\text{Cost per Down Time hour}]. \qquad (16)$$

Total corrective maintenance cost for the dispenser area was provided by the gas company as 23,170 KD/year. The total number of failures in the dispenser area for the 41 stations

was 6869 failures/year. Thus, cost per failure, which resulted in corrective maintenance, was calculated by 23170/6869 = 3.4 KD/failure. This cost is the same for all categories of dispensers in all stations. Similarly, the PM cost for the dispenser area was provided by the gas station company as 1704 KD/year. Currently PM is scheduled to be performed once every 4 months, or 3 times per year, in each of the 41 stations which resulted in a total of $3 \times 41 = 123$ PM per year. By dividing the PM cost by the total number of preventive maintenances, cost per preventive maintenance was found to be 1704/123 = 13.85 KD/PM. This cost is also the same for all categories of dispensers in all stations. Downtime cost was not considered in the total maintenance cost based on the assumption that the sale will not be lost due to a dispenser failure as the customer will use another dispenser instead of leaving the system. Economic analyses are performed for each category of stations under the current (old) and proposed (new) PM policies.

4.1. High Failure Stations. As it was mentioned before, under the current PM schedule a PM is performed once every 4 months. As a result, high failure stations have 3281 failures/year, which results in 3281 CM operations/year, and 3 PMs per station per year. Equation (13) is utilized in cost calculations for dispensers in each category of stations. Note that 3 stations from the medium failure category are added to the high category due to same frequency of PM as obtained from LP model. Thus, using (13), maintenance costs for the current schedule for dispensers in high failure stations are calculated as CMC = 3281 failures/year*3.4 KD/failure = 11156 KD/year; PMC = 3 PM/station/year*10 stations*13.85 KD/PM = 416 KD/year; and TMC = 11.572 KD/year.

Under the proposed PM schedule, where PM is to be performed once every 2 months, the number of failures is expected to be 3047 failures/year and 6 PMs are performed per station/year. Thus, the maintenance costs are calculated by (13) as CMC = 3047 failure/year*3.4 KD/failure = 10360 KD/year; PMC = 6 PM/station/year*10 stations*13.85 KD/PM = 831 KD/year; and TMC = 11.191 KD/year. These maintenance cost for the current and proposed PM schedules are compared in Table 5, where system parameters are also listed. It can be seen that extra costs of PM is justified since the total costs are reduced from 11.572 KD to 11.191 KD for this category of stations.

4.2. Medium Failure Stations. For the medium failure stations, it was proposed to have a PM once every 3 months, that is, 4 times per year. Currently, PM is performed once every 4 months or 3 times per year. Thus, the maintenance costs for the current and proposed schedules are calculated as for high failure stations using (13) based on the parameters given in Table 5. The final cost results are also listed in the table. Note that 3.4 KD/failure and 13.85 KD/PM are common parameters for all cases. The cost results are compared in Table 5. PM costs are justified for this category also since total maintenance costs are reduced from 7.394 KD/year to 5.076 KD/year.

4.3. Low Failure Stations. As it was mentioned earlier, PM was proposed to be performed once every 4 months, which is the same as the current schedule. Therefore, the costs would be the same for the current and the proposed schedules.

5. Conclusions

Maintenance is one of the major operational activities in industrial systems. In particular, implementation of a PM program requires detailed and careful analysis to justify the related costs since maintenance costs are a significant part of operational costs. Implementing a PM has to be justified by analyzing its effect on reducing CM costs. In this paper, a gas station company with a chain of 41 stations is considered and detailed procedures are presented for maintenance analysis. After analyzing the current system, a linear programming model is used to determine optimum PM activities to be performed for each category of gas stations. Furthermore, detailed cost analyses are performed for the dispensers to establish cost-saving PM schedules. LP model and the economic cost analysis procedures proved to be effective for the company considered. The models and procedures utilized in this paper could be used by operational engineers in maintenance analysis in other industrial settings in order to improve productivity and to reduce related operational costs.

References

[1] R. K. Mobley, *An Introduction to Predictive Maintenance*, Van Nostrand Reinhold, New York, NY, USA, 1990.

[2] M. Savsar, "Analysis and modeling of maintenance operations in the context of an oil filling plant," *Journal of Manufacturing Technology Management*, vol. 22, no. 5, pp. 679–697, 2011.

[3] D. I. Cho and M. Parlar, "A survey of maintenance models for multi-unit systems," *European Journal of Operational Research*, vol. 51, no. 1, pp. 1–23, 1991.

[4] R. Dekker, "Applications of maintenance optimization models: a review and analysis," *Reliability Engineering and System Safety*, vol. 51, no. 3, pp. 229–240, 1996.

[5] C. Sheu and L. J. Krajewski, "Decision model for corrective maintenance management," *International Journal of Production Research*, vol. 32, no. 6, pp. 1365–1382, 1994.

[6] Z. Simeu-Abazi, O. Daniel, and B. Descotes-Genon, "Analytical method to evaluate the dependability of manufacturing systems," *Reliability Engineering and System Safety*, vol. 55, no. 2, pp. 125–130, 1997.

[7] G. Waeyenbergh, L. Pintelon, and L. Gelders, "A stepping stone towards knowledge based maintenance," *South African Journal of Industrial Engineering*, vol. 12, no. 2, pp. 61–61, 2001.

[8] G. Waeyenbergh and L. Pintelon, "Maintenance concept development: a case study," *International Journal of Production Economics*, vol. 89, no. 3, pp. 395–405, 2004.

[9] K. Komonen, "A cost model of industrial maintenance for profitability analysis and benchmarking," *International Journal of Production Economics*, vol. 79, no. 1, pp. 15–31, 2002.

[10] F. Abdulmalek, M. Savsar, and M. Aldaihani, "Simulation of tool change policies in a flexible manufacturing cell," *WSEAS Transactions on Systems*, vol. 7, no. 3, pp. 2546–2552, 2004.

[11] F. T. S. Chan, H. C. W. Lau, R. W. L. Ip, H. K. Chan, and S. Kong, "Implementation of total productive maintenance: a case study,"

International Journal of Production Economics, vol. 95, no. 1, pp. 71–94, 2005.

[12] E. G. Kyriakidis and T. D. Dimitrakos, "Optimal preventive maintenance of a production system with an intermediate buffer," *European Journal of Operational Research*, vol. 168, no. 1, pp. 86–99, 2006.

[13] F. C. Gómez de León Hijes and J. J. R. Cartagena, "Maintenance strategy based on a multicriterion classification of equipments," *Reliability Engineering and System Safety*, vol. 91, no. 4, pp. 444–451, 2006.

[14] M. Carnero, "An evaluation system of the setting up of predictive maintenance programmes," *Reliability Engineering and System Safety*, vol. 91, no. 8, pp. 945–963, 2006.

[15] M. Savsar, "Reliability analysis of a flexible manufacturing cell," *Reliability Engineering and System Safety*, vol. 67, no. 2, pp. 147–152, 2000.

[16] M. Savsar and M. Aldaihani, "Modeling of machine failures in a flexible manufacturing cell with two machines served by a robot," *Reliability Engineering and System Safety*, vol. 93, no. 10, pp. 1551–1562, 2008.

[17] M. Savsar, "Performance analysis of an FMS operating under different failure rates and maintenance policies," *International Journal of Flexible Manufacturing Systems*, vol. 16, no. 3, pp. 229–249, 2005.

[18] M. Savsar, "Effects of maintenance policies on the productivity of flexible manufacturing cells," *Omega*, vol. 34, no. 3, pp. 274–282, 2006.

[19] M. Savsar, "Buffer allocation in serial production lines with preventive and corrective maintenance operations," *Kuwait Journal of Science and Engineering*, vol. 33, no. 2, pp. 253–266, 2006.

4

Chip Attach Scheduling in Semiconductor Assembly

Zhicong Zhang, Kaishun Hu, Shuai Li, Huiyu Huang, and Shaoyong Zhao

Department of Industrial Engineering, School of Mechanical Engineering, Dongguan University of Technology, Songshan Lake District, Dongguan, Guangdong 523808, China

Correspondence should be addressed to Zhicong Zhang; stephen1998@gmail.com

Academic Editor: Josefa Mula

Chip attach is the bottleneck operation in semiconductor assembly. Chip attach scheduling is in nature unrelated parallel machine scheduling considering practical issues, for example, machine-job qualification, sequence-dependant setup times, initial machine status, and engineering time. The major scheduling objective is to minimize the total weighted unsatisfied Target Production Volume in the schedule horizon. To apply Q-learning algorithm, the scheduling problem is converted into reinforcement learning problem by constructing elaborate system state representation, actions, and reward function. We select five heuristics as actions and prove the equivalence of reward function and the scheduling objective function. We also conduct experiments with industrial datasets to compare the Q-learning algorithm, five action heuristics, and Largest Weight First (LWF) heuristics used in industry. Experiment results show that Q-learning is remarkably superior to the six heuristics. Compared with LWF, Q-learning reduces three performance measures, objective function value, unsatisfied Target Production Volume index, and unsatisfied job type index, by considerable amounts of 80.92%, 52.20%, and 31.81%, respectively.

1. Introduction

Semiconductor manufacturing consists of four basic steps: wafer fabrication, wafer sort, assembly, and test. Assembly and test are back-end steps. Semiconductor assembly contains many operations, such as reflow, wafer mount, saw, chip attach, deflux, EPOXY, cure, and PEVI. IS factory is a site for back-end semiconductor manufacturing where chip attach is the bottleneck operation in the assembly line. In terms of Theory of Constraints (TOC), the capacity of a shop floor depends on the capacity of the bottleneck, and a bottleneck operation gives a tremendous impact upon the performance of the whole shop floor. Consequently, scheduling of chip attach station has a significant effect on the performance of the assembly line. Chip attach is performed in a station which consists of ten parallel machines; thus, chip attach scheduling in nature is some form of unrelated parallel machine scheduling under certain realistic restrictions.

Research on unrelated parallel machine scheduling focuses on two sorts of criteria: completion time or flow time related criteria and due date related criteria. Weng et al. [1]

proposed a heuristic algorithm called "Algorithm 9" to minimize the total weighted completion time with setup consideration. Algorithm 9 was demonstrated to be superior to six heuristic algorithms. Gairing et al. [2] presented an effective combinatorial approximate algorithm for makespan objective. Mosheiov [3] and Mosheiov and Sidney [4] converted an unrelated parallel machine scheduling problem with total flow time objective into polynomial number of assignment problems. The scheduling problem was tackled by solving the derived assignment problems. Yu et al. [5] formulated unrelated parallel machine scheduling problems as mixed integer programming and dealt with them using Lagrangian Relaxation. They examined six measures such as makespan and mean flow time. Promising results were achieved compared with a modified FIFO method.

Besides completion time or flow time related criteria, tardiness objectives are also employed frequently. Dispatching rules are widely applied to production scheduling with a tardiness objective, such as Earliest Due Date (EDD), Shortest Processing Time (SPT), Critical Ratio (CR), Minimal Slack (MS), Modified Due Date (MDD) [6, 7], Apparent Tardiness

Cost (ATC) [8, 9], and COVERT [10–12]. More complicated heuristic algorithms and local search methods are also developed. Bank and Werner [13] addressed the problem of minimizing the weighted sum of linear earliness and tardiness penalties in unrelated parallel machine scheduling. They derived some structural properties useful to searching for an approximate solution and proposed various constructive and iterative heuristic algorithms. Liaw et al. [14] found the efficient lower and upper bounds of minimizing the total weighted tardiness by a two-phase heuristics based on the solution to an assignment problem. They also presented a branch-and-bound algorithm incorporating various dominance rules. Kim et al. [15] studied batch scheduling of unrelated parallel machines with a total weighted tardiness objective and setup times consideration. They examined four search heuristics for this problem: the earliest weighted due date, the shortest weighted processing time, the two-level batch scheduling heuristic, and the simulated annealing method.

We are concerned in the paper about a particular Target Production Volume (TPV) oriented optimization objective. In real production in IS factory, the planning department figures out the TPV of each job type on chip attach operation in a schedule horizon. Thus, the major objective of chip attach scheduling is to meet TPVs to the fullest extent (see Section 2.1 for details). We apply reinforcement learning (RL), an artificial intelligence method, for this study. We first present a brief concept of reinforcement learning.

1.1. Q-Learning. Reinforcement learning is a machine learning method proposed to approximately solve large-scale Markov Decision Process (MDP) or Semi-Markov Decision Process (SMDP) problems. Reinforcement learning problem is a model in which an agent learns to select optimal or near-optimal actions for achieving its long-term goals (to maximize the total or average reward) through trial-and-error interactions with dynamic environment. In this paper, we address RL problems of episodic task, that is, problems with a terminal state. Sutton and Barto [16] defined four key elements of RL algorithms: policy, reward function, value function, and model of the environment. A policy determines the agent's action at each state. A reward function determines the payment on transition from one state to another. A value function specifies the value of a state or a state-action pair in the long run, the expected total reward for an episode. By learning from interaction between the agent and its environment, value-based RL algorithms aim to approximate the optimal state or action value function through iteration and thus find a near-optimal policy. Compared with dynamic programming, RL algorithms do not need to know the transition probability and reduce the computational effort.

Q-learning is one of the most widely applied RL algorithms based on value iteration. Q-learning was first proposed by Watkins [17]. Convergence results of tabular Q-learning were obtained by Watkins and Dayan [18], Jaakkola et al. [19], and Tsitsiklis [20]. Bertsekas and Tsitsiklis [21] demonstrated that Q-learning produces the optimal policy

in discounted reward problems under certain conditions. Q-learning uses $Q(s, a)$, called Q-value, to represent the value of a state-action pair. $Q(s, a)$ is defined as follows:

$$Q(s,a) = \sum_{s' \in S} p(s,a,s') \left[r(s,a,s') + \gamma V^*(s') \right], \quad (1)$$

where S denotes the state space, $p(s, a, s')$ denotes the transition probability from s to s' taking action a, $r(s, a, s')$ denotes the reward on transition from s to s' taking action a, γ ($0 < \gamma \le 1$) is a discounted factor, and $V^*(\cdot)$ is the optimal state value function.

In terms of Bellman optimality function, the following holds for arbitrary $s \in S$, where $A(s)$ denotes the set of actions available for state s:

$$V^*(s) = \max_{a \in A(s)} Q(s,a). \quad (2)$$

From (1) and (2), the following equation holds:

$$Q(s,a) = \sum_{s' \in S} p(s,a,s') \left[r(s,a,s') + \gamma \max_{a' \in A(s')} Q(s',a') \right]$$
$$\forall (s,a). \quad (3)$$

Equation (3) is the basic transformation of Q-learning algorithm. The step-size version of Q-learning is

$$Q(s,a) = Q(s,a) + \alpha \left[r(s,a,s') + \gamma \max_{a' \in A(s')} Q(s',a') \right.$$
$$\left. - Q(s,a) \right], \quad \forall (s,a), \quad (4)$$

where α ($0 < \alpha \le 1$) is learning rate. Using historical samples or simulation experiments, Q-learning obtains a near-optimal policy by driving action-value function, $Q(s, a)$, towards the optimal action-value function, $Q^*(s, a)$, through iteration based on formula (4).

Recently, RL has drawn attention from production scheduling. S. Riedmiller and M. Riedmiller [22] used Q-learning to solve stochastic and dynamic job shop scheduling problem with the overall tardiness objective. Some typical heuristic dispatching rules, SPT, LPT, EDD, and MS, were chosen as actions and compared with the Q-learning method. Aydin and Öztemel [23] applied a Q-learning algorithm to minimize the mean tardiness of dynamic job shop scheduling. Their results showed that the RL-scheduling system outperformed the use of each of the three rules (SPT, COVERT, and CR) individually with mean tardiness objective in most of the testing cases. Hong and Prabhu [24] formulated setup minimization problem (minimizing the sum of due date deviation and setup cost) in JIT manufacturing systems as an SMDP and solved it by tabular Q-learning method. Experiment results showed that Q-learning algorithms achieved significant performance improvement over usual dispatching rules such as EDD in complex real-time shop floor control problems for JIT production. Wang and Usher [25] applied Q-learning to select dispatching rules for the single machine scheduling problem. Csáji et al. [26] proposed an adaptive iterative distributed scheduling algorithm operated in a market-based production control system, where every machine and job is associated with its own software agent. Singh et al. [27] proposed an online reinforcement learning algorithm for call admission control. The approach optimized the SMDP performance criterion with respect to a family of parameterized policies. Multi-agent reinforcement learning system has also been applied to scheduling or control problems, for example, Kaya and Alhajj [28], Paternina-Arboleda and Das [29], Mariano-Romero et al. [30], Vengerov [31], Iwamura et al. [32].

Applications of RL algorithms to scheduling problems have not been thoroughly explored in the prior studies. In this study, we employ Q-learning algorithm to resolve chip attach scheduling problem and achieve overwhelming experimental results compared with six heuristic algorithms. The remainder of this paper is organized as follows. We describe the problem and convert it into RL problem explicitly in Section 2, present the RL algorithm in Section 3, conduct the computational experiments and analysis in Section 4, and draw conclusions in Section 5.

2. RL Formulation

2.1. Problem Statement. The scheduling problem concerned in this paper is described as follows. The work station for chip attach operation consists of m parallel machines and processes n types of jobs. The bigger the weight of a job type is, the more important it is. Each job needs to be processed on one machine only and one machine processes at most one job at a time. Any job type (say, j) is only allowed to be processed on subset M_j of the m parallel machines. The jobs of the same type j have a deterministic processing time $p_{i,j}$ ($1 \le i \le m$; $1 \le j \le n$) if they are processed on machine i. The machines are unrelated; that is, $p_{i,j}$ is independent of $p_{k,j}$ for all jobs j and all machines $i \ne k$. The production is lot based. Normally, one lot contains more than 1000 units. Thus, the processing time is the time for processing one lot and processing is nonpreemptive (i.e., once a machine starts processing one lot, it cannot process another one until it completely processes this lot). Setup time between job type $j1$ and $j2$ is $s_{j1,j2}$ ($1 \le j1, j2 \le n$). The setup times are deterministic and sequence dependant. Trivially, $s_{j,j} = 0$ holds for arbitrary j ($1 \le j \le n$) and $s_{j,x} + s_{x,q} > s_{j,q}$ holds for arbitrary j, x, q ($1 \le j, x, q \le n$).

The usage of a machine is considered to be in one of two categories: engineering time (e.g., maintenance time) and production time. We only need to schedule the production in production time, the total available time in a schedule horizon deducting the engineering time. Production time is divided into initial production time and normal production time. We consider the initial machine status in the schedule horizon. If a machine is processing a lot, called "initial lot," at the beginning of a schedule horizon, it is not allowed to process any other lot until it completely processes the remaining units in the initial lot (called initial volume).

The time for processing the unprocessed initial volume in the initial lot is called "initial production time." Since the production of nonbottleneck operations is determined by the bottleneck operation, we assume that the jobs are always available for processing on chip attach operation when they are needed.

The primary objective of chip attach scheduling is to minimize the total weighted unsatisfied TPV of a schedule horizon. Since equipment of semiconductor manufacturing is very expensive, machine utilization should be kept in a high level. Hence, on the premise that TPVs of all job types are entirely satisfied, the secondary objective of chip attach scheduling is to process as much as weighted excess volume to relieve the burden of the next schedule horizon. The objective function is formulated as follows:

$$\min \sum_{j=1}^{n} w_j \left(D_j - Y_j \right)^+ - \sum_{j=1}^{n} \frac{w_j}{M} \left(Y_j - D_j \right)^+, \quad (5)$$

where w_j ($1 \le j \le n$) is the weight per unit of job type j, D_j ($1 \le j \le n$) is the predetermined TPV of job type j (including the initial volume in the initial lots), and Y_j ($1 \le j \le n$) is the processed volume of job type j. D_j can be represented as follows:

$$D_j = \sum_{i=1}^{m} \omega(i, j) I_i + k_j L \quad \left(k_j = 0, 1, \ldots \right), \quad (6)$$

where I_i denotes the initial volume in the initial lot processed by machine i at the beginning of the schedule horizon, L is lot size, and

$$\omega(i, j) = \begin{cases} 1, & \text{if machine } i \text{ is processing job type } j \\ & \text{in the beginning of the schedule horizon,} \\ 0, & \text{otherwise.} \end{cases}$$
$$(7)$$

Calculation of Y_j is rate based, interpreted as follows. Suppose machine i processes lot LQ (belonging to job type q), proceeding lot LJ (belonging to job type j). Let ts_{LJ} denote the start time of setup for LJ; then, the completion time of LJ is $ts_{LJ} + s_{q,j} + p_{i,j}$. Let $\Delta Y_{i,j}(t)$ denote the increase of processed volume of job type j because of processing LJ on machine i from time ts_{LJ} to t, defined as follows:

$$\Delta Y_{i,j}(t) = \frac{(t - ts_{LJ}) L}{s_{q,j} + p_{i,j}} \quad \left(ts_{LJ} \le t \le ts_{LJ} + s_{q,j} + p_{i,j} \right). \quad (8)$$

M is a positive number which is large enough. M is set following the next inequality:

$$M > \max \left\{ \frac{\left(w_q + w_x \right) \left(s_{v,j} + p_{i,j} \right)}{w_j p_{i,q}}, \right.$$

$$\left. \frac{w_q \left(s_{v,x} + p_{i,x} \right) \left(s_{x,j} + p_{i,j} \right)}{p_{i,j} \min_{1 \le c \le m, 1 \le a,b,k \le n} \left\{ w_k \left(s_{a,b} + p_{c,b} \right) \right\}} \right\}$$

$$\left(\forall 1 \le i \le m, 1 \le j, q, v, x \le n \right). \quad (9)$$

For an optimal schedule minimizing objective function (5), if (9) holds and there exists j $(1 \le j \le n)$ such that $Y_j > D_j$, then

$$Y_j + \sum_{i=1}^{m} \beta(i,j) U_i \ge D_j \quad (\forall 1 \le j \le n), \quad (10)$$

where U_i denotes the unprocessed volume in the last lot processed by machine i at the end of this schedule horizon (i.e., the initial volume of the next schedule horizon) and

$$\beta(i,j)$$

$$= \begin{cases} 1, & \text{if machine } i \text{ is processing job type } j \text{ at} \\ & \text{the end of the schedule horizon,} \\ 0, & \text{otherwise.} \end{cases} \quad (11)$$

According to inequality (9), in any schedule minimizing objective function (5), any machine will not process a lot belonging to a job type whose TPV has been satisfied until TPV of any other job types is also fully satisfied. In other words, inequality (9) guarantees that the objective function takes minimization of the total weighted unsatisfied TPV (the first item of objective function (5)) as the first priority. The fundamental problem in applying reinforcement learning to scheduling is to convert scheduling problems into RL problems, including representation of state, construction of actions, and definition of reward function.

2.2. State Representation and Transition Probability. We first define the state variables. State variables describe the major characteristics of the system and are capable of tracking the change of the system status. The system state can be represented by the vector

$$\varphi = \left[T_i^0 (1 \le i \le m); T_i (1 \le i \le m); t_i (1 \le i \le m); \right. \\ \left. d_j (1 \le j \le n); e_i (1 \le i \le m) \right], \quad (12)$$

where T_i^0 $(1 \le i \le m)$ denotes the job type of which the latest lot completely processed on machine i, T_i $(1 \le i \le m)$ denotes the job type of which the lot being processed on machine i (T_i equals zero if machine i is idle), t_i $(1 \le i \le m)$, denotes the time starting from the beginning of the latest setup on machine i (for convenience, we assume that there is a zero-time setup if $T_i^0 = T_i$), d_j $(1 \le j \le n)$ is unsatisfied TPV (i.e., $(D_j - Y_j)^+$), and e_i $(1 \le i \le m)$ represents the unscheduled normal production time of machine i.

Considering the initial status of machines, the initial system state of the schedule horizon is

$$s_0 = \left[T_{i,0}^0 (1 \le i \le m); T_{i,0} (1 \le i \le m); t_{i,0} (1 \le i \le m); \right. \\ \left. D_j (1 \le j \le n); \text{TH} - \sigma_i - \text{TE}_i (1 \le i \le m) \right], \quad (13)$$

where TH denotes the overall available time in the schedule horizon, σ_i denotes the initial production time of machine i, and TE_i denotes the engineering time of machine i.

There are two kinds of events triggering state transitions: (1) completion of processing a lot on one or more machines; (2) any machine's normal production time is entirely scheduled. If the triggering event is completion of processing, the state at the decision-making epoch is represented as

$$s_d = \left[T_{i,d}^0 (1 \le i \le m); T_{i,d} (1 \le i \le m); t_{i,d} (1 \le i \le m); \right. \\ \left. d_{j,d} (1 \le j \le n); e_{i,d} (1 \le i \le m) \right], \quad (14)$$

where $\{i \mid T_{i,d} = 0, 1 \le i \le m\} \ne \Phi$. If $T_{i,d} = 0$ (machine i is idle), then $t_{i,d} = 0$. If the triggering event is using up a machine's normal production time, then $\{i \mid e_{i,d} = 0, 1 \le i \le m\} \ne \Phi$.

Assume that after taking action a, the system state immediately transfers form s_d to an interim state, s, as follows:

$$s = \left[T_i^0 (1 \le i \le m); T_i (1 \le i \le m); t_i (1 \le i \le m); \right. \\ \left. d_j (1 \le j \le n); e_i (1 \le i \le m) \right], \quad (15)$$

where $T_i > 0$ for all i $(1 \le i \le m)$; that is, all machines are busy.

Let Δt denote the sojourn time at state s; then, $\Delta t = \min\{\min_{1 \le i \le m}\{s_{T_i^0, T_i} + p_{i,T_i} - t_i\}, \min_{1 \le i \le m}\{e_i \mid e_i > 0\}\}$. Let $\Lambda = \{i \mid s_{T_i^0, T_i} + p_{i,T_i} - t_i = \Delta t\}$; then, the state at the next decision-making epoch is represented as

$$s' = \left[T_i (i \in \Lambda), T_i^0 (i \notin \Lambda); T_i = 0 (i \in \Lambda), T_i (i \notin \Lambda); \right. \\ 0 (i \in \Lambda), t_i + \Delta t (i \notin \Lambda); \\ d_j - \frac{\Delta t L \sum_{i=1}^{m} \delta_Y(T_i, j)}{s_{T_i^0, j} + p_{i,j}} (1 \le j \le n); \\ \left. \max\{e_i - \Delta t, 0\} (1 \le i \le m) \right], \quad (16)$$

where

$$\delta_Y(T_i, j) = \begin{cases} 1, & \text{if } T_i = j, \\ 0, & \text{if } T_i \ne j. \end{cases} \quad (17)$$

Apparently we have $P_{s_d, s'}^a = 1$, where $P_{s_d, s'}^a$ denotes the one-step transition probability from state s_d to state s' under action a. Let s_u and τ_u denote the system state and time, respectively, at the uth decision-making epoch. It is easy to show that

$$P\{s_{u+1} = X, \tau_{u+1} - \tau_u \le t \mid s_0, s_1, \dots, s_u; \tau_0, \tau_1, \dots, \tau_u\} \\ = P\{s_{u+1} = X, \tau_{u+1} - \tau_u \le t \mid s_u; \tau_u\}, \quad (18)$$

where $\tau_{u+1} - \tau_u$ is the sojourn time at state s_u. That is, the decision process associated with (s, τ) is a Semi-Markov Decision Process with particular transition probability and sojourn times. The terminal state of an episode is

$$s_e = \left[T_{i,e}^0 (1 \le i \le m); T_{i,e} (1 \le i \le m); t_{i,e} (1 \le i \le m); \right. \\ \left. d_{j,e} (1 \le j \le n); 0 (1 \le i \le m) \right]. \quad (19)$$

2.3. Action. Prior domain knowledge can be utilized to fully exploit the agent's learning ability. Apparently, an optimal schedule must be nonidle (i.e., any machine has no idle time during the whole schedule). It may happen that more than one machine are free at the same decision-making epoch. An action determines which lot to be processed on which machine. In the following, we define seven actions using heuristic algorithms.

Action 1. Select jobs by WSPT heuristics as follows.

Algorithm 1. WSPT heuristics.

Step 1. Let SM denote the set of free machines at a decision-making epoch.

Step 2. Choose machine k to process job type q, with $(k, q) =$ $\text{argmin}_{(i,j)}\{(s_{T_i^0, j} + p_{i,j})/w_j \mid 1 \leq j \leq n, i \in M_j$ and $i \in \text{SM}\}$.

Step 3. Remove k from SM. If $\text{SM} \neq \Phi$, go to Step 2; otherwise, the algorithm halts.

Action 2. Select jobs by MWSPT (modified WSPT) heuristics as follows.

Algorithm 2. MWSPT heuristics.

Step 1. Define SM as Step 1 in Algorithm 1, and let SJ denote the set of job types whose TPVs have not been satisfied at a decision-making epoch; that is, $\text{SJ} = \{j \mid Y_j < D_j, 1 \leq j \leq n\}$. If $\text{SJ} = \Phi$, go to Step 4.

Step 2. Choose job type q to process on machine k, with $(k, q) = \text{argmin}_{(i,j)}\{(s_{T_i^0, j} + p_{i,j})/w_j \mid j \in \text{SJ}, i \in M_j$ and $i \in \text{SM}\}$.

Step 3. Remove k from SM. Set $Y_q = Y_q + L$ and update SJ. If $\text{SJ} \neq \Phi$ and $\text{SM} \neq \Phi$, go to Step 2; if $\text{SJ} = \Phi$ and $\text{SM} \neq \Phi$, go to Step 4; otherwise, the algorithm halts.

Step 4. Choose machine k to process job type q, with $(k, q) = \text{argmin}_{(i,j)}\{(s_{T_i^0, j} + p_{i,j})/w_j \mid 1 \leq j \leq n, i \in M_j$ and $i \in \text{SM}\}$.

Step 5. Remove k from SM. If $\text{SM} \neq \Phi$, go to Step 4; otherwise, the algorithm halts.

Action 3. Select jobs by Ranking Algorithm (RA) as follows.

Algorithm 3. Ranking Algorithm.

Step 1. Define SM and SJ as Step 1 in Algorithm 2. If $\text{SJ} = \Phi$, go to Step 5.

Step 2. For each job type j ($j \in \text{SJ}$), sort the machines in increasing order of $(s_{V_i, j} + p_{i,j})$ ($1 \leq i \leq m$), where V_i is defined as follows.

$$V_i = \begin{cases} T_i, & \text{if machine } i \text{ is busy} \\ T_i^0, & \text{if machine } i \text{ is free} \end{cases} \quad (1 \leq i \leq m). \quad (20)$$

Let $g_{i,j}$ ($1 \leq g_{i,j} \leq m$) denote the order of machine i ($1 \leq i \leq m$) for job type j ($1 \leq j \leq n$).

Step 3. Choose job q to process on machine k, with $(k, q) = \text{argmin}_{(i,j)}\{g_{i,j} \mid j \in \text{SJ}, i \in M_j$ and $i \in \text{SM}\}$. If there exist two or more machine-job combinations (say, machine-job combination $(i_1, j_1), (i_2, j_2), \ldots, (i_h, j_h)$) with the same minimal order; that is, $(i_e, j_e) = \text{argmin}_{(i,j)}\{g_{i,j} \mid j \in \text{SJ}, i \in M_j$ and $i \in \text{SM}\}$ holds for e ($1 \leq e \leq h$), then choose job type j_e to process on machine i_e, with $(i_e, j_e) = \text{argmin}_{(i,j)}\{(s_{V_{i_e}, j_e} + p_{i_e, j_e})/w_{j_e} \mid 1 \leq e \leq h\}$.

Step 4. Remove k or i_e from SM. Set $Y_q = Y_q + L$ or $Y_{j_e} = Y_{j_e} + L$ and update SJ. If $\text{SJ} \neq \Phi$ and $\text{SM} \neq \Phi$, go to Step 3; if $\text{SJ} = \Phi$ and $\text{SM} \neq \Phi$, go to Step 5; otherwise, the algorithm halts.

Step 5. Choose job q to process on machine k, with $(k, q) = \text{argmin}_{(i,j)}\{g_{i,j} \mid 1 \leq j \leq n, i \in M_j$ and $i \in \text{SM}\}$. If there exist two or more machine-job combinations (say, machine-job combinations $(i_1, j_1), (i_2, j_2), \ldots, (i_h, j_h)$) with the same minimal order, choose job type j_e to process on machine i_e, with $(i_e, j_e) = \text{argmin}_{(i,j)}\{(s_{V_{i_e}, j_e} + p_{i_e, j_e})/w_{j_e} \mid 1 \leq e \leq h\}$.

Step 6. Remove k or i_e from SM. If $\text{SM} \neq \Phi$, go to Step 5; otherwise, the algorithm halts.

Action 4. Select jobs by LFM-MWSPT heuristics as follows.

Algorithm 4. LFM-MWSPT heuristics.

Step 1. Define SM and SJ as Step 1 in Algorithm 2.

Step 2. Select a free machine (say, k) from SM by LFM (Least Flexible Machine; see [33]) rule and choose a job type to process on machine k following MWSPT heuristics.

Step 3. Remove k from SM. If $\text{SM} \neq \Phi$, go to Step 2; otherwise, the algorithm halts.

Action 5. Select jobs by LFM-RA heuristics as follows.

Algorithm 5. LFM-RA heuristics.

Step 1. Define SM and SJ as Step 1 in Algorithm 2.

Step 2. Select a free machine (say, k) from SM by LFM rule and choose a job type to process on machine k following Ranking Algorithm.

Step 3. Remove k from SM. If $\text{SM} \neq \Phi$, go to Step 2; otherwise, the algorithm halts.

Action 6. Each free machine selects the same job type as the latest one it processed.

Action 7. Select no job.

At the start of a schedule horizon, the system is at initial state s_0. If there are free machines, they select jobs to process by taking one of Actions 1–6; otherwise, Action 7 is chosen. Afterwards, when any machine completes processing a lot or any machine's normal production time is completely scheduled, the system transfers into a new state, s_u. The agent selects an action at this decision-making epoch and the system state transfers into an interim state, s. When, again, any machine completes processing a lot or any machine's normal production time used is up, the system transfers into the next decision-making state s_{u+1} and the agent receive reward r_{u+1}, which is computed due to s_u and the sojourn time between the two transitions into s_u and s_{u+1} (as shown in Section 2.4). The previous procedure is repeated until a terminal state is attained. An episode is a trajectory from the initial state to a terminal state of a schedule horizon. Action 7 is available only at the decision-making states when all machines are busy.

2.4. Reward Function. A reward function follows several disciplines. It indicates the instant impact of an action on the schedule, that is, to link the action with immediate reward. Moreover, the accumulated reward indicates the objective function value; that is, the agent receives large total reward for small objective function value.

Definition 6 (reward function). Let K denote the number of decision-making epoch during an episode, t_u ($0 \leq u < K$) the time at the uth decision-making epoch, $T_{i,u}$ ($1 \leq i \leq m$, $1 \leq u \leq K$) the job type of the lot which machine i processes during time interval $(t_{u-1}, t_u]$, $T_{i,u}^0$ the job type of the lot which precedes the lot machine i processes during time interval $(t_{u-1}, t_u]$, and $Y_j(t_u)$ the processed volume of job type j by time t_u. It follows that

$$Y_j(t_u) - Y_j(t_{u-1}) = \sum_{i=1}^{m} \frac{(t_u - t_{u-1})\,\delta(i,j)\,L}{s_{T_{i,u}^0, T_{i,u}} + p_{i,T_{i,u}}}, \quad (21)$$

where $\delta(i,j)$ is an indicator function defined as

$$\delta(i,j) = \begin{cases} 1, & T_{i,u} = j, \\ 0, & T_{i,u} \neq j. \end{cases} \quad (22)$$

Let r_u ($u = 1, 2, \ldots, K$) denote the reward function at the uth decision-making epoch. r_u is defined as

$$r_u = \sum_{j=1}^{n} \min \left\{ \sum_{i=1}^{m} \frac{(t_u - t_{u-1})\,\delta(i,j)\,L}{s_{T_{i,u}^0, T_{i,u}} + p_{i,T_{i,u}}}, \left[D_j - Y_j(t_{u-1})\right]^+ \right\} w_j$$

$$+ \max \left\{ \sum_{i=1}^{m} \frac{(t_u - t_{u-1})\,\delta(i,j)\,L}{s_{T_{i,u}^0, T_{i,u}} + p_{i,T_{i,u}}} - \left[D_j - Y_j(t_{u-1})\right]^+, 0 \right\}$$

$$\times \frac{w_j}{M}. \quad (23)$$

The reward function has the following property.

Theorem 7. *Maximization of the total reward R in an episode is equivalent to minimization of objective function (5).*

Proof. The total reward in an episode is

$$R = \sum_{u=1}^{K} r_u$$

$$= \sum_{u=1}^{K} \sum_{j=1}^{n} \min \left\{ \sum_{i=1}^{m} \frac{(t_u - t_{u-1})\,\delta(i,j)\,L}{s_{T_{i,u}^0, T_{i,u}} + p_{i,T_{i,u}}}, \right.$$

$$\left. \left[D_j - Y_j(t_{u-1})\right]^+ \right\} w_j$$

$$+ \max \left\{ \sum_{i=1}^{m} \frac{(t_u - t_{u-1})\,\delta(i,j)\,L}{s_{T_{i,u}^0, T_{i,u}} + p_{i,T_{i,u}}} - \left[D_j - Y_j(t_{u-1})\right]^+, 0 \right\}$$

$$\times \frac{w_j}{M}$$

$$= \sum_{j=1}^{n} \sum_{u=1}^{K} \min \left\{ \sum_{i=1}^{m} \frac{(t_u - t_{u-1})\,\delta(i,j)\,L}{s_{T_{i,u}^0, T_{i,u}} + p_{i,T_{i,u}}}, \left[D_j - Y_j(t_{u-1})\right]^+ \right\}$$

$$\times w_j$$

$$+ \max \left\{ \sum_{i=1}^{m} \frac{(t_u - t_{u-1})\,\delta(i,j)\,L}{s_{T_{i,u}^0, T_{i,u}} + p_{i,T_{i,u}}} - \left[D_j - Y_j(t_{u-1})\right]^+, 0 \right\}$$

$$\times \frac{w_j}{M}. \quad (24)$$

It is easy to show that

$$Y_j = \sum_{u=1}^{K} \sum_{i=1}^{m} \frac{(t_u - t_{u-1})\,\delta(i,j)\,L}{s_{T_{i,u}^0, T_{i,u}} + p_{i,T_{i,u}}}. \quad (25)$$

It follows that

$$R = \sum_{j=1}^{n} \left[w_j \min\{D_j, Y_j\} + \frac{w_j}{M} \max\{0, Y_j - D_j\} \right]$$

$$= \sum_{j \in \Omega_1} \left[w_j D_j + \frac{w_j}{M}(Y_j - D_j) \right] + \sum_{j \in \Omega_2} w_j Y_j$$

$$= \sum_{j=1}^{n} w_j D_j - \left\{ \sum_{j \in \Omega_1} \left[-\frac{w_j}{M}(Y_j - D_j) \right] \right. \quad (26)$$

$$\left. + \sum_{j \in \Omega_2} w_j(D_j - Y_j) \right\}$$

$$= \sum_{j=1}^{n} w_j D_j - \sum_{j=1}^{n} \left[w_j(D_j - Y_j)^+ - \frac{w_j}{M}(Y_j - D_j)^+ \right],$$

where $\Omega_1 = \{j \mid Y_j > D_j\}$ and $\Omega_2 = \{j \mid Y_j \leq D_j\}$. Since $\sum_{j=1}^{n} w_j D_j$ is a constant, it follows that

$$\max R \iff \min \sum_{j=1}^{n} \left[w_j(D_j - Y_j)^+ - \frac{w_j}{M}(Y_j - D_j)^+ \right]. \quad (27)$$

$\square$

3. The Reinforcement Learning Algorithm

The chip attach scheduling problem is converted into an RL problem with terminal state in Section 2. To apply Q-learning to solve this RL problem, another issue arises, that is, how to tailor Q-learning algorithm in this particular context. Since some state variables are continuous, the state space is infinite. This RL system is not in tabular form, and it is impossible to maintain Q-values for all state-action pairs. Thus, we use linear function with gradient-descent method to approximate the Q-value function. Q-values are represented as linear combination of a set of basis functions, $\Phi_k(s)$ ($1 \leq k \leq 4m + n$), as shown in the next formula:

$$Q(s, a) = \sum_{k=1}^{4m+n} c_k^a \Phi_k(s), \qquad (28)$$

where c_k^a ($1 \leq a \leq 6, 1 \leq k \leq 4m + n$) are the weights of basis functions. Each state variable corresponds to a basis function. The following basis functions are defined to normalize the state variables:

$$\Phi_k(s)$$

$$= \begin{cases} \dfrac{T_k^0}{n} & (1 \leq k \leq m), \\[2mm] \dfrac{T_{k-m}}{n} & (m + 1 \leq k \leq 2m), \\[2mm] \dfrac{t_{k-2m}}{\max\left\{s_{j1,j2} + p_{j2} \mid 1 \leq j1 \leq n, 1 \leq j2 \leq n\right\}} \\ \hspace{3cm} (2m + 1 \leq k \leq 3m), \\[2mm] \dfrac{d_{k-3m}}{D_{k-3m}} & (3m + 1 \leq k \leq 3m + n), \\[2mm] \dfrac{e_{k-3m-n}}{\text{TH}} & (3m + n + 1 \leq k \leq 4m + n). \end{cases}$$

$$(29)$$

Let C^a denote the vector of weights of basis functions as follows:

$$C^a = \left(c_1^a, c_2^a, \ldots, c_{4m+n}^a\right)^T. \qquad (30)$$

The RL algorithm is presented as Algorithm 8, where α is learning rate, γ is a discount factor, $E(a)$ is the vector of eligibility traces for action a, $\delta(a)$ is an error variable for action a, and λ is a factor for updating eligibility traces.

Algorithm 8. Q-learning with linear gradient-descent function approximation for chip attach scheduling.

Initialize C^a and $E(a)$ randomly. Set parameters α, γ, and λ.

Let num_episode denote the number of episodes having been run. Set num_episode = 0.

While num_episode < MAX_EPISODE do

Set the current decision-making state $s \leftarrow s_0$.

While at least one of state variables e_i ($1 \leq i \leq m$) is larger than zero do

Select action a for state s by ε-greedy policy.

Implement action a. Determine the next event for triggering state transition and the sojourn time. Once any machine completes processing a lot or any machine's normal production time is completely scheduled, the system transfers into a new decision-making state, s' (e_i' ($1 \leq i \leq m$) is a component of s').

Compute reward $r_{s,s'}^a$.

Update the vector of weights in the approximate Q-value function of action a:

$$\delta(a) \longleftarrow r_{s,s'}^a + \gamma \max_{a'} Q\left(s', a'\right) - Q(s, a),$$

$$E(a) \longleftarrow \lambda E(a) + \nabla_{C^a} Q(s, a), \qquad (31)$$

$$C^a \longleftarrow C^a + \alpha \delta(a) E(a).$$

Set $s \leftarrow s'$.

If $e_i = 0$ holds for all i ($1 \leq i \leq m$), set num_episode = num_episode + 1.

4. Experiment Results

In the past, the company used a manual process to conduct chip attach scheduling. A heuristic algorithm called Largest Weight First (LWF) was used as follows.

Algorithm 9 (Largest Weight First (LWF) heuristics). Initialize SM with the set of all machines (i.e., SM = $\{i \mid 1 \leq i \leq m\}$) and define SJ as Step 1 in Algorithm 2. Initialize e_i ($1 \leq i \leq m$) with each machine's normal production time. Set $Y_j = I_j$, where I_j is the initial production volume of job type j.

Step 1. Schedule the job types in decreasing order of weights in order to meet their TPVs.

While SJ $\neq \Phi$ and SM $\neq \Phi$ do

Choose job q with $q = \text{argmax}\{w_j \mid j \in \text{SJ}\}$.

While SM $\cap M_q \neq \Phi$ and $Y_q < D_q$ do

Choose machine i to process job q, with $i = \text{argmin}\{p_{k,q}/w_q \mid k \in \text{SM} \cap M_q\}$.

If $e_i - s_{T_i^0,q} < (D_q - Y_q)p_{i,q}$, then

set $Y_q \leftarrow Y_q + L(e_i - s_{T_i^0,q})/p_{i,q}$, $e_i = 0$, and remove i from SM;

else, set $e_i \leftarrow e_i - s_{T_i^0,q} - (D_q - Y_q)p_{i,q}/L$, and $Y_q = D_q$.

$T_i^0 = q$

Step 2. Allocate the excess production capacity.

If SM $\neq \Phi$, then

For each machine i ($i \in$ SM),

Choose job j with $j = \text{argmax}\{(e_i - s_{T_i^0,q})w_q/p_{i,q} \mid 1 \leq q \leq n\}$, set $Y_j \leftarrow Y_j + L(e_i - s_{T_i^0,j})/p_{i,j}$, $e_i = 0$.

TABLE 1: Comparison of objective function values using heuristics and Q-learning.

Dataset no.	WSPT	MWSPT	RA	LFM-MWSPT	LFM-RA	LWF	Q-Learning
1	88.867	78.116	59.758	87.689	57.582	42.253	−3.8613
2	138.44	135.86	110.747	126.69	109.07	95.926	7.6657
3	119.01	108.75	124.09	104.25	121.90	83.332	23.775
4	83.681	60.797	39.073	69.405	45.575	33.920	−4.4275
5	129.38	128.47	96.960	109.17	99.827	89.863	21.414
6	70.840	55.692	51.108	66.213	51.041	16.930	−5.4467
7	120.90	100.60	95.399	109.33	90.754	76.422	27.374
8	102.42	107.80	116.56	103.33	107.62	93.663	11.840
9	94.606	87.914	81.763	88.812	80.331	60.164	33.036
10	90.803	88.164	90.773	87.926	88.56293	56.307	22.798
11	111.13	88.287	82.916	97.882	85.605	60.160	16.493
12	100.29	89.005	86.692	95.836	78.342	60.744	−3.8617
Average	104.19	94.123	86.321	95.547	84.685	64.147	12.233

TABLE 2: Comparison of unsatisfied TPV index using heuristics and Q-learning.

Dataset no.	WSPT	MWSPT	RA	LFM-MWSPT	LFM-RA	LWF	Q-learning
1	0.1179	0.1025	0.0789	0.1170	0.0754	0.0554	0.0081
2	0.1651	0.1611	0.1497	0.1499	0.1482	0.1421	0.0137
3	0.1421	0.1289	0.1475	0.1227	0.1455	0.0987	0.0691
4	0.1540	0.1104	0.0716	0.1258	0.0854	0.0614	0.0088
5	0.1588	0.1564	0.1186	0.1303	0.1215	0.1094	0.0571
6	0.1053	0.0819	0.0757	0.1006	0.0762	0.0248	0.0137
7	0.1462	0.1209	0.1150	0.1292	0.1082	0.0917	0.0582
8	0.1266	0.1324	0.1437	0.1272	0.1309	0.1150	0.0381
9	0.1315	0.1211	0.1133	0.1249	0.1127	0.0828	0.0815
10	0.1154	0.1112	0.1151	0.1105	0.1110	0.0709	0.0536
11	0.1544	0.1215	0.1146	0.1387	0.1204	0.0827	0.0690
12	0.1262	0.1112	0.1088	0.1194	0.1002	0.0758	0.0118
Average	0.1370	0.1216	0.1112	0.1247	0.1113	0.0842	0.0402

The chip attach station consists of 10 machines and normally processes more than ten job types. We selected 12 sets of industrial data for experiments comparing the Q-learning algorithm (Algorithm 8) and the six heuristics (Algorithms 1–5, 9): WSPT, MWSPT, RA, LFM-MWSPT, LFM-RA, and LWF. For each dataset, Q-learning repeatedly solves the scheduling problem 1000 times and selects the optimal schedule of the 1000 solutions. Table 1 shows the objective function values of all datasets using the seven algorithms. Individually, any of WSPT, MWSPT, RA, LFM-MWSPT, and LFM-RA obtains larger objective function values than LWF for every dataset. Nevertheless, taking WSPT, MWSPT, RA, LFM-MWSPT, and LFM-RA as actions, Q-learning algorithm achieves an objective function value much smaller than LWF for each dataset. In Tables 1–4, the bottom row presents the average value over all datasets. As shown in Table 1, the average objective function value of Q-learning is only 12.233, less than that of LWF, 66.147, by a large amount of 80.92%.

Besides objective function value, we propose two indices, unsatisfied TPV index and unsatisfied job type index, to measure the performance of the seven algorithms. Unsatisfied TPV index (UPI) is defined as formula (32) and indicates the weighted proportion of unfinished Target Production Volume. Table 2 compares UPIs of all datasets using seven algorithms. Also, any of WSPT, MWSPT, RA, LFM-MWSPT, and LFM-RA individually obtains larger UPI than LWF for each dataset. However, Q-learning algorithm achieves smaller UPI than LWF does for each dataset. The average UPI of Q-learning is only 0.0402, less than that of LWF, 0.0842, by a large amount of 52.20%. Let J denote the set $\{j \mid 1 \le j \le n, Y_j < D_j\}$. Unsatisfied job type index (UJTI) is defined as formula (33) and indicates the weighted proportion of the job types whose TPVs are not completely satisfied. Table 3 compares UJTIs of all datasets using seven algorithms. With most datasets, Q-learning algorithm achieves smaller UJTIs than LWF. The average UJTI of Q-learning is 0.0802, which is remarkably less than that of LWF, 0.1176, by 31.81%. Consider

$$\mathrm{UPI} = \frac{\sum_{j=1}^{n} w_j \left(D_j - Y_j \right)^+}{\sum_{j=1}^{n} w_j D_j}, \tag{32}$$

$$\mathrm{UJTI} = \frac{\sum_{j=J} w_j}{\sum_{j=1}^{n} w_j}. \tag{33}$$

TABLE 3: Comparison of unsatisfied job type index using heuristics and Q-learning.

Dataset no.	WSPT	MWSPT	RA	LFM-MWSPT	LFM-RA	LWF	Q-learning
1	0.1290	0.2615	0.1667	0.1650	0.1793	0.0921	0.0678
2	0.1924	0.2953	0.2302	0.2650	0.2097	0.1320	0.0278
3	0.2250	0.3287	0.2126	0.2564	0.2278	0.0921	0.0921
4	0.0781	0.2987	0.0278	0.1290	0.0828	0.0278	0.0278
5	0.2924	0.3169	0.3002	0.2224	0.2632	0.2055	0.0571
6	0.2290	0.3062	0.1817	0.2290	0.1632	0.0571	0.0278
7	0.2650	0.2987	0.2160	0.2529	0.2075	0.1320	0.1647
8	0.1924	0.3225	0.2067	0.1813	0.1696	0.1320	0.1320
9	0.2221	0.3250	0.1403	0.1892	0.2073	0.1320	0.0749
10	0.2621	0.3304	0.2667	0.2859	0.2708	0.1781	0.0678
11	0.2029	0.2896	0.2578	0.2194	0.2220	0.1381	0.1542
12	0.1924	0.3271	0.2302	0.1838	0.2182	0.0921	0.0678
Average	0.2069	0.3084	0.2031	0.2149	0.2018	0.1176	0.0802

TABLE 4: Comparison of the total setup time using heuristics and Q-learning.

Dataset no.	WSPT	MWSPT	RA	LFM-MWSPT	LFM-RA	LWF	Q-learning
1	0.8133	0.8497	0.3283	0.7231	0.3884	0.3895	1.0000
2	0.8333	1.3000	0.4358	0.7564	0.5263	0.4094	1.0000
3	0.8712	1.1633	0.4207	0.6361	0.4937	0.4298	1.0000
4	1.1280	0.7123	0.4629	0.8516	0.5139	0.4318	1.0000
5	0.9629	1.3121	0.4179	0.8597	0.5115	0.3873	1.0000
6	0.7489	1.0393	0.4104	0.7489	0.4542	0.4074	1.0000
7	1.7868	2.2182	0.8223	1.4069	1.0125	0.4174	1.0000
8	0.6456	0.8508	0.4055	0.6694	0.5053	0.3795	1.0000
9	0.9245	0.9946	0.5013	0.7821	0.6694	0.4163	1.0000
10	1.1025	1.7875	0.6703	1.0371	0.9079	0.4894	1.0000
11	0.9973	1.3655	0.3994	0.9686	0.5129	0.4066	1.0000
12	0.7904	1.1111	0.4419	0.6195	0.5081	0.4258	1.0000
Average	0.9671	1.2254	0.4764	0.8383	0.5837	0.4158	1.0000

Table 4 shows the total setup time of all datasets using seven algorithms. For the reason of commercial confidentiality, we used the normalized data with the setup time of a dataset divided by the result of this dataset using Q-learning. Thus, the total setup times of all datasets by Q-learning are converted into one and the data of the six heuristics are adjusted accordingly. Q-learning algorithm requires more than twice of setup time than LWF does for each dataset. The average accumulated setup time of LWF is only 41.58 percents of that of Q-learning.

The previous experimental results reveal that for the whole scheduling tasks, any individual one of the five action heuristics (WSPT, MWSPT, RA, LFM-MWSPT, and LFM-RA) for Q-learning performs worse than LWF heuristics. However, Q-learning greatly outperforms LWF in terms of the three performance measures, the objective function value, UPI, and UJTI. This demonstrates that some action heuristics provide better actions than LWF heuristics at some states. During repeatedly solving the scheduling problem, Q-learning system perceives the insights of the scheduling problem automatically and adjusts its actions towards the optimal ones facing different system states. The actions at all states form a new optimized policy which is different from any policies following any individual action heuristics or LWF heuristics. That is, Q-learning incorporates the merit of five alternative heuristics, uses them to schedule jobs flexibly, and obtains results much better than any individual action heuristics and LWF heuristics. In the experiments, Q-learning achieves high-quality schedules at the cost of inducing more setup time. In other words, Q-learning utilizes the machines more efficiently by increasing conversions among a variety of job types.

5. Conclusions

We apply Q-learning to study lot-based chip attach scheduling in back-end semiconductor manufacturing. To apply reinforcement learning to scheduling, the critical issue being conversion of scheduling problems into RL problems. We convert chip attach scheduling problem into a particular SMDP problem by Markovian state representation. Five heuristic algorithms, WSPT, MWSPT, RA, LFM-MWSPT, and LFM-RA, are selected as actions so as to utilize prior domain knowledge. Reward function is directly related to scheduling objective function, and we prove that maximizing the accumulated reward is equivalent to minimizing the objective function. Gradient-descent linear function approximation is combined with Q-learning algorithm.

Q-learning exploits the insight structure of the scheduling problem by solving it repeatedly. It learns a domain-specific policy from the experienced episodes through interaction and then applies it to latter episodes. We define two indices, unsatisfied TPV index and unsatisfied job type index, together with objective function value to measure the performance of Q-learning and the heuristics. Experiments with industrial datasets show that Q-learning apparently outperforms six heuristic algorithms: WSPT, MWSPT, RA, LFM-MWSPT, LFM-RA, and LWF. Compared with LWF, Q-learning achieves reduction of the three performance measures, respectively, by an average level of 52.20%, 31.81%, and 80.92%. With Q-learning, chip attach scheduling is optimized through increasing effective job type conversions.

Disclosure

Given the sensitive and proprietary nature of the semiconductor manufacturing environment, we use normalized data in this paper.

Acknowledgments

This project is supported by the National Natural Science Foundation of China (Grant no. 71201026), Science and Technological Program for Dongguan's Higher Education, Science and Research, and Health Care Institutions (no. 2011108102017), and Humanities and Social Sciences Program of Ministry of Education of China (no. 10YJC630405).

References

[1] M. X. Weng, J. Lu, and H. Ren, "Unrelated parallel machine scheduling with setup consideration and a total weighted completion time objective," *International Journal of Production Economics*, vol. 70, no. 3, pp. 215–226, 2001.

[2] M. Gairing, B. Monien, and A. Woclaw, "A faster combinatorial approximation algorithm for scheduling unrelated parallel machines," in *Automata, Languages and Programming*, vol. 3580 of *Lecture Notes in Computer Science*, pp. 828–839, 2005.

[3] G. Mosheiov, "Parallel machine scheduling with a learning effect," *Journal of the Operational Research Society*, vol. 52, no. 10, pp. 1–5, 2001.

[4] G. Mosheiov and J. B. Sidney, "Scheduling with general job-dependent learning curves," *European Journal of Operational Research*, vol. 147, no. 3, pp. 665–670, 2003.

[5] L. Yu, H. M. Shih, M. Pfund, W. M. Carlyle, and J. W. Fowler, "Scheduling of unrelated parallel machines: an application to PWB manufacturing," *IIE Transactions*, vol. 34, no. 11, pp. 921–931, 2002.

[6] K. R. Baker and J. W. M. Bertrand, "A dynamic priority rule for scheduling against due-dates," *Journal of Operations Management*, vol. 3, no. 1, pp. 37–42, 1982.

[7] J. J. Kanet and X. Li, "A weighted modified due date rule for sequencing to minimize weighted tardiness," *Journal of Scheduling*, vol. 7, no. 4, pp. 261–276, 2004.

[8] R. V. Rachamadugu and T. E. Morton, "Myopic heuristics for the single machine weighted tardiness problem," Working Paper 28-81-82, Graduate School of Industrial Administration, Garnegie-Mellon University, 1981.

[9] A. Volgenant and E. Teerhuis, "Improved heuristics for the n-job single-machine weighted tardiness problem," *Computers and Operations Research*, vol. 26, no. 1, pp. 35–44, 1999.

[10] D. C. Carroll, *Heuristic sequencing of jobs with single and multiple components [Ph.D. thesis]*, Sloan School of Management, MIT, 1965.

[11] A. P. J. Vepsalainen and T. E. Morton, "Priority rules for job shops with weighted tardiness costs," *Management Science*, vol. 33, no. 8, pp. 1035–1047, 1987.

[12] R. S. Russell, E. M. Dar-El, and B. W. Taylor, "A comparative analysis of the COVERT job sequencing rule using various shop performance measures," *International Journal of Production Research*, vol. 25, no. 10, pp. 1523–1540, 1987.

[13] J. Bank and F. Werner, "Heuristic algorithms for unrelated parallel machine scheduling with a common due date, release dates, and linear earliness and tardiness penalties," *Mathematical and Computer Modelling*, vol. 33, no. 4-5, pp. 363–383, 2001.

[14] C. F. Liaw, Y. K. Lin, C. Y. Cheng, and M. Chen, "Scheduling unrelated parallel machines to minimize total weighted tardiness," *Computers and Operations Research*, vol. 30, no. 12, pp. 1777–1789, 2003.

[15] D. W. Kim, D. G. Na, and F. F. Chen, "Unrelated parallel machine scheduling with setup times and a total weighted tardiness objective," *Robotics and Computer-Integrated Manufacturing*, vol. 19, no. 1-2, pp. 173–181, 2003.

[16] R. S. Sutton and A. G. Barto, *Reinforcement Learning: An Introduction*, MIT Press, Cambridge, Mass, USA, 1998.

[17] C. J. C. H. Watkins, *Learning from delayed rewards [Ph.D. thesis]*, Cambridge University, 1989.

[18] C. J. C. H. Watkins and P. Dayan, "Q-learning," *Machine Learning*, vol. 8, no. 3-4, pp. 279–292, 1992.

[19] T. Jaakkola, M. I. Jordan, and S. P. Singh, "On the convergence of stochastic iterative dynamic programming algorithms," *Neural Computation*, vol. 6, pp. 1185–1201, 1994.

[20] J. N. Tsitsiklis, "Asynchronous stochastic approximation and Q-learning," *Machine Learning*, vol. 16, no. 3, pp. 185–202, 1994.

[21] D. P. Bertsekas and J. N. Tsitsiklis, *Neuro-Dynamic Programming*, Athena Scientific, Belmont, Mass, USA, 1996.

[22] S. Riedmiller and M. Riedmiller, "A neural reinforcement learning approach to learn local dispatching policies in production scheduling," in *Proceedings of the 16th International Joint Conference on Artificial Intelligence*, Stockholm, Sweden, 1999.

[23] M. E. Aydin and E. Öztemel, "Dynamic job-shop scheduling using reinforcement learning agents," *Robotics and Autonomous Systems*, vol. 33, no. 2, pp. 169–178, 2000.

[24] J. Hong and V. V. Prabhu, "Distributed reinforcement learning control for batch sequencing and sizing in just-in-time manufacturing systems," *Applied Intelligence*, vol. 20, no. 1, pp. 71–87, 2004.

[25] Y. C. Wang and J. M. Usher, "Application of reinforcement learning for agent-based production scheduling," *Engineering Applications of Artificial Intelligence*, vol. 18, no. 1, pp. 73–82, 2005.

[26] B. C. Csáji, L. Monostori, and B. Kádár, "Reinforcement learning in a distributed market-based production control system," *Advanced Engineering Informatics*, vol. 20, no. 3, pp. 279–288, 2006.

[27] S. S. Singh, V. B. Tadić, and A. Doucet, "A policy gradient method for semi-Markov decision processes with application to call admission control," *European Journal of Operational Research*, vol. 178, no. 3, pp. 808–818, 2007.

[28] M. Kaya and R. Alhajj, "A novel approach to multiagent reinforcement learning: utilizing OLAP mining in the learning process," *IEEE Transactions on Systems, Man and Cybernetics Part C*, vol. 35, no. 4, pp. 582–590, 2005.

[29] C. D. Paternina-Arboleda and T. K. Das, "A multi-agent reinforcement learning approach to obtaining dynamic control policies for stochastic lot scheduling problem," *Simulation Modelling Practice and Theory*, vol. 13, no. 5, pp. 389–406, 2005.

[30] C. E. Mariano-Romero, V. H. Alcocer-Yamanaka, and E. F. Morales, "Multi-objective optimization of water-using systems," *European Journal of Operational Research*, vol. 181, no. 3, pp. 1691–1707, 2007.

[31] D. Vengerov, "A reinforcement learning framework for utility-based scheduling in resource-constrained systems," *Future Generation Computer Systems*, vol. 25, no. 7, pp. 728–736, 2009.

[32] K. Iwamura, N. Mayumi, Y. Tanimizu, and N. Sugimura, "A study on real-time scheduling for holonic manufacturing systems—determination of utility values based on multi-agent reinforcement learning," in *Proceedings of the 4th International Conference on Industrial Applications of Holonic and Multi-Agent Systems*, pp. 135–144, Linz, Austria, 2009.

[33] M. Pinedo, *Scheduling: Theory, Algorithms, and Systems*, Prentice Hall, Englewoods Cliffs, NJ, USA, 2nd edition, 2002.

5

Environmentally Lean Production: The Development and Incorporation of an Environmental Impact Index into Value Stream Mapping

T. J. Roosen and D. J. Pons

Department of Mechanical Engineering, University of Canterbury, Private Bag 4800, Christchurch 8140, New Zealand

Correspondence should be addressed to D. J. Pons; dirk.pons@canterbury.ac.nz

Academic Editor: Gabor Szederkenyi

There is a need to include environmental waste alongside other lean wastes. Current concepts of environmental waste focus on the total production of waste from a plant. However waste is generated by individual processes within the production. Therefore focused management of waste requires engineers to know what and where waste is being generated. This is often simply not known with any accuracy. This work offer a solution by developing a method to integrate environmental waste into the lean method of Value Stream Mapping (VSM). Specifically it integrate corporate environmental standards with the VSM process, thereby permitting established lean improvement process to be focused at specific environmental improvement actions. Application of the method is demonstrated in a manufacturing setting, representing a variety of environmental impacts. The deployment is capable of being generalised to any number of environmental factors. It is able to represent a customised waste index for a particular industry. Several ways to represent the multidimensional environmental wastes were explored via industry focus group. The resulting method can be used by production staff to quantify environmental impacts at the level of the individual process and aggregated to report wastes for the whole value stream.

1. Introduction

Lean seeks to reduce waste in a production process. One of the more common lean management tools is the use of value stream mapping (VSM). This analyses and represents the time taken to complete a process, with a particular emphasis on time that does not add value to the product, hence nonvalue-added time. VSM is used to reduce task time and subsequently reduce company monetary overheads.

VSM focuses on *time* as a wasted consumable. However lean as a whole is concerned with many other types of waste. Consequently organisations that seek to implement lean are typically required to use different lean tools to cover the various waste dimensions of their processes. This invariably means multiple systems, with their own implementation, culture, and reporting processes. There is ongoing interest in developing integrated lean systems that avoid this duplication.

One of these areas where better integration is desirable is between the time dimension as covered by VSM and the environmental waste dimension. Environmental waste is only weakly represented in current lean thinking, which tends to simply perceive waste as merely cost of the raw materials or decrements to the productivity of the production system. However, from the environmental perspective, the type of waste is important because of the different toxicity and effect on the environment. There are also problems in getting the environmental waste considerations embedded in the production activities. Collecting data on environmental waste and its impact on the environment is the easier part. The more difficult problem to overcome is the lack of vertical integration between the organisational data on environmental waste and the processes that originally created the waste (see Figure 1). It is difficult to attribute environmental waste back to its source in the production process and consequently difficult to apply the continuous improvement methods.

Environmentally Lean Production: The Development and Incorporation of an Environmental Impact
Index into Value Stream Mapping

41

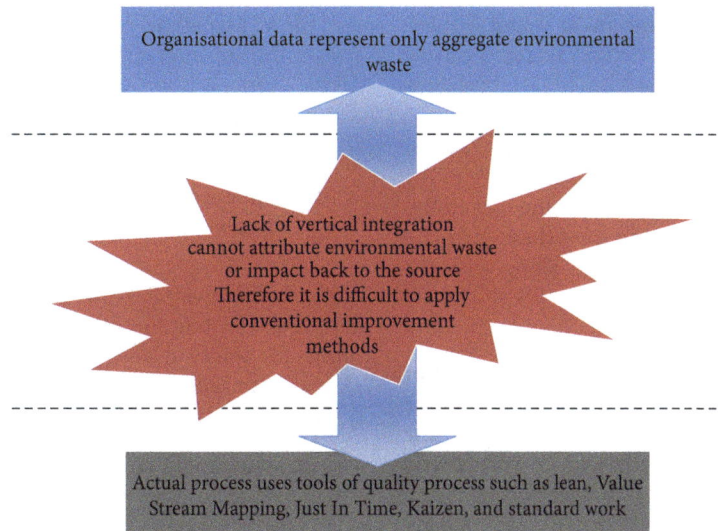

FIGURE 1: There is a lack of vertical integration between hard organisational data and process from which waste originated. This hinders the deployment of sustainability measures through the production system and down to the level of individual processes and operator work teams.

This paper provides a method for the integration of environmental waste into VSM processes. The particular area under examination is manufacturing, and representative data from a case are provided.

2. Literature Review

2.1. Lean Manufacturing and the Waste Principle. The perception of waste reduction primarily focuses on the diminution of environmental impact (ENI) through the use of traditional waste management programmes. Waste management is most often associated with objects disposed or recycled.

Lean manufacturing aims to reduce costs of production by eliminating waste and nonvalue-added activities and is a common underlying principle in many major businesses and production facilities around the world. Lean itself developed as a generalisation of the Toyota Production System (TPS), which itself was an embodiment of previous production quality systems [1–5].

In essence, lean manufacture seeks to preserve value within an organisation with overall less work and thus maximising efficiency through the reduction of waste. Though all these systems started in the manufacturing industry, the concept of "production process" can readily be applied to any other set of processes, even those that do not produce physical products. Consequently lean manufacturing has been greatly influential as a way of thinking in many industries beyond its automotive roots [6].

The TPS focused on pinpointing and eliminating waste [2, 4]. A series of tools were developed to help map and consequently eliminate three areas. These were "Muda", also known as the seven wastes, "Muri" the overburdening of people or equipment, and "Mura" the unevenness or irregular production [3, 7, 8]. The categories developed to describe the seven primary wastes (Muda), plus the eighth waste of underutilisation of people added later in development, are shown in Figure 2.

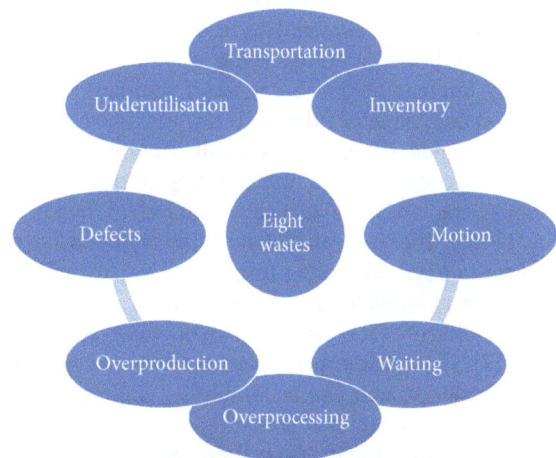

FIGURE 2: The eight wastes to be eliminated in a lean manufacturing system.

The lean methodology also subsumes many of the ideas from total quality systems, particularly the problem-solving approach. This may be summarised as investigating a problem > identification of the impediments > application of an improvement process > ongoing cycles of continuous improvement. The concept of empowerment of operators to make suggestions and arrange their own work is also common. Indeed both quality and lean are reliant on a culture that welcomes operator engagement in the production processes beyond merely the provision of labour. The burst of activity that creates the incremental improvement is a Kaizen activity. (Kaizen refers to the lean philosophy that no process can ever be perfect, so operations must be improved continuously through waste elimination events.) In this context the term means that adequate is never good enough and that no process can ever be thought to be perfect, so therefore each process must be continually evolved and improved.

The lean production paradigm can be accomplished by applying a wide variety of lean manufacturing tools such as Heijunka, Six Sigma, Kanbans, First In-First Out (FIFO), Value Stream Mapping (VSM), Takt (from Taktzeit meaning cycle) time, Just In Time (JIT), Single Minute Dye Exchange (SMDE), and 5 S principles [4].

There have been many attempts to explore the effectiveness of different techniques used to implement lean thinking in a real practice along with examining why some techniques might be preferential to others [9]. The tools themselves are a vital component of lean implementation along with the defining culture of lean.

2.2. Value Stream Mapping. VSM is a functional method or visual flow chart by which the production process can be represented as a set of processes connected in time. The method excels at showing the time dimension, particularly the nonvalue-added or waste time. It is therefore the lean method of choice for industries where costs are mostly determined by time or where a shorter production cycle confers competitive advantage. VSM can map an entire process, supply chain network, or the subtasks within a single process. It therefore readily scales hierarchically. In addition it maps both the material flow and the information that controls production [10]. The method, being a type of flow chart, is typically implemented using a set of standard icons for information and material flow [3, 11–13].

A given value stream includes all activities that contributed to a product, that is, value adding, nonvalue adding and supporting activities that are required to render the service [14–16]. The concept of waste within a manufacturing or information system can be further expanded through a categorisation of nonvalue adding (NVA) work, necessary but nonvalue adding (NNVA) work, and finally value-added (VA) work [3, 17]. Using these principles, the baseline processes within the value stream can be established and categorised. Once the value stream has been mapped, it becomes the baseline for improvement which then can be used to help create a future state map, which represents the desired future state including process improvements and reduction of NVA and NVA waste.

VSM is widely recognised in many different organisations irrespective of the type of system under examination. Research has mostly been focused on push/pull, Kanbans, inventory control, and mixed model assembly implementation. There has been less research into adapting concepts such as JIT, continuous improvement, cycle time reduction, visual management, automation, and floor space reduction into VSM simulation [18]. Another commonly recognised flaw in VSM is the inability to map value streams other than cycle time or cost. A limited number of modified VSM concepts have been developed to cope with complex value streams primarily network value mapping and critical path Value Stream Mapping.

2.2.1. Strengths of VSM. Some of the primary strengths of VSM are [9, 18].

(i) VSMs are able to easily identify waste (time and cost) from the values stream;

(ii) VSMs allow organisations to guide and visualise future information and material flow with iterative process improvements;

(iii) They map more than just waste and allow source and root cause to be examined,

(iv) They provide simple and objective analysis of complex systems.

2.2.2. Limitations of VSM. As with all processes, VSM has associated weaknesses inherent within the system design that limits the ability of VSM to be applied in every circumstance. A variety of limitations inherent in VSM are described below [9, 16, 18, 19].

(i) Static tool that captures snapshot-in-time not continuous flow,

(ii) Future state assumes every Kaizen will be fully completed,

(iii) Editing VSMs drawn by hand is cumbersome,

(iv) Detail capture of value stream is limited, especially in more complex multi stream systems,

(v) VSM does not represent spatial layout and consequent impacts of distance.

2.3. Methods of Environmental Waste Management. Waste management is the processing, collection, and transportation of waste (defined as "non-wanted things, that are perceived to have no purpose or value") to recover residual value or reduce consequences for the natural environment. Arguably the most widely used and universal waste management focused system is the ISO 14000 which is a family of standards relating to environmental waste management [20, 21]. They assist an organisation in minimising how their operations or processes can negatively affect the environment (i.e., cause adverse changes to air, water, or land). The ISO 14031 standard (from the ISO family) relates to Environmental Performance Evaluation (EPE) and is a management system which aims to assist organisations in identifying their environmental impacts by determining which aspects will be treated as significant, setting criteria for environmental performance, and assessing their environmental performance against these criteria. As part of the IS0 14000 family, another approach is found within the ISO 14040 set of standards, described as the Environmental Management—Life Cycle Assessment—Principles and Framework. The principal definition of the "Life Cycle Assessment" (LCA) is the assemblage and evaluation of the inputs, outputs, and potential environmental impacts of a product system throughout the product's life cycle. The LCA model is a more focused approach to waste management than ISO 14031.

Cradle to Cradle (C2C) is a methodology that uses biomimicry to compare and analyse the human resource system as a biological organism where materials and resources are modelled as nutrients in a health metabolism. The initial coining of the term was by Walter R. Stahel in the 1970s, but it was not until a modification of the Life Cycle Assessment saw the birth of the C2C ideology through the publication

Environmentally Lean Production: The Development and Incorporation of an Environmental Impact
Index into Value Stream Mapping

43

of Cradle to Cradle: *Remaking the Way We Make Things* [22]. The primary theory of the C2C principle is the idea of regenerative design in which every product is produced in a way in which it ensures recyclability of the resource.

The polluter pays principle (PPP), also known as extended producer responsibility (EPR), emphasises that the summation of all environmental costs throughout the life cycle of any product should be reflected in the market price of that product. PPP aims to change the waste paradigm from a governmental focus on waste and environmental initiatives to corporate or manufacturing entities which produce the waste and thus should also deal with waste impacts and disposal. This would mean that manufacturers would absorb greater responsibility in the cleaning, storing, recycling, and reuse of waste produced. This type of thinking has increasingly affected national policy formulation. Therefore it is becoming increasingly important for manufacturers to develop systems to better manage their environmental waste. The preferential method of waste management would be prevention and minimisation of waste at point-of-generation, as opposed to disposal and energy recovery. Hence it is desirable to include environmental waste into lean thinking.

2.3.1. Waste Management Indices.

Once an overall waste management framework is determined, it is crucial to then decide on an appropriate index in which specific environmental performance factors can be evaluated against. There exist several methods in which the environmental consequences can be measured or evaluated directly. It should be noted that a majority of the indices do not directly account for the principles of a Lean Manufacturing Programme. The ISO 14031 standards highlight the development of specific metrics through indicators. The process of choosing the indicator may include choosing from existing indicators or developing new indicators. (This standard describes the two general categories for Environmental Performance Evaluation (EPE) as Environmental Performance Indicators (EPI) or Environmental Condition Indicators (ECI). EPI can be further broken down into Management Performance Indicators (MPI) and Operation Performance Indicators (OPI). MPI is a type of EPI that provide information about management's efforts to influence the overall environmental performance of the organisation. On the other hand, OPI provides information about the environmental performance of an organisation's operations. Examples of how these three indicators interrelate are given in ISO 14031:2000.)

The US Environmental Protection Agency [20] environmental toolkit provides assistance in developing an environmentally conscious organisation. The most relevant features of the EPA toolkit relate to identification of environmental wastes and Environmental Value Stream Mapping (EVSM) adaptation. This discussion is primarily interested in the identification of wastes. Initially the toolkit describes links between the "seven wastes" and environmental wastes in identifying critical ENI. The EPA toolkit further explores the ability of targeting environmental waste in an organisation.

Environmental Management Accounting (EMA) (As described by United Nations Division for Sustainable Development UNDSD 2001) is a combined process that provides a method to translate data from financial accounting, cost accounting, and mass balance to improve material efficiency and reduce environmental impacts [23]. The primary focus of EMA is an assessment of the total annual environmental expenditure on emissions' treatment, disposal, and environmental protection and management. In essence EMA sets up procedures for internal decision making which include both physical procedures for material and energy consumption, flows, and final disposal and monetarized procedures for costs, savings, and revenues related to activities with a potential ENI. The total emissions method seeks to determine (through empirical analysis) evidence of a link between lean production practices and environmental performance [24]. The method explores three interrelated hypotheses. The hypotheses state that the more an organisation establishes lean principles, the more likely it will adopt formal environmental management systems, the less likely it will generate waste, and finally, the lower its emissions will be. In other words, an organisation's environmental performance could be defined by the degree it emits toxic pollutants [25]. The systematic (or strategic) environmental assessment (SEA) incorporates environmental considerations into policies, plans, programmes, and strategies of an organisation [26, 27]. Life Cycle Assessment (LCA) is a core concept in the development of environmentally conscious design and cleaner practices in industry and involves the evaluation of environmental burdens associated with product, process, service, or practice. Volvo along with the Federation of Swedish Industries jointly developed an Environmental Priorities Strategies (EPS) system to select appropriate materials to use during construction of its products [28, 29]. This method is based on environmental indices calculated for specific materials.

Another possible cumulative measurement for wastes is the use of a "carbon footprint" analysis in which waste of a very specific form can be aggregated and measured. The "carbon footprint" analysis is a method in which the total emissions of greenhouse gasses (GHG) are estimated in terms of the carbon equivalence (tCO_2e-tonnes of carbon dioxide equivalent or grams of CO_2 equivalent per kilowatt hour of generation (gCO_2eq/kWh)) from a specific product. The measurement is taken across a product's life cycle from raw materials used in manufacturing to the disposal of the final product. Its purpose is to measure the individual gas emissions from each activity within a supply chain process and framework and attribute these to each output product [30]. A carbon footprint, in other words, is a measure of the total amount of greenhouse gas (GHG) emissions. Carbon dioxide, methane, nitrous oxide, hydrofluorocarbons, perfluorocarbons, sulphur hexafluoride, and ozone are examples. These GHG emissions are either directly or indirectly caused by an activity or are accumulated over the life stages of a product. Toxicity was another possible measure of environmental impact, particularly the impact of a set process with respect to human health. Initial investigation of the use of toxicity as a potential EIF, particularly LD_{50}, was discarded due to the high degree in variability of data available for any substance measured. High use of estimated data along with large uncertainties and safety factors did not promote the use of this particular EIF as a contribution to the total Environmental Impact Index (EII).

The Global Report Initiative [31] promotes economic, environmental, and social sustainability. GRI provides companies and organisations with a sustainability reporting framework. The framework includes identification of a variety of aspects oriented towards long-term sustainability for the often described economic, environmental, and social categories. Within the environmental dimension is a section with a number of aspects concerning emissions, effluents, with both core and additional performance. Other performance indicators of the GRI (environment) include the aspects of materials, energy, water, and biodiversity making a total of 30 performance indicators. The GRI has become a widely used methodology for companies to measure and report on their sustainability practices with specific measurements identified.

3. Purpose: A Need to Integrate Environmental Factors with Lean

Current concepts of environmental waste focus on the total production of waste from a plant. They are interested in quantifying the amount of waste and its consequences on the natural environment. Hence there is an emphasis on containing the waste within the plan boundary then applying a postproduction process to neutralise the environmental impact, and finally releasing it across the plant boundary into the environment.

There is a growing awareness of the importance of incorporating environmental factors into lean methods. There have been a number of initiatives in this direction. One was to use the Integrated Definition for Function Modelling (IDEF0) as a modelling notation to incorporate an existing waste index [32]. That work at least showed that it was conceptually possible but did not implement environmental factors into operational practices in the real industrial setting. The United States Environmental Protection Agency (EPA) has developed an environmental value stream map (EVSM) method which examines natural resource flow by expanding the mapping process to include environmental waste streams [20]. This method has been applied to reduce water consumption in an alcohol and sugar industry case study [33]. This method easily focuses on one particular form of waste but lacks the ability to focus on environmental waste as a whole or even multiple environmental waste streams.

However, clean environmental identification practices will also require reduction of waste at its point-of-generation. Waste is not generated by a plant but by individual processes within the production. Therefore focussed management of environmental waste requires that production engineers first know what the waste is and where it is being generated. This is the crux of the problem, because this is often simply not known with any accuracy. In addition, production plants are controlled and improved by lean methods, and if some waste is not visible to the lean methods, then it will not be included in the continuous improvement cycles. It is therefore imperative to identify and embed the environmental issues into the lean tools.

There have been only minor developments in creating an overall value stream environmental index and an encompassing methodology. What is needed is a way to include environmental waste alongside the other lean wastes. If this can be achieved, then the organisational momentum and culture that sustain the lean initiatives will automatically ensure that environmental waste is included in the decision making.

4. Approach Taken

This project was contextualised in a research collaboration with a local industry partner. This firm provides remanufacture services for a high-value precision engineering product. The firm already had an established process for implementing VSM. What was missing was the incorporation of the environmental impact of each process. This was important for the firm for two major reasons: first, that the processes can involve toxic materials and secondly that the reduction of environmental waste was seen as a strategic competitive advantage.

We approached this problem in the following way. First, we created a composite environmental waste index. We used a variety of environmental impact factors, which were then integrated to form a single new impact index that was relevant to the operational purpose of the firm. We created several different concepts for how such an index might be visually represented within the VSM framework.

Second, we tested the relevance of these concepts within the firm. Focus groups within the industry were used to identify the waste types and index factors that were most applicable to the situation. They also selected, from among the multiple concepts, which visual representation was the best for them. The focus group was comprised of several people with a variety of roles within the firm, including engineering managers, Environment Health and Safety officers, and quality control engineers. This part of the method ensures that the results are relevant to the industrial perspective and provides a degree of confidence in the applicability. We did this with awareness that adoption within an organisational culture is important for the success of any new initiative, hence the special care to engage stakeholders in the design process.

Third, from the results of the focus group we then designed the details of an integrated environmental waste-VSM (EW-VSM) method. We shaped this around VSM as that is the dominant lean tool used in this type of industry. We found a way to represent multiple dimensions of environmental waste (in this case five) for each process in the value stream. We also found a way to represent the aggregated environmental waste for the whole value stream. This permits the methodology to scale with the production hierarchy.

The fourth part of our method was to deploy this EW-VSM in the firm, on actual production lines. An environmental value stream map was conducted on a process that was identified to incorporate a large amount of environmental impacts such as high energy use, carbon footprint, high cost of waste removal, and toxic materials. A current state map of the process was constructed by a team including a quality engineer, VSM specialist, production workers, and

Environmentally Lean Production: The Development and Incorporation of an Environmental Impact
Index into Value Stream Mapping

45

technical manager. This exercise was conducted over a three-day period. The implementation began with a tutorial of how the environmental impact analysis methodology worked and how it was integrated with VSM use. The selected practitioners were informed of the new methodology through the use of standard operating procedures (SOP) that had been specially written.

After informing the users, the index implementation and evaluation began. The evaluation started with a review of a particular process (Annulus Filler). Once all participants were informed of the overall approach of the environmental index method and its relationship to VSM, the first stage of the analysis was instigated. The data acquisition begins with setting the initial percent target waste reduction (in this application 80% was chosen) followed by the capture of all five impact factor components. The data capture included calculation of all carbon footprint data by hand, perceived impact, determining cost to remove waste, volume of waste removed and remaining, and finally the site based Risk Register values for each process. The environmental impact factors were then aggregated into the single environmental impact index for each of the nine stages of the VSM. The VSM with added environmental impact index data bar and summary system radar chart was then analysed along with the process radar charts to determine which process had the highest environmental impact. Finally, after all information was captured as required, the environmental value stream ladder was added to the VSM, as well as Kaizen events identified.

The fifth and final part of our approach was to survey users for their responses to the method. We did this by a survey. We were interested in the relevance and ease-of-use from the perspective of industry practitioners. This part of the method was therefore a check on the applicability of the EW-VSM construct. The survey questions are included in the results. The respondents were from those who had participated in the EW-VSM as well as other roles within the plant. Ethics approval was obtained for the survey from the University of Canterbury.

5. Results

5.1. Environmental Impact Index (EII). Several factors relating to the use of an index at the local industry based sponsor were required to be taken into consideration when developing the appropriate aggregated composite EII scale. The first key factor for aggregated scales is the need for an index that can consider the broader definitions of waste and environmental impact and accommodate the specific operational characteristics and strategic purposes of the organisation. A design with a multileveled weighting scale can accommodate a wide variety of EIF.

A series of nine possible environmental waste impact indices were initially examined. The EPA toolkit, EMA method, Emissions index, SEA index, and ISO 14000 have all been omitted at this stage of the project due to several limitations. Of these, the EPA toolkit, EMA method, and ISO 14000 were eliminated due to their low scores for ease of use, ease of integration, and adaptability. The Emissions index and

SEA index suffer from being too specific and inflexible in accommodating different forms of waste or environmental impact scenarios.

The surviving candidate indices were an adapted Volvo environmental priority system, simple carbon footprint index, GRI index, simplified risk and consequence index, and a custom scale. Benefits of these indices include the following:

(i) ability for some indices to accommodate multiple environmental factors (custom scale),

(ii) some proposed indices are widely used and recognized (GRI and ISO),

(iii) ability to adapt the index is recognized as a key benefit (custom scale),

(iv) ability to quickly and effectively reflect poor performing processes,

(v) ability for practitioners of various skill levels to use and operate.

Detriments of these selected indices include:

(i) some indices are based on single environmental factors (Volvo and carbon footprint),

(ii) some indices (including custom scale) are not recognized or officially vetted,

(iii) overly complex index creation (GRI and ISO).

5.2. Conceptual Design of an Index for Environmental Waste. We applied a conceptual design process to the development of the Environmental Impact Index (EII) and its visual representation. We did this because representation is an important factor in usability, and we were specifically interested in a scale design that would be easy to implement. Thus, we were also designing for change management. For this reason the process of design specifically included focus groups from within the industry under examination.

The study examined possible visual displays to represent the chosen index. We also needed the representation to be easily integrated into current VSM maps. To consolidate the disparity gap between overall site waste data and process level information, two main design criteria were required to be met. The first element required to consolidate the disparity was to create or modify an appropriate waste index and encompass this index into an overall evaluation methodology that could be used to determine specific environmental impacts at the process level. The second criteria required to be fulfilled was to create a robust visual representation method that would effectively highlight high environmental impact processes that required Kaizen (waste reduction) to initiate.

Several concepts were explored through focus group review sessions. These concepts included a bar graph display, representative symbols, and simple process flow charts, as shown in Figure 3. Participants selected a coloured flow process chart, for clarity of communication and ease of integration with VSMs. The summary of the EII may then be displayed as an environmental waste impact ladder below the current lead time ladder as shown in Figure 3.

(a) Bar graph concept

(b) Waste pipe concept

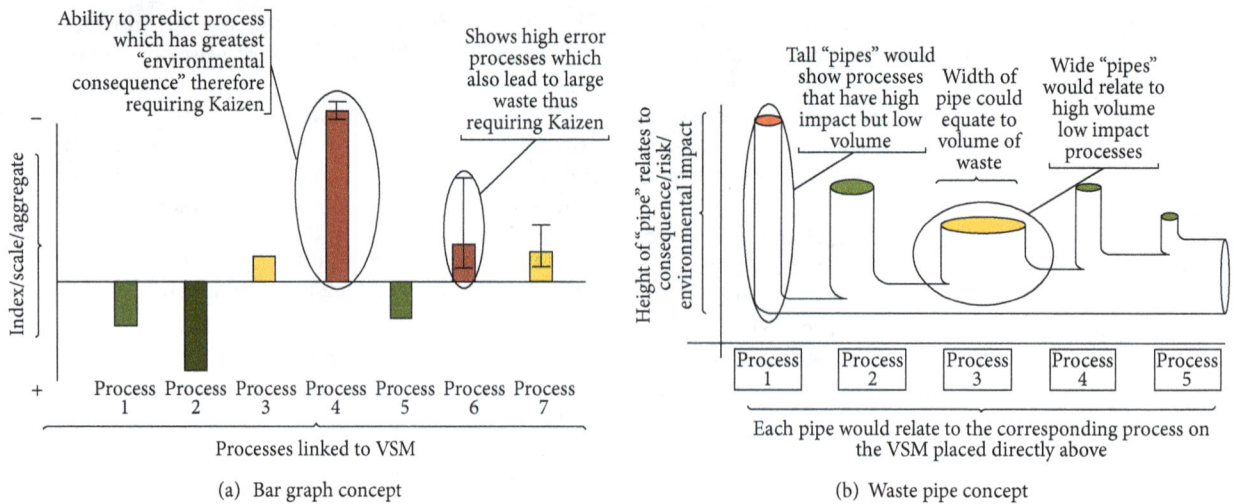

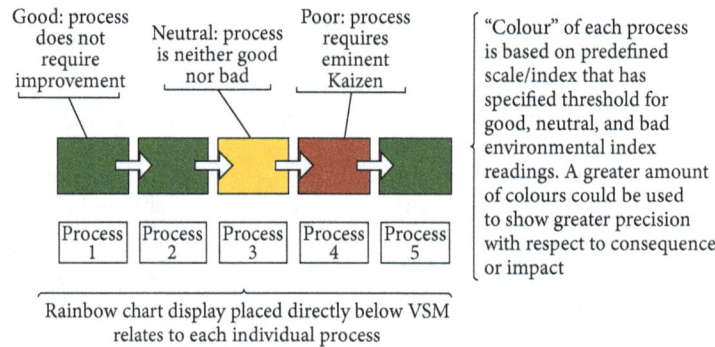

(c) Process chart concept

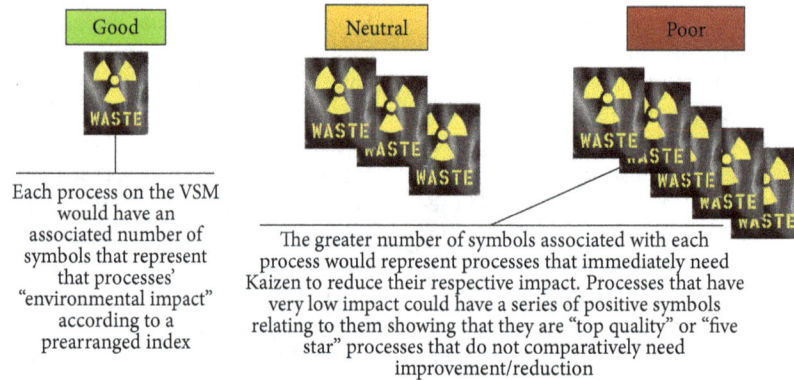

(d) Symbolic representation concept

FIGURE 3: Summary of initial visual display concepts.

Following further industrial practitioner based focus-group review sessions with leaders in the Environment Health and Safety (EH and S), lean, and VSM (Value Stream Mapping) groups, a final customised index was chosen which incorporated various aspects of the previously described standards and indices. The most favourable index by general consensus was a customizable index that would allow the organisation to modify the index based on current site objectives and organisational purpose. A custom scale was also deemed the most preferred option because it allowed

Environmentally Lean Production: The Development and Incorporation of an Environmental Impact
Index into Value Stream Mapping

47

a balance to be created between accuracy of results, adjustability of index, and adaptability of the applied method to highlight high environmental waste impact process. An important specification identified by the focus group was to develop a composite index to be customisable to allow for future modifications as a result of changes to the organisational purpose of site goals, essentially future proofing of the methodology and index. Five EIF were chosen, reflecting the current strategic goals and organisational purpose of the particular industrial application. Descriptions of the chosen set of EIF for this application were as follows:

(i) Carbon footprint.

(ii) Perceived impact of waste (levels 1–10):

   (a) *Level 1*: relates to near zero or minimal perceived human impact such as paper or storm.

   (b) *Level 5*: relates to medium level of perceived human impact such as sewage.

   (c) *Level 10*: relates to very high perceived human impact such as anthrax, radiation, or asbestos.

(iii) Cost of cleanup/remediation per kg.

(iv) Removed waste volume × Site Environmental Risk Register value (based on ISO 14001 standards),

(v) Remaining waste volume × Site Environmental Risk Register value (based on ISO 14001 standards),

These cover all the factors that the focus group deemed pertinent to the site. However we note that the method is able to accommodate different factors and different numbers thereof, and we recommend that practitioners give thought to the wastes appropriate in their own situation rather than unthinkingly adopting the above list.

5.3. Creation of Composite Index for Environmental Waste. Creating a composite index consists of several key stages, the initial EIF estimation, determination of an average EIF, and aggregation of the final EII. This overall process is shown in Figure 7. The aggregation of the composite index starts with the definition of the chosen environmental impact factors (EIF) shown in Figure 4. These interchangeable factors are the foundation for which the final EII will be based on and must be selected carefully to reflect the organisational purpose, goals, and environmental aims of the organisation in question. The chosen factors used in this particular application were decided through a series of focus group discussion as discussed previously.

The second aspect that must be defined is the scaling factor (SF). This element allows a layered system approach to be undertaken when determining which EIF is most important from a customer, practitioner, or manufacturing perspective. This preweighting also allows compensations to be made for low numerical valued EIF. At the outset, the SF would remain one unless a specific EIF needs to be highlighted or targeted. If a larger SF is required the practitioners have been advised to increase the SF in increments of 10 until a suitable value is reached, reducing the complexity

of determining an appropriate number. This SF is used as an alignment modification factor to reduce or enlarge the importance of any of the chosen EIF. This might be useful to reflect a changing organisational strategic purpose, for example placing greater importance on, say, carbon footprint. By increasing the SF of the carbon footprint aspect, the company would effectively increase the percent contribution of that EIF to the overall index. Importantly the production improvement processes inherent in the lean systems will automatically refocus to reduce this particular waste.

Once the appropriate EIF have been confirmed, the data collection for each EIF begins. To compensate for inaccurate, limited, or estimated data collection of EIF, a project evaluation and review technique (PERT) analysis was used to determine an average EIF value. This proceeds from fitting a beta probability distribution to three estimates, shown in (1) and Figure 5. The EIF values are separated into Pessimistic (P), Expected (E), and Optimistic (O) values. The distribution is weighted towards the expected EIF value, as per the function for the mean of the beta distribution, and this also minimises extreme data outliers such as an overly optimistic or pessimistic evaluation:

$$\text{EIF estimated} = \frac{(O + 4E + P)}{6}. \tag{1}$$

After the mean EIF value is determined, the EIF is then multiplied by a scaling factor (SF), as determined above. The next stage of the aggregated composite index is to assimilate the various EIF into a single index. This is determined by adding the vector magnitude of each EIF together, as shown in (2):

EII (vector magnitude)

$$= \sqrt{(\text{EIF}_1 * \text{SF}_1) + (\text{EIF}_2 * \text{SF}_2) + \cdots + (\text{EIF}_n * \text{SF}_n)} \tag{2}$$

There are several reasons for using a vector magnitude to determine the final EII. The first is that this permits any number of waste dimensions to be consolidated to a single value; that is, it makes the method scalable. A representation of a 3-dimensional waste problem is shown in Figure 6, and although a graphic representation is unavailable for the general nth dimensional problem, the vector magnitude still works. (This reason relates to the theoretical modelling used to address the problem and create a suitable solution. The approach was used to examine if the application of risk maps and consequence scales, representing environmental risk, could be used to provide a single valued solution. This concept of a risk map was replaced by a model in which the x and y axes described EIF characteristic of carbon footprint and volume of waste for a specific process. This model was further expanded to include a greater number of axes that represented different appropriate EIF. The end result was the creation of an nth dimensional model that could be used to describe any number of EIF. Finally a five-dimensional model was chosen, with each EIF being represented by a separate axis. Each process could then be mapped in accordance to the contribution of EIF, represented by a separate axis. This

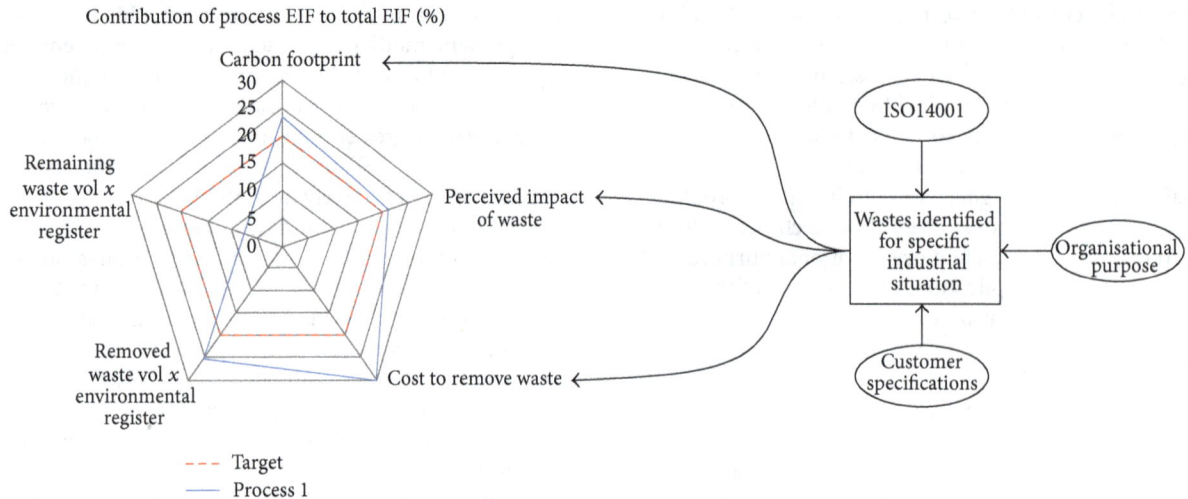

FIGURE 4: Initial environment impact factors consideration and subsequent radar chart for each process in value stream.

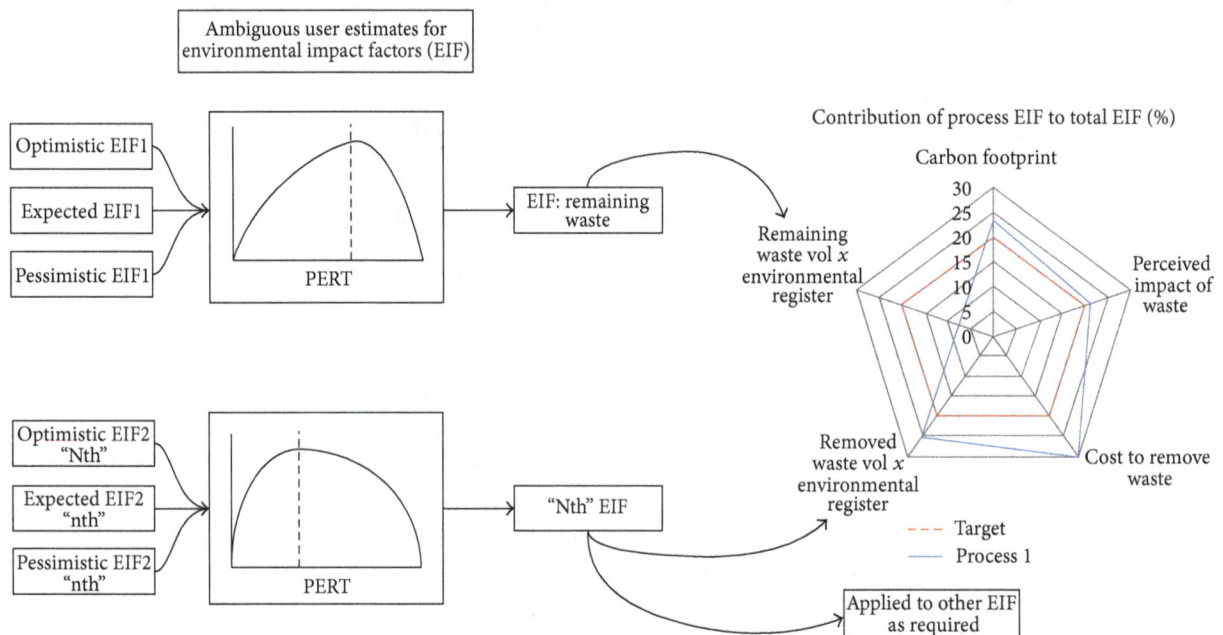

FIGURE 5: Method of determination for each environmental impact factor value using the PERT three-point estimation.

resulted in a representative 5-dimensional vector for each process, shown by a simplified illustration in Figure 6. The vectors describing each process could then be consolidated into a single valued unit through the use of the vector magnitude equation. This also means that with the addition of any extra "dimensions" describing a different EIF, final solution can be easily adjusted by adding in another vector component.)

The second reason for using a vector magnitude relates to the inability to simply multiply or add the EIF together. Direct multiplication or addition of the chosen EIF is not recommended as this could often result in large number valued solutions for specific processes as a result of one particularly high EIF that could skew the results. This problem is solved by using the vector magnitude equation as well as incorporating

a scaling factor in the magnitude equation to ensure no single EIF or process dominates the overall analysis. Thirdly the vector magnitude approach allows for the likely event of a specific process having a zero valued EIF. If multiplication was used then the final value representing a process with a zero valued EIF would be reduced to zero, reflecting inaccurate result. The vector approach allows for any number of EIF to be zero values and still results in a final indicative EII.

Finally addition of EIF was considered a possible aggregation method but due to both large number dominance of some EIF compared to others and unit mismatch, this was discarded in favour of the vector approach. The methodology created is able to accommodate any number of types of waste, as discussed above, and we refer to this as an nth dimensional concept. The current model uses 5 waste dimensions. Each of

Environmentally Lean Production: The Development and Incorporation of an Environmental Impact
Index into Value Stream Mapping

49

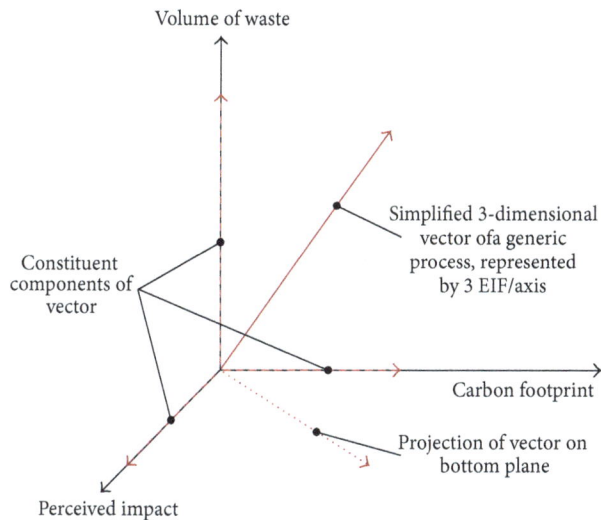

FIGURE 6: Simplified three-dimensional vector representation of generic system process.

these is represented on one axis, and additional axes may be added and further wastes are included. The vector magnitude then reduces the nth dimensional representation to a single value. This is useful for the ability to report summary data to managers and corporate staff and hence indicates how well the plant is meeting the strategic objectives. Thus the method integrates well with strategic management initiatives at the one level and lean improvement (via VSM specifically) at the operational level.

5.4. Identifying Environmental Kaizen Opportunities for Improvement. The purpose of lean initiatives is to identify areas for continuous improvement. These improvement foci are termed Kaizen events. Also important in lean is the realisation that not everything can be improved, because of finite resources, and it is therefore important to be able to identify which deficiencies are most worth targeting. In the case of value stream mapping, it is usual to use a burst symbol to represent the Kaizen events on the VSM chart. Also, VSM uses the concept of future state to identify the target reduction in nonvalue adding times. In the case of the environmental VSM approach described here, the Kaizen concept is directly applicable. It is straightforward to identify where to apply the environmental Kaizen, based on the process activities with the highest waste scores. Contextual knowledge of the plant may then be used to further identify which processes are likely to be more or less amenable to change. Note that the environmental Kaizen are not necessarily at the same location as those for the standard VSM. This is because the one set of process improvements are focussed on the environmental issues and the other on the temporal. (We use a green burst symbol to show the environmental Kaizen and yellow for the temporal.)

In application, the selected environmental impacts are integrated into a single EII, and a series of radar charts are created. These display the performance of individual processes and the overall system. Radar charts and conditional formatting are then used to identify the processes

which required environmental Kaizen initiatives. The first set of radar charts used are at the process level and break down each individual processes performance compared to an overall threshold value (see Figure 7). This threshold may be determined by creating an overall "target" percent based value of the maximum calculated index. The highest index would be multiplied by the high and low percent targets. These percents are then used across the entire system to determine good, neutral, and bad performing processes.

The radar charts are used in two ways. The first radar chart ((B) in Figure 7) is a summary figure which displays overall process performance of each process compared to the percent thresholds. The high percent bound is determined by reducing the highest calculated EII by the top percent target, whilst the low bound level is determined by multiplying the highest calculated EII by the low bounded percent target. Any process above the maximum bound in the summary radar chart can be described as a critical process requiring Kaizen activities to reduce the overall EII value. Conditional formatting has been used to set the displayed summary process EII to red to reflect a poor EII performance if above the maximum bound. Processes that are between the bounds are ones that do not require immediate attention but have the potential to have a large EII over the next few EVSM iterations and are set to display yellow. Finally, processes below the minimum threshold are set to a green showing that they will most likely not require intervention.

The second use of the radar chart is to display a breakdown of each processes performance with respect to the chosen EIF. The first step is to determine the total sum of the total system EIFs Each process radar chart is then created by determining the percent contribution of that processes EIF to the total system EIF. The practitioner can easily compare and identify which environmental factor of what particular process requires Kaizen implementation (as shown in Figure 8). The final aspect of the index incorporation is the inclusion and transfer of the summary EII data onto the standard VSM templates creating the final EVSM product. An example representative EVSM is shown in Figure 7. The figure shows a representative standard value stream (yellow data boxes), standard "time" domain Kaizen, and associated lead time ladder. Below the lead time ladder is the main contribution of this paper, the inclusion of an integrated environmental impact ladder and associated environmental Kaizen linking key lean VSM use with environmental considerations.

5.5. Application to Industrial Case Study. The industrial case study under examination is an organisation that remanufactures aviation turbines. Quality of work is of utmost importance, due to the safety and reliability considerations. In addition, a rapid turnaround of the product is important for the client's utilisation of expensive airframes. The minimisation of environmental waste is important for both the client (the aviation industry is sensitive to carbon footprint) and the remanufacturer (toxicity of plating processes in particular). The environmental VSM approach was applied in this environment and results follow.

First, the firm identified the EIF to which it was sensitive (see Section 5.1). These were carbon footprint, perceived

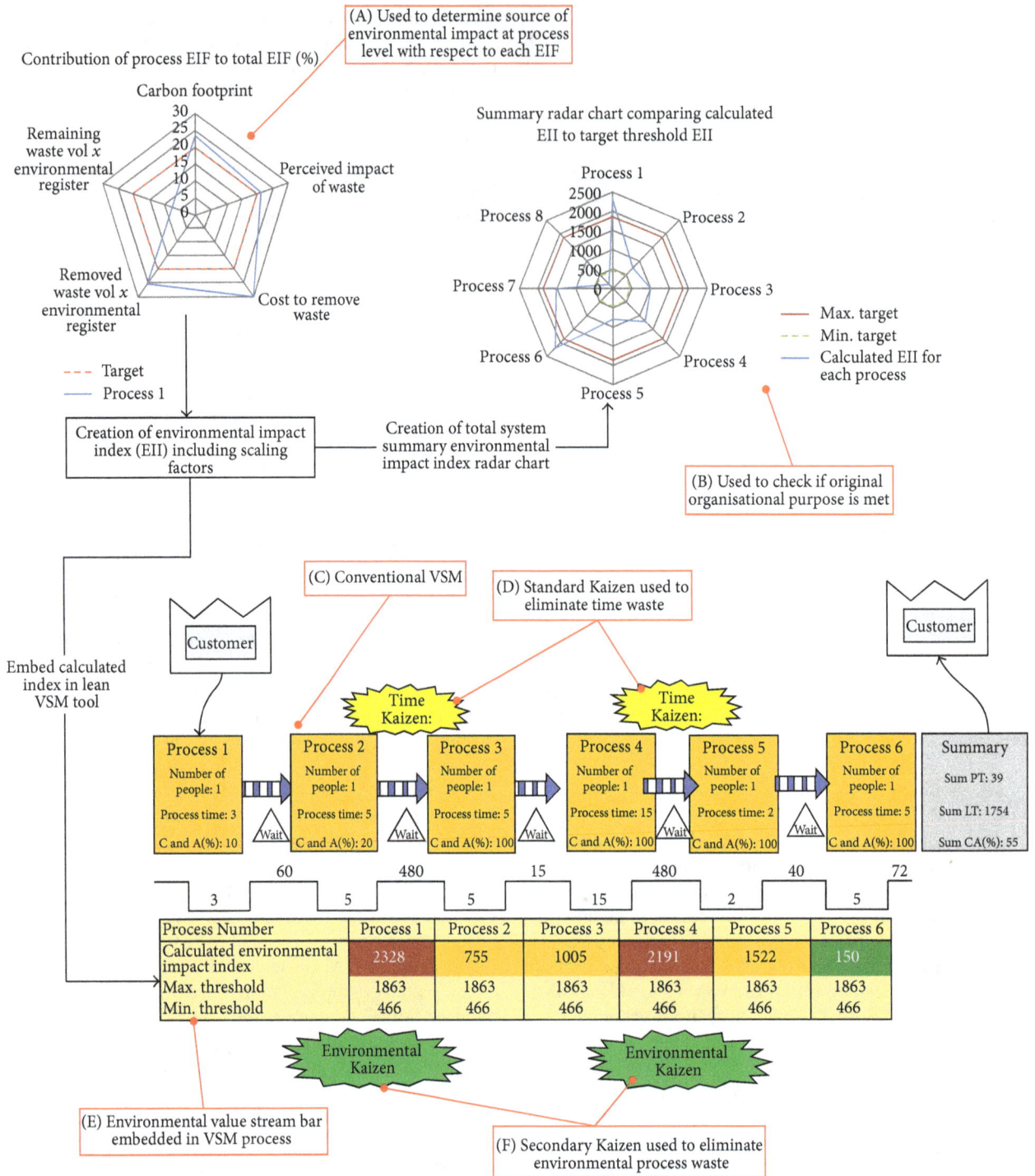

FIGURE 7: Illustration displaying the method in which final EII is aggregated from EIF data ((A) and (B)), the conventional time value stream ((C) and (D)), how the EII are incorporated into VSM (E), and the resulting environmental Kaizen created (F).

impact of waste, cost of cleanup/remediation, removed waste volume (weighted according to Site Environmental Risk Register), and remaining waste volume (likewise weighted). This was made for a total of five impacts (the methodology accepts any number). The environmental impacts were then assessed as part of a real VSM development.

5.5.1. *Current State Environmental VSM.* The EVSM method described was applied to a production process value stream within the industrial setting. A typical process stream might consist of between seven and a hundred activities depending on the level of detail required for analysis. The chosen value stream consisted of nine process stages that contained a large

Environmentally Lean Production: The Development and Incorporation of an Environmental Impact
Index into Value Stream Mapping

51

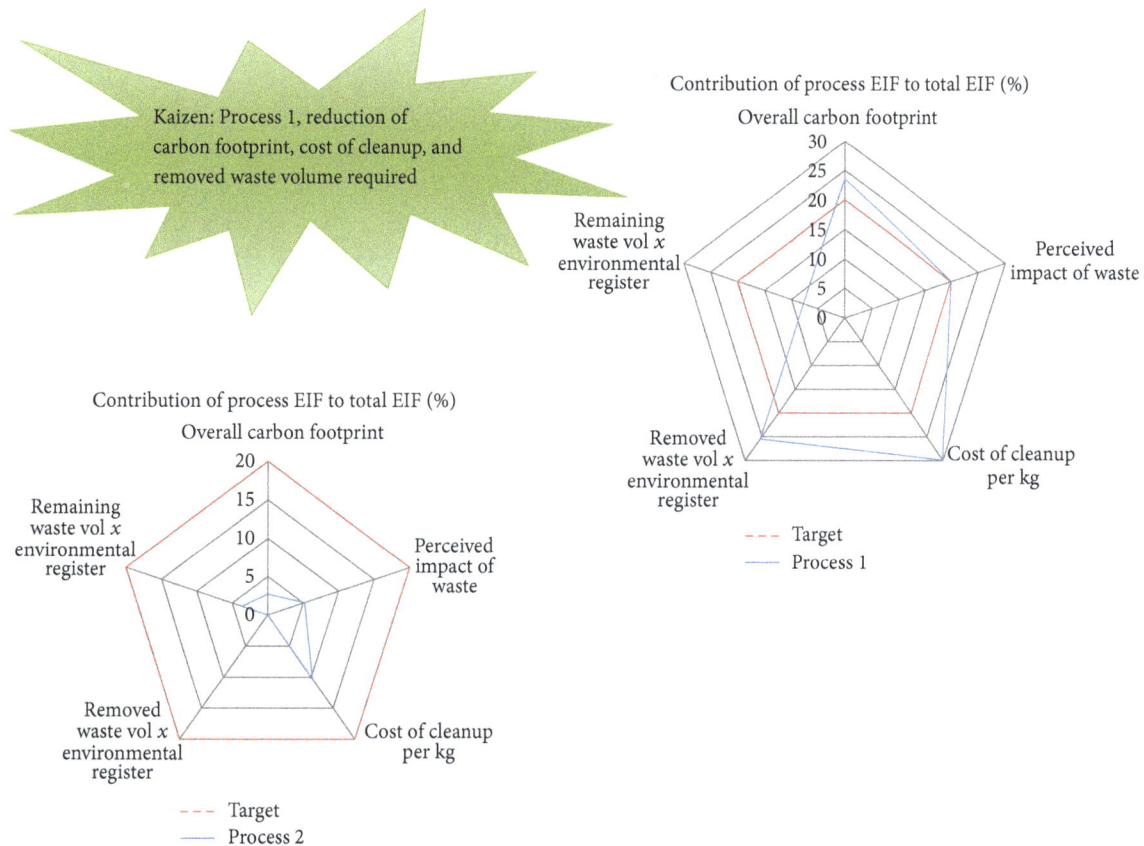

FIGURE 8: Representative sample and comparison of two process EIF radar charts and example of Kaizen for high environmental impact Process.

variety of environmental waste impacts. The method was applied to the VSM chosen and the EVSM created as shown in Figure 9.

5.5.2. Environmental Kaizen. This process resulted in a key Kaizen being created, for the high environmental impact Process 6, shown in Figure 10.

There is always a balance between economic and environmental goals during the continuous improvement process, and for this reason it is useful to have managerial representation in the Kaizen event. In principle the target future environmental waste levels can be included in the future state map.

5.6. User Survey of Applicability. To validate the effectiveness of the created EVSM method and associated index, a survey was conducted of industry participants. The questions relevant to the present study are as follows.

  (i) Question 1: to what extent is it important to measure environmental waste impacts?

  (ii) Question 3: to what extent does the practitioner feel the tool was successful in promoting new thinking and continuous improvement?

  (iii) Question 4: to what extent does the practitioner feel the method was effective at identifying environmental waste impacts?

Responses are shown in Figure 11.

These results show that practitioners understood the importance of measuring the impact of environmental waste (Q1). They felt the tool was successful in promoting new thinking and continuous improvement (Q3) and effective at identifying environmental waste (Q4). The practitioners also felt the tool helped sensitise the user to the environmental impact of processes as well as show actual process level data attributing overall site wide data to source of environmental waste impact. We therefore conclude that the tool was successful in achieving the primary purpose.

6. Discussion

6.1. Outcomes: What Has Been Achieved? This work has made several contributions to the body of knowledge regarding environmental Value Stream Mapping and lean manufacturing principles. The first contribution is the creation of a method to integrate environmental and lean methods. Specifically we have shown integration from the generic *environmental standard* ISO 14001 through an *organisational environmental risk register*, onwards to integration within the *VSM process*, and thus finally permitting the established lean *improvement process* (e.g., kaizen) to be focussed at specific environmental improvement actions. Thus we have found a way to take the abstract concepts of environmental waste and make them concrete. Specifically, we have developed a

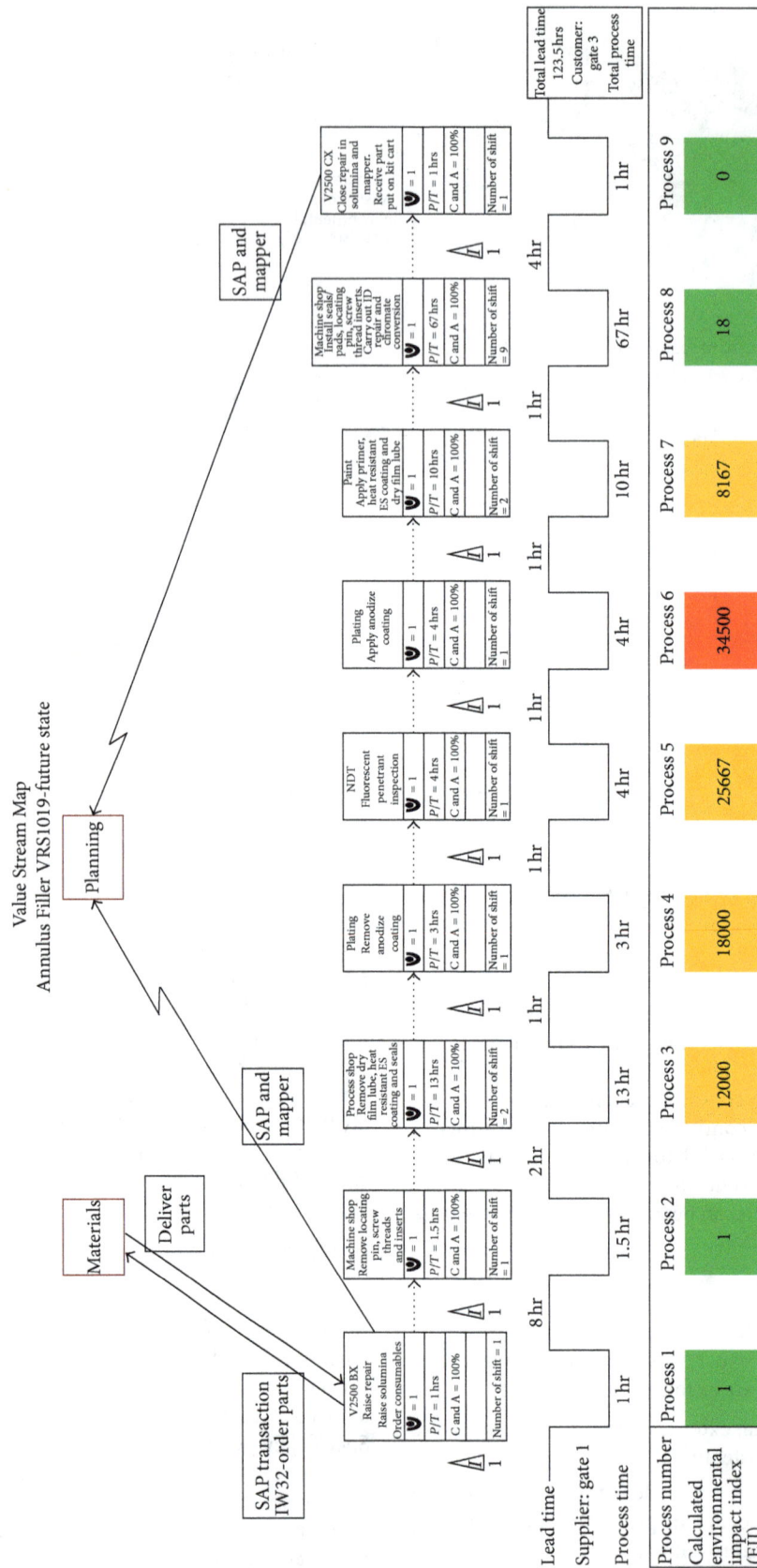

FIGURE 9: Implemented Environmental Impact Index incorporated with VSM for the chosen industrial value stream.

Environmentally Lean Production: The Development and Incorporation of an Environmental Impact
Index into Value Stream Mapping

53

Contribution of process EIF to total EIF (%)

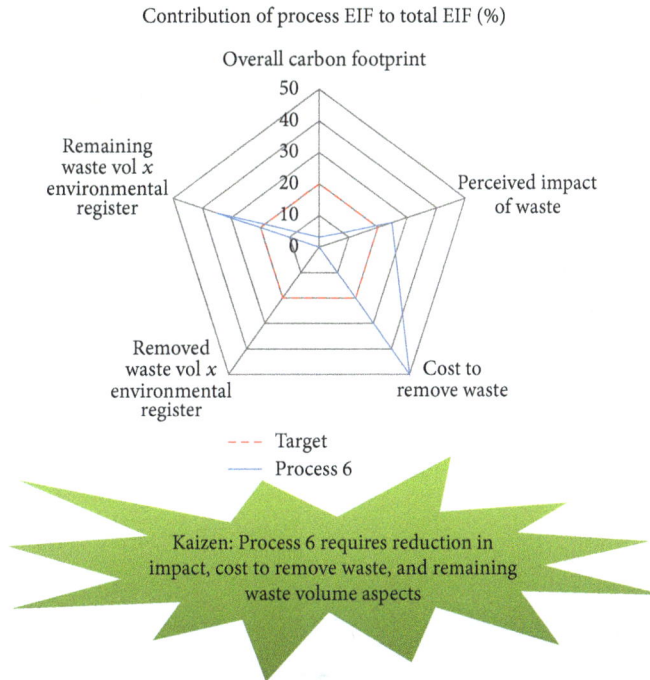

FIGURE 10: Process level radar chart providing key output: environmental Kaizen created in response to high impact process and waste identification through use of applied method and radar charts (Process 6-Plating: apply anodize coating).

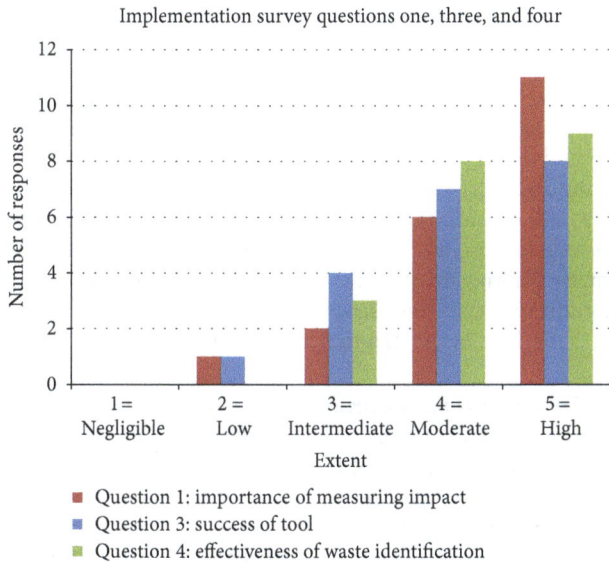

FIGURE 11: Results of survey questions one, three and four. These results demonstrate effectiveness of the method.

method to operationalise environmental waste within the VSM lean method. This methodology is shown in Figure 12.

A second contribution is the development of an nth dimension environmental factor methodology to create a *customised* environmental waste index for a particular industry. While the index created for this specific case used carbon footprint, perceived impact, cost to remediate, and waste volumes (removed and residual), the method is capable of being generalised to use different and any number of factors.

A third advancement is that we developed a way to use *ambiguous user estimates* of the quantity of each type of waste. This is important because it provides a basis for estimating values that are imprecise and otherwise difficult for operators to commit to a single deterministic value. Thus the method is capable of identifying areas for improvement (which is the overall purpose) despite ambiguous and imperfect information. To achieve this we used the PERT beta distribution, which already has acceptance in the project-management field.

A fourth contribution is the design of a way to represent the multidimensional environmental wastes that are relevant to diverse industry situations. Specifically we have used radar charts to help attribute process level environmental impact to overall system data and attribute environmental impacts to the source of the problem. This concept also allows the practitioner to drill down or up from a process to an overall system level (or vice versa) for the required information.

6.2. Implications for Practitioners. Industry practitioners at the production level now have a method to identify specific improvement activities (e.g., Kaizen) for environmental waste consistent with the organisational priorities. Thus environmental waste can be considered alongside other forms of waste during the VSM process (see Figure 12). To implement this, production engineers and supervisors would thus apply the environmental waste considerations as part of VSM (see action 2.2 in Figure 12). Then Kaizen solutions are developed Kaizen in the normal manner. Optionally, they can also report back to senior management against objectives for environmental waste and can do so at the level of whole value streams or individual processes.

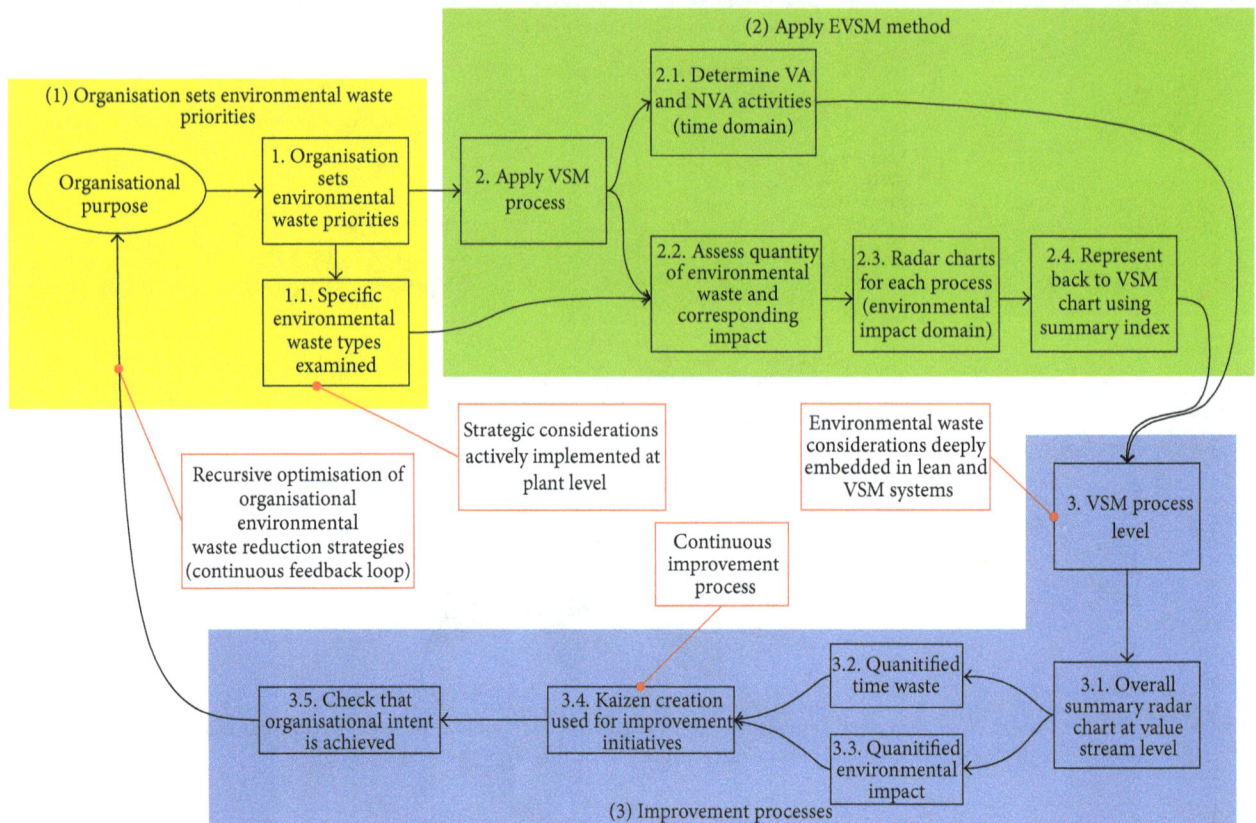

FIGURE 12: Summary of the overall method as implemented in industry.

Complementary to that, senior management now have a method to take the external environmental standard ISO 14001 and develop a customised construct for environmental waste for their particular organisation. They can then align the production processes, particularly the priorities going into the continuous improvement processes, to achieve those organisational objectives. Thus the method provides a strategic tool for firms that seek to improve their environmental position as summarised in Figure 12. A further implication from a management perspective is that the method has been developed with implementation and change management in mind. It has been specifically designed to be easy to implement and to fit in with existing organisational cultures. It achieves this by being complementary to the established practices of lean and VSM in particular: it takes advantage of practices and ideas with which the organisation is already familiar.

To implement this method, executives and production managers would decide on *which* environmental wastes to include and *set* the priorities for each (see activity 1 in Figure 12). Production staff would then implement this alongside the usual VSM processes. One of the quality staff would need training to use the method, but widespread training of other staff is not necessary (providing they already know how to implement VSM). Executives can then request summary information on overall environmental waste burden and efficacy of improvement measures. They can then use this information to further refine the strategic approach to manage the environmental waste.

There are many other industries that use lean principles, such as service organisations and project management. The method derived here is generic and not limited to manufacturing and therefore has potential applicability to these other areas. All organisations, including service firms, can identify waste priorities, assess their waste impacts, and implement Kaizen improvements. The concept of time is particularly relevant to service industries, so Value Stream Mapping is a particularly relevant lean tool in these situations.

6.3. Limitations and Opportunities for Further Research. During the implementation of the described method, several bottlenecks in usability of the system were discovered. The most notable bottleneck was the calculation component when determining the carbon footprint for each stage of the EVSM. This was remedied through the inclusion of an excel spreadsheet that determined carbon footprint for any process. The application of the method was also limited by the level of understanding of the practitioner with respect to environmental impacts and actual process level data instead of overall site level data.

An obvious limitation is that although we have integrated *environmental waste* with *lean manufacturing* practices, the integration is only for value stream mapping. There are many other lean methods, and not all organisations use VSM. Where *time* is the main driver of cost or quality, then VSM is appropriate, but this is not relevant to the production economics of all organisations. At present the integration has only been demonstrated for the manufacturing industry.

Environmentally Lean Production: The Development and Incorporation of an Environmental Impact
Index into Value Stream Mapping

55

Consequently there are opportunities for future research to extend and adapt the method to other situations.

7. Conclusion

This work develops a method to integrate evaluation of environmental impacts and lean methods. The method has been developed and tested in a manufacturing setting, and is able to represent a variety of environmental wastes within the Value Stream Mapping (VSM) method. Specifically it integrates from generic *environmental standard* ISO 14001, through *organisational environmental risk register*, onwards to integration within the *VSM process*, and thus finally permitting the established lean *improvement process* (e.g., Kaizen) to be focussed at specific environmental improvement actions. The deployment used carbon footprint, perceived impact, cost to remediate, and waste volumes (removed and residual), but the method is capable of being generalised to nth dimension environmental factors. It is thereby able to represent a *customised* environmental waste index for a particular industry. *Ambiguous user estimates* of waste quantities are accommodated through the PERT beta distribution. Several ways to represent the multi-dimensional environmental wastes were explored via industry focus group and the preferred representation designed to completion. The resulting method can be used by production staff and represents environmental impacts at the level of the individual process and aggregated to the whole value stream. The method may also be used by executives to align organisational practices with strategic objectives for waste reduction.

Appendix

A. Implementation of VSM

A VSM requires five steps that can then be applied to information, material, or process flow. A brief summary of the five steps is provided [3, 13].

A.1. Identify Target Product, Family, or Service. This stage requires the translation of customer requirements into process requirements. The customer base can be both external and internal and is described as those who accept, evaluate, install/inspect, own and use products or services.

A.2. Map Current State. Creating a current state VSM requires a team of people (who both manage and support various parts of the value stream) and who have been closely associated or involved with the process or information flow. Once the critical value stream has been chosen, every task or component is noted in the order that it is required to complete the service or product, starting at the shipping process and working backwards in the value stream to the raw materials or suppliers, while collecting information at each stage [10, 15].

A.3. Asses Current VSM in terms of Creating a Better Flow by Eliminating Waste. Once the current state map has been completed, an assessment should be carried out to determine which processes add value. This step requires the identification of all value-added (VA) and nonvalue -dded (NVA) activities, as well as necessary but nonvalue-adding (NNVA). A common exercise used during this operation is the lean implementation tool called a "Kaizen burst" in which areas that represent large amounts of NV-added time (lead time) are targeted and reduced or eliminated. In this circumstance, a Kaizen event is one in which a process is critically reviewed to determine areas which could be improved.

A.4. Draw Future State VSM. Once the target waste (Kaizen initiatives) areas are identified, an ideal future state map (FSM) should be determined. This map should represent how the value stream will look after the identified waste has been eliminated and all Kaizen implemented. The FSM should be indicative of a situation in which all the individual processes produce only what its customer/process needs (or as close as possible) and only when required.

A.5. Work toward the Future State Condition. The final stage in VSM analysis is the creation and implementation of a work plan to accomplish the waste reduction goals identified whilst determining the FSM. The implementation plan describes how the goals set whilst creating the FSM are going to be achieved. Waste identification is a crucial element of any VSM as it is indicative of the Kaizen events held to reduce NVA activities. Some common reasons for waste within an information or manufacturing system are as follows [34]:

  (i) push rather than pull based specifications and requirements,

  (ii) nonoptimal use of human resource (e.g., using the wrong staff to do the wrong job such as a manager level or high engineering level staff doing NVA or NNVA work),

  (iii) lack of detail, lack of organisation in planning, and lack of leadership and management,

  (iv) use of obsolete two-dimensional drawings instead of single point release database with three-dimensional data.

Acknowledgment

The authors acknowledge the support of the Christchurch Engine Centre, New Zealand, particularly Tim Coslett for hosting and supporting this project.

References

[1] W. E. Deming, *Out of the Crisis*, MIT Press, Cambridge, Mass, USA, 1986.

[2] J. Womack, D. Jones, and D. Roos, *The Machine that Changed the World : The Story of Lean Production*, Harper Perennial, 1991.

[3] J. P. Womack and D. T. Jones, *Lean Thinking: Banish Waste and Create Wealth in your Corporation*, Simon and Schuster, London, UK, 1996.

[4] F. A. Abdulmalek and J. Rajgopal, "Analyzing the benefits of lean manufacturing and value stream mapping via simulation:

a process sector case study," *International Journal of Production Economics*, vol. 107, no. 1, pp. 223–236, 2007.

[5] T. Melton, "The benefits of lean manufacturing: what lean thinking has to offer the process industries," *Chemical Engineering Research and Design*, vol. 83, no. 6, pp. 662–673, 2005.

[6] P. Hines, M. Holwe, and N. Rich, "Learning to evolve: a review of contemporary lean thinking," *International Journal of Operations and Production Management*, vol. 24, no. 10, pp. 994–1011, 2004.

[7] B. J. Hicks, "Lean information management: understanding and eliminating waste," *International Journal of Information Management*, vol. 27, pp. 233–249, 2007.

[8] P. Hines and N. Rich, "The seven value stream mapping tools," *International Journal of Operations and Production Management*, vol. 17, no. 1, pp. 46–64, 1997.

[9] I. Serrano Lasa, R. D. Castro, and C. O. Laburu, "Extent of the use of Lean concepts proposed for a value stream mapping application," *Production Planning and Control*, vol. 20, no. 1, pp. 82–98, 2009.

[10] M. Braglia, G. Carmignani, and F. Zammori, *A New Value Stream Mapping Approach for Complex Production Systems*, Taylor and Francis, 2006.

[11] D. Tapping, T. Luyster, and T. Shuker, *Value Stream Management*, Productivity Press, A Division of Kraus Productivity Organization, 2002.

[12] Y. H. Lian and H. Van Landeghem, "Analysing the effects of Lean manufacturing using a value stream mapping-based simulation generator," *International Journal of Production Research*, vol. 45, no. 13, pp. 3037–3058, 2007.

[13] M. Rother and J. Shook, Eds., *Learning to See: Value Stream Mapping to Add Value and Eliminate MUDA*, The Lean Enterprise Institue, Brookline, Mass, USA, 1999.

[14] P. Kuhlang, T. Edtmayr, and W. Sihn, "Methodical approach to increase productivity and reduce lead time in assembly and production-logistic processes," *CIRP Journal of Manufacturing Science and Technology*, vol. 4, no. 1, pp. 24–32, 2011.

[15] D. Seth and V. Gupta, "Application of value stream mapping for lean operations and cycle time reduction: an Indian case study," *Production Planning and Control*, vol. 16, no. 1, pp. 44–59, 2005.

[16] B. Singh, S. K. Garg, and S. K. Sharma, "Value stream mapping: literature review and implications for Indian industry," *International Journal of Advanced Manufacturing Technology*, vol. 53, pp. 799–809, 2011.

[17] Y. Monden, *Toyota Production System: An Integrated Approach to Just-in-Time*, Industrial Engineering and Management Press, Norcross, Ga, USA, 1993.

[18] A. Gurumurthy and R. Kodali, "Design of lean manufacturing systems using value stream mapping with simulation A case study," *Journal of Manufacturing Technology Management*, vol. 22, no. 4, pp. 444–473, 2011.

[19] S. Irani, "Value Network Mapping (VNM): visualization and analysis of multiple interacting value streams in jobshops," in *Proceedings of the 4th Annual Lean Management Solutions Conference*, Institute of Industrial Engineers, Los Angeles, Calif, USA, September 2004.

[20] EPA, "EPA, The lean and Environmental Toolkit," Contract EP-W-04-23, 2011, http://www.epa.gov/lean/environment/toolkits/index.htm.

[21] ISO:14031, "Environmental Managment—Environmental Performance Evaluation Guidelines," ISO 14031:2000, International Organization for Standards, Geneva, Switzerland, 2000.

[22] W. McDonough and M. Braungart, *Cradle to Cradle: Remaking the Way we Make Things*, North Point Press, 2002.

[23] C. Jasch, "The use of Environmental Management Accounting (EMA) for identifying environmental costs," *Journal of Cleaner Production*, vol. 11, no. 6, pp. 667–676, 2003.

[24] A. A. King and M. J. Lenox, "Lean and green? An empirical examination of the relationship between lean production and environmental performance," *Production and Operations Management*, vol. 10, no. 3, pp. 244–256, 2001.

[25] S. Hart and G. Ahuja, "A natural resourced based view of the firm," *Academy of Management Review*, vol. 20, no. 4, pp. 985–1014, 1995.

[26] A. Brinkley, M. Karlsson, J. R. Kirby, and D. Pitts, "Systematic environmental assessment: a decision methodology for strategic environmental issues," in *Proceedings of the IEEE International Symposium on Electronics and the Environment (ISEE '00)*, pp. 178–183, San Francisco, Calif, USA, October 2000.

[27] S. Salhofer, G. Wassermann, and E. Binner, "Strategic environmental assessment as an approach to assess waste management systems. Experiences from an Austrian case study," *Environmental Modelling and Software*, vol. 22, no. 5, pp. 610–618, 2007.

[28] I. Hokerby, "Environmentally compatible product and process development," in *Proceedings of the NAE Workshop on Corporate Environmental Stewardship*, 1993.

[29] D. J. Richards, "Environmentally conscious manufacturing," *World Class Design to Manufacture*, vol. 1, no. 3, pp. 15–22, 1994.

[30] T. Wiedmann and J. Minx, "A defintion of, 'carbon footprint'," in *Ecological Economics Reserach Trends*, C. C. Pertsova, Ed., chapter 1, Nova Science Publishers, New York, NY, USA, 2008.

[31] GRI, *G3 Stustainability Reporting Guidelines*, Global Reporting Initiative, Amsterdam, The Netherlands, 2006.

[32] A. S. Patil, "Incorporating environmetnal index as waste into value stream mapping," in *Mechanical Eningeering, Masters of Science, Major of Industrial Engineering*, p. 93, Wichita State University, Wichita, Kan, USA, 2002.

[33] A. S. Torres and A. M. Gati, "Environmental value stream mapping (EVSM) as sustainability management tool," in *Proceedings of the Portland International Conference on Management of Engineering and Technology (PICMET '09)*, pp. 1689–1698, Portland, Ore, USA, August 2009.

[34] B. W. Oppenheim, "Lean product development flow," *Systems Engineering*, vol. 7, pp. 352–376, 2004.

6

A Novel Metaheuristic for Travelling Salesman Problem

Vahid Zharfi and Abolfazl Mirzazadeh

Industrial Engineering Department, Faculty of Engineering, Kharazmi University, Tehran 31979-37551, Iran

Correspondence should be addressed to Vahid Zharfi; v.zharfi@tmu.ac.ir

Academic Editor: Alan Chan

One of the well-known combinatorial optimization problems is travelling salesman problem (TSP). This problem is in the fields of logistics, transportation, and distribution. TSP is among the NP-hard problems, and many different metaheuristics are used to solve this problem in an acceptable time especially when the number of cities is high. In this paper, a new meta-heuristic is proposed to solve TSP which is based on new insight into network routing problems.

1. Introduction

Although there are various classic methods for solving optimization problems, but they are not always able to solve real and applied optimization problems. It is generally believed that these problems are known as difficult and complicated problems. TSP is a typical example of a very hard combinatorial optimization problem. TSP can be modelled as an undirected weighted graph, such that cities are the graph's vertices, paths are the graph's edges, and a path's distance is the edge's length. It is a minimization problem starting and finishing at a specified vertex after having visited each other's vertex exactly once. Metaheuristics can solve high-dimensional problems quickly. Heuristics may be classified into two families: specific heuristics and metaheuristics. Specific heuristics are tailored and designed to solve a specific problem and/or instance. Metaheuristics are general-purpose algorithms that can be applied to solve almost any optimization problem. They may be viewed as upper level general methodologies that can be used as a guiding strategy in designing underlying heuristics to solve specific optimization problems [1]. The word metaheuristic was first used by Glover [2] during introducing tabu search concepts.

Metaheuristics gained more popularity in few recent decades, and various algorithms are proposed to solve the problem such as Genetic Algorithm [3–5], Ant Colony [6–10], Particle Swarm [11], and Simulated Annealing [12, 13].

In this paper, a novel metaheuristic is proposed for solving TSP. This algorithm is used to foresee the best next city so as to gain shorter rout. The 2-Opt and 3-Opt algorithms are also used in local search. They are probably the most basic local search heuristics for the TSP.

The paper is organized as follows: Section 2 illustrates the background theory of suggested algorithm. In Section 3, formulating algorithms are presented. Algorithm improvement by using k-Opt is discussed in Section 4. The algorithm is employed into several standard TSP instances in the next section and the results are reported. According to the obtained results, it is noticeable that the algorithm achieves proper answers, quickly so that the answer of the first iteration will be a good and acceptable answer.

Finally, Section 6 concludes the paper.

2. Background Theory of the Algorithm

In this model, it is supposed that every node in the network is a vibration source and has a special weight which vibrates the edges, so that the nodes with the heavier weight release more vibration. This emitted force by a node distributes in edges that are connected to. Shared amount of each node is dependent on the edges distances. Shorter edges have more quota of this force. In other words, if the length of an edge connected to the node is shorter, less vibration falls down along the edge and it is vibrated more intensely. There is a simple network (Figure 1) to illustrate the theory. The

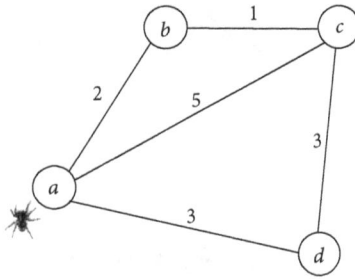

FIGURE 1: Routing in a network.

searcher is located at the node a in the following figure and wants to move to the next node after a.

Signs which lead to a are amount of vibrations that each edge indicates. Therefore, vibration of the edges can be considered as tendency that each neighbour node has in order to be selected by the searcher.

Node c as a sample has relationship with nodes a, b, and d. The distance for each edge is determined, and the sum of the inverted distances equals $(1/5 + 1 + 1/3) = 23/15$. Weight of this node is w_c. Therefore, the vibration that this node sends to the node a equals:

$$\frac{(1/5)}{(23/15)} \times w_c = 0.13 \times w_c. \tag{1}$$

Also, about nodes b and d, the amount of vibration which is sent to node a is, respectively,

$$\frac{(1/2)}{(3/2)} \times w_b = 0.33 \times w_b,$$
$$\frac{(1/3)}{(2/3)} \times w_d = 0.05 \times w_d. \tag{2}$$

If all weights are considered equal to 1, the searcher receives a total vibration with an amount of $(0.13 + 0.33 + 0.50) = 0.96$ in node a. As a result, the searcher at the node a selects c with 14% possibility $(0.13/0.96)$, b with 34% possibility $(0.33/0.96)$, and d with 52% possibility $(0.50/0.96)$.

Movement from node c in addition to direct route $c \rightarrow a$ can be transferred to node a of other routes such as "$c \rightarrow d \rightarrow a$" and "$c \rightarrow b \rightarrow a$." If we also consider binary routes, calculations would be changed. Thus, by considering routes with maximum two edges, there are three different routes from c to a (Table 1). Also, about nodes b and d, similar calculations will be performed (Tables 2 and 3).

According to the tables, it is possible to obtain an amount of vibration in each edge. For example, edge "$c \rightarrow a$" receives a force with an amount of $(0.20/2.28) \times w_c$ unit of c, $(0.16/2.25) \times w_b$ unit of b, and a force with an amount of $(0.12/1.35) \times w_d$ unit of node d. Therefore, "$c \rightarrow a$" edge vibrates as follows:

$$\left(\frac{0.20}{2.28}\right) \times w_c + \left(\frac{0.16}{2.25}\right) \times w_b + \left(\frac{0.12}{1.35}\right) \times w_d. \tag{3}$$

It is supposed that all weights are 1. Therefore, the amount of vibration in "$c \rightarrow a$" is 0.25. Similarly, "$b \rightarrow a$" and

"$d \rightarrow a$" are 0.37 and 0.31, respectively. It should be mentioned that the total amount of these three values is not necessarily equal to 1. In the investigated example final total amount of vibration force of the three edges which end with a is 0.93.

Now, the searcher is faced with vibrant edges in node a Therefore, it selects "$b \rightarrow a$" with 40% possibility $(0.40/0.93)$ and moves toward "b," selects "$c \rightarrow a$" with 27% possibility $(0.25/0.93)$ and moves toward "c," or also selects "$d \rightarrow a$" with 33% possibility $(0.31/0.93)$ and moves toward node "d." Total amount of these possibilities equals 1.

3. Model Formulation

In the previous section we got acquainted with the method of routing by the algorithm. As mentioned, edges with more vibration are selected with more possibility.

The vibration of a supposed edge "$j \rightarrow i$" is derived from two factors:

(i) direct force which is emitted from jth node;

(ii) indirect forces that are emitted from other nodes and distribute up to "$j \rightarrow i$." In other words, it is released from node k ($k \neq i, j$) to node i after passing node j:

$$D_{kji} = \begin{cases} d_{kj} + d_{ji} & \text{if } k \neq i, j, \\ d_{ji} & \text{if } k = j \neq i, \end{cases}$$
$$D_{kji} = \frac{1}{d_{kji}}. \tag{4}$$

In this relation d_{kji} is the distance of the route initiated from k that ends in i after passing j. Through definition D_k is sum of inverted distances of routes connected to k which has up to two edges (see (5)); it is possible to gain the force which is applied from node k to edge $j \rightarrow i$ by:

$$D_k = \sum_{i \neq k} \sum_{j \neq k} D_{kji}, \tag{5}$$

$$r_{kji} = \left(\frac{D_{kji}}{D_k}\right) \times w_k, \tag{6}$$

so that w_k represents weight of kth node. r_{kji} is the applied force for "$j \rightarrow i$" resulting from k. Vibration of "$j \rightarrow i$" equals

$$v_{ji} = \sum_k r_{kji}. \tag{7}$$

The final result of the previous calculations is matrix $V = (v_{ji})$ in which v_{ji} represents vibration degree in the edge "$j \rightarrow i$." It is necessary to use another element in order to increase ability of this matrix. To do so we can use vision power matrix.

In distance matrix, d_{ji} represents the length of an edge which is connector of nodes i and j. The vision matrix $\Gamma = (\gamma_{ij})$ is defined as follows:

$$\gamma_{ij} = \frac{1}{d_{ij}}. \tag{8}$$

TABLE 1: Emitted force by node "c" in the network.

Origin	Destination	Route	Route distance	Inversion of route distance	Final edge of the route	Amount of vibration in the final edge
c	a	$c \to b \to a$	3	0.33	$b \to a$	$(0.33/2.28) \times w_c$
c	a	$c \to a$	5	0.20	$c \to a$	$(0.20/2.28) \times w_c$
c	a	$c \to d \to a$	6	0.16	$d \to a$	$(0.16/2.28) \times w_c$
c	b	$c \to a \to b$	7	0.14	$a \to b$	$(0.14/2.28) \times w_c$
c	b	$c \to b$	1	1.00	$c \to b$	$(1.00/2.28) \times w_c$
c	d	$c \to a \to d$	8	0.12	$a \to d$	$(0.12/2.28) \times w_c$
c	d	$c \to d$	3	0.33	$c \to d$	$(0.33/2.28) \times w_c$
	Total			2.28	w_c	

TABLE 2: Emitted force by node "b" in the network.

Origin	Destination	Route	Route distance	Inversion of route distance	Final edge of the route	Vibration force in the final edge
b	a	$b \to a$	2	0.50	$b \to a$	$(0.50/2.25) \times w_b$
b	a	$b \to c \to a$	6	0.16	$c \to a$	$(0.16/2.25) \times w_b$
b	c	$b \to a \to c$	7	0.14	$a \to c$	$(0.14/2.25) \times w_b$
b	c	$b \to c$	1	1.00	$b \to c$	$(1.00/2.25) \times w_b$
b	d	$b \to a \to d$	5	0.20	$a \to d$	$(0.20/2.25) \times w_b$
b	d	$b \to c \to d$	4	0.25	$c \to d$	$(0.25/2.25) \times w_b$
	Total			2.25		w_b

By combination of two matrixes Γ and V, it is possible to use both visual sense and mechanical sensor of the searcher for detection and recognition of the best route.

The result of performed calculations in this section is possibility matrix $P = (p_{ij})$:

$$P_{ij} = \frac{v_{ij}^{\alpha} \cdot \gamma_{ij}^{\beta}}{\sum_i \sum_j v_{ij}^{\alpha} \cdot \gamma_{ij}^{\beta}}. \tag{9}$$

The parameters α and β are investigated from the obtained results. A proper amount of these parameters is dependent on nature and structure of the investigated problem. Adjustment of α and β has an important role in algorithm efficiency. Investigation of the obtained results shows that $\alpha = 8$ and $\beta = 2$ are appropriate for most of the instances.

4. Algorithm Improvement by Local Search

The suggested algorithm searches to get to the best route in the network globally. But it makes less chance to envelop all answers. Consequently, local search is necessary in order to cover the solution space more entirely. A local search algorithm starts from a candidate solution and then iteratively moves to a neighbour solution.

In optimization, 2-Opt and 3-Opt are simple local search algorithms for solving the travelling salesman problem [14, 15]. In this paper, we propose to combine the 2-Opt, 3-Opt, and the suggested algorithm to improve the answers. The algorithm steps are as follows:

Initialize matrix P
For (i < Iteration Max) do:
    While the searcher complete a solution
        Choose probabilistically the next state to move;
        Add that move to the tabu list;
    end
    Implementation of the 2-Opt and 3-Opt;
    If (local best solution better than global solution)
        Save local best solution as global solution;
    end
    Update matrix P for each edge that the searcher chose
    (Increase elements of matrix P related to the selected route up to q percent);
end

5. Computational Results

In this section we are going to solve different standard examples by using the proposed algorithm. A large number of standard samples are available in public references. These references include famous samples for combinatorial optimization problems. One of the valid references in travelling salesman problems is TSPLIB [16].

5.1. Quality of Solutions. In order to estimate the algorithm function from absolute difference obtained by optimized answer, the following relation is used:

$$\frac{|f(s) - f(s^*)|}{f(s^*)}, \tag{10}$$

so that s is the obtained answer and s^* is the optimized answer.

TABLE 3: Emitted force by node "d" in the network.

Origin	Destination	Route	Route distance	Inversion of route distance	Final edge of the route	Vibration force in the final edge
d	a	$d \to c \to a$	8	0.12	$c \to a$	$(0.12/1.35) \times w_d$
d	a	$d \to a$	3	0.33	$d \to a$	$(0.33/1.35) \times w_d$
d	b	$d \to a \to b$	5	0.20	$a \to b$	$(0.20/1.35) \times w_d$
d	b	$d \to c \to b$	4	0.25	$c \to b$	$(0.25/1.35) \times w_d$
d	c	$d \to a \to c$	8	0.12	$a \to c$	$(0.12/1.35) \times w_d$
d	c	$d \to c$	3	0.33	$d \to c$	$(0.33/1.35) \times w_d$
	Total			1.35		w_d

TABLE 4: Searching for optimized solution of different travelling salesman problems.

Instance	Best known	First sol.	Best sol.	Average sol.	Worst sol.	Error
ftv33	1286	1462	1286	1286	1286	0.0000000
ftv38	1530	1591	1530	1535	1546	0.0032679
swiss42	1273	1364	1273	1273	1273	0.0000000
ftv44	1613	1742	1615	1619	1625	0.0037197
ry48p	14422	15341	14466	14493	14507	0.0049230
berlin52	7542	7951	7542	7542	7542	0.0000000
ftv64	1839	2022	1909	1925	1940	0.0467645
eil76	538	559	538	538	538	0.0000000
eil101	629	717	630	638	650	0.0143084

5.2. Robustness. Metaheuristics must be able to act properly in an extended range of samples or problems with the same parameters. It is possible to adjust metaheuristic parameters for a collection of samples in a good way which has less efficiency for other samples.

Different problems of travelling salesman have been solved with different sizes and structures by the algorithm. For all instances, 9, 2, and 0.2% are allocated for α, β, and q, respectively (Table 4).

The optimal path length of the algorithm's iterations for berlin52 is shown in Figure 2.

According to the obtained results, it is noticeable that the algorithm achieves proper answers quickly so that the answer of the first iteration will be a good and acceptable answer. Accordingly, although the suggested algorithm shows its success alone in this issue, it can be used as a powerful initial solution generator for other solution-based metaheuristics in order to use advantages of other algorithms and getting more close to the optimized answer.

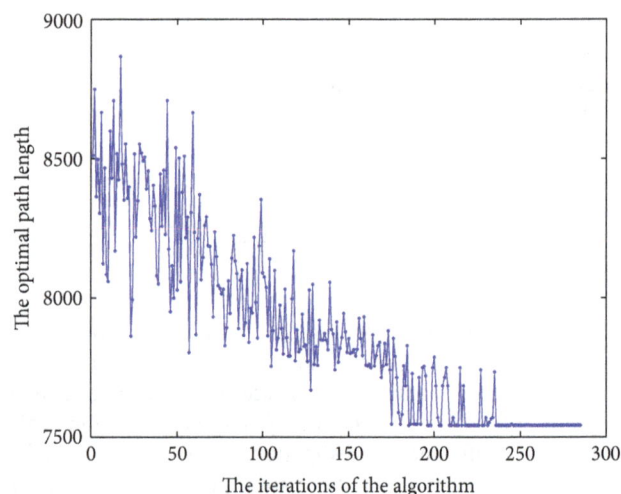

FIGURE 2: The optimal path length of proposed algorithm for berlin52.

6. Conclusion

In this paper we presented a new metaheuristic that has new insight into network problems. At the end of every iteration, the algorithm also tries to improve both the performance of the algorithm and the quality of solutions for the Tsp by using local search algorithms in the name of 2-Opt and 3-Opt. As a result, the proposed algorithm can be used for variety network problems due to its insight to these kinds of problems. Therefore, through extending this behaviour we can move toward solving different optimization problems by this algorithm in a way that each position of problem solutions represents a node and each algorithm step is passing from one state to another state. Future work will be conducted to improve the proposed algorithm and examine other local search algorithms to enhance the performance of the method for improving solutions. Additional improvements might lie on the combination of our approach with the other metaheuristics like genetic algorithm, tabu search, simulated annealing, and so on.

References

[1] T. El-ghazali, *Meta-Heuristic from Design to Implementation*, Wiley, 2009.

[2] F. Glover, "Future paths for integer programming and links to artificial intelligence," *Computers and Operations Research*, vol. 13, no. 5, pp. 533–549, 1986.

[3] S. Yuan, B. Skinner, S. Huang, and D. Liu, "A new crossover approach for solving the multiple travelling salesmen problem using genetic algorithms," *European Journal of Operational Research*, vol. 228, no. 1, pp. 72–82, 2013.

[4] Y. Nagata and D. Soler, "A new genetic algorithm for the asymmetric traveling salesman problem," *Expert Systems with Applications*, vol. 39, no. 10, pp. 8947–8953, 2012.

[5] W. Hui, "Comparison of several intelligent algorithms for solving TSP problem in industrial engineering," *Systems Engineering Procedia*, vol. 4, pp. 226–2235, 2012.

[6] S. M. Chen and C. Y. Chien, "Solving the traveling salesman problem based on the genetic simulated annealing ant colony system with particle swarm optimization techniques," *Expert Systems with Applications*, vol. 38, no. 12, pp. 14439–14450, 2011.

[7] J. Yang, X. Shi, M. Marchese, and Y. Liang, "Ant colony optimization method for generalized TSP problem," *Progress in Natural Science*, vol. 18, no. 11, pp. 1417–1422, 2008.

[8] M. Dorigo and L. M. Gambardella, "Ant colonies for the travelling salesman problem," *BioSystems*, vol. 43, no. 2, pp. 73–81, 1997.

[9] J. Bai, G. K. Yang, Y. W. Chen, L. S. Hu, and C. C. Pan, "A model induced max-min ant colony optimization for asymmetric traveling salesman problem," *Applied Soft Computing*, vol. 13, no. 3, pp. 1365–1375, 2013.

[10] A. Colorni, M. Dorigo, F. Maffioli, V. Maniezzo, G. Righini, and M. Trubian, "Heuristics from nature for hard combinatorial optimization problems," *International Transactions in Operational Research*, vol. 3, no. 1, pp. 1–21, 1996.

[11] X. H. Shi, Y. C. Liang, H. P. Lee, C. Lu, and Q. X. Wang, "Particle swarm optimization-based algorithms for TSP and generalized TSP," *Information Processing Letters*, vol. 103, no. 5, pp. 169–176, 2007.

[12] K. Meer, "Simulated annealing versus metropolis for a TSP instance," *Information Processing Letters*, vol. 104, no. 6, pp. 216–219, 2007.

[13] X. Geng, Z. Chen, W. Yang, D. Shi, and K. Zhao, "Solving the traveling salesman problem based on an adaptive simulated annealing algorithm with greedy search," *Applied Soft Computing Journal*, vol. 11, no. 4, pp. 3680–3689, 2011.

[14] C. Engels and B. Manthey, "Average-case approximation ratio of the 2-opt algorithm for the TSP," *Operations Research Letters*, vol. 37, no. 2, pp. 83–84, 2009.

[15] K. H. Hsieh, "Fourier descriptors for 2-Opt and 3-Opt heuristics for traveling salesman problem," *Journal of the Chinese Institute of Industrial Engineers*, vol. 28, no. 3, pp. 237–246, 2011.

[16] http://elib.zib.de/pub/Packages/mp-testdata/tsp/tsplib/tsplib.html.

7

Ergonomic Fuzzy Evaluation of Firefighting Operation Motion

Lifang Yang, Tianjiao Zhao, and Fanyu Meng

Department of Industrial Design, Harbin Institute of Technology, Harbin 150001, China

Correspondence should be addressed to Lifang Yang; yanglifang@hit.edu.cn

Academic Editor: Durga Rao Karanki

The firefighting operation motion has an important impact on the safety and comfort of firefighting operation. As a judgment criterion of the firefighting efficiency, the comfort level is hard to judge in that it is completely decided by human feeling, so the comprehensive fuzzy evaluation is utilized for evaluation of comfort level. In this paper, firstly the factor and judgment set of firefighting operation comfort level are determined, and the fuzzy weight evaluation is obtained by questionnaires and analytic hierarchy process. Secondly, the joint angles of some particular motions are determined by motion capture equipment, the moment is obtained by ergonomic engineering software, and then the comprehensive comfort evaluation on firefighting operation motion is completed. Finally, the objective evaluation system of firefighting operation comfort is established.

1. Introduction

Firefighting grows more and more difficult owing to many types of fire accidents nowadays. The operation efficiency becomes the important factor on the firefighting according to statistics of fire accidents. Therefore, the rationality and reliability of firefighting operation directly affect the rescue work.

Motion analysis can be used to optimize and standardize human operation motion by means of detecting and tracking human operation. The motion analysis of firefighting is based on the collection, classification, and evaluation of particular rescue motion of fire men and the research results will be the fundamentals for firefighting product design and fire man training optimization. The motion analysis can be classified into two methods: visual motion observation method and image motion observation method [1].

The ergonomics analysis of firefighting operation and fire men mainly focuses on firefighting training, fire extinguisher, and fire man uniforms. Jiao et al. applied BP neural network to firefighting training evaluation. The comprehensive evaluation on the basic information, training methods, training contents, training management, and achievement of the tested groups are discussed [2]. Based on SAQ+B (Scenario Animation Question and Browse), Chen and Li proposed an evaluation method for simulation of firefighting training.

The training process is first divided into several units, and then the training problems are abstracted. Through experts grading, integrating calculation method of AHP by Matlab, the firefighting training evaluation is divided into two parts of knowledge acquisition and capability test. Therefore the quantitative estimation for the training subject is obtained [3]. AHP is short for analytic hierarchy, which is to process a structured technique for organizing and analyzing complex decisions. Comprehensive fuzzy evaluation is applied in plenty of areas. For ergonomic evaluation, Hongzhe and Damin utilized this method in evaluating the comfort level of fighter plane cockpit arrangement. Park et al. conducted evaluation on the comfort level of cab in engineering machines [4].

In the paper, evaluation model for firefighting operation motion will be built by fuzzy integrating method. The human body modeling and motion simulation will be conducted in virtual environment. Through motion capture equipment, the human biomechanics data will be obtained and then an ergonomic software is used for motion simulation, the comfort level judgment for typical firefighting motions. The paper connects the subjective perception and the objective evaluation of firefighters and implements the comfort judgment based on human factors, which play a quite significant role in modern industry [5].

2. Basic Ideas of Comprehensive Fuzzy Evaluation

2.1. Definition of Comfort Level. The comfort level can be defined as follows from the judgment of joint:

$$U = \frac{\tau(\alpha)_{\max,i} - \tau_{ci}}{\tau(\alpha)_{\max,i}}, \tag{1}$$

where τ_{ci} is actual moment of joint, $\tau(\alpha)_{\max,i}$ is joint, and i is maximum moment angel α.

From above, it is clear to see that what influences human comfort the most is the relationship between joint maximum moment and actual moment, and the less the actual moment the better the comfort. The comfort judgment set of firefighting operation is established as follow:

$$V = \{v_1, v_2, v_3\}, \tag{2}$$

where V_i $(i = 1, 2, 3)$ represents all possible judgment results and $\{v_1, v_2, v_3\}$ = {Comfortable, Medium, Uncomfortable}.

2.2. Determination of Membership Function. Membership function is used to describe fuzzy set. Research indicates that the energy utilization rate comes to maximum if half used [6], thus human will not feel tired even when working longer. The membership function is described in Figure 1.

In this figure, the membership function of U to fuzzy "comfort" is shown in

$$\mu_A(U) = \begin{cases} 0, & U < 0.5, \\ \dfrac{U - 0.5}{0.2}, & 0.5 \leq U \leq 0.7, \\ 1, & U > 0.7. \end{cases} \tag{3}$$

2.3. Determination of Factors Set That Influences Motion Comfort. This paper mainly takes body movement into account. Human body consists of joints, which move in different degrees of freedom and combine into motions. Therefore, the judgment factors set is established based on joint movements in the comprehensive comfort evaluation for firefighters (Figure 2).

This evaluation model is a multilevel one, to begin with, conducting first-order evaluation on joint degree of freedom, then the second-order and stepping up, and finally obtaining comprehensive comfort evaluation [7]. The single factor evaluation matrix R is established based on the definition of comfort level and fuzzy membership function. One has

$$R = \begin{bmatrix} r_{11} & r_{12} & r_{13} \\ r_{21} & r_{22} & r_{23} \\ r_{31} & r_{32} & r_{33} \\ \cdots & \cdots & \cdots \\ r_{n1} & r_{n2} & r_{n3} \end{bmatrix}. \tag{4}$$

Taking the first-order shoulder joint judgment, for instance, r_{ij} $(j = 1, 2, 3)$ represents i $(i = 1, 2, 3)$ in factor set U_{11} and membership to j in evaluation set V, and then making comprehensive fuzzy evaluation,

$$B = w \cdot R \tag{5}$$

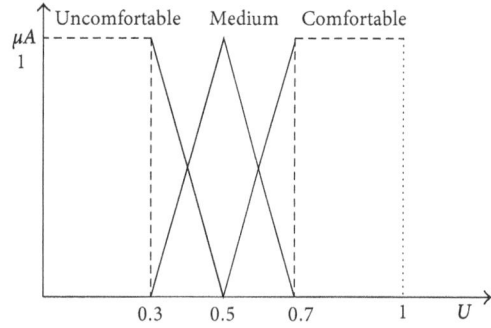

FIGURE 1: Membership of comfort level in different evaluation indexes.

B is comprehensive fuzzy evaluation set in V; w is the weight set for factor set U versus V. In the evaluation system, the importance level for the evaluation objectives is different, so fuzzy evaluation is applied to make clear the important parts when establishing evaluation objectives [8]. Weight coefficient is used to represent the importance of evaluation objectives. Three-level evaluation is applied taking comfort evaluation into consideration as shown in the following equations [9].

First-level comprehensive fuzzy evaluation method:

$$B_{\text{Shoulder}} = w_{111} R_{\text{UASS}} + w_{112} R_{\text{UAFE}} + w_{113} R_{\text{UAR}},$$
$$B_{\text{Elbow}} = w_{121} R_{\text{FLE}}. \tag{6}$$

Second-level comprehensive fuzzy evaluation method:

$$B_{\text{Upper Limbs}} = w_{11} R_{\text{Shoulder}} + w_{12} R_{\text{Elbow}},$$
$$B_{\text{Lower Limbs}} = w_{21} R_{\text{Hip}} + w_{22} R_{\text{Foot}} + w_{23} R_{\text{Knee}}. \tag{7}$$

Third-level comprehensive fuzzy evaluation method:

$$B = w_1 R_{\text{Upper Limbs}} + w_2 R_{\text{Lower Limbs}}. \tag{8}$$

To obtain the membership of operation motion to comfort level, subjective survey is requisite to acquire weight w, and the maximum moment and actual moment of joints are needed according to the comfort level formula. Thus, the first step is to conduct weight acquisition experiment, and the second is to calculate the joint motion angle and actual moment.

3. Weight Index Acquisition

3.1. Subjective Weight Investigation Experiment. When it comes to judge how much joints influence human comfort, joint movement is usually used. The questionnaire is completed through judging the body condition after finishing some movement, taking two joints into comparison. If the body comfort level is lower, which means that this very joint plays an important role in evaluating human body condition, this joint will score high.

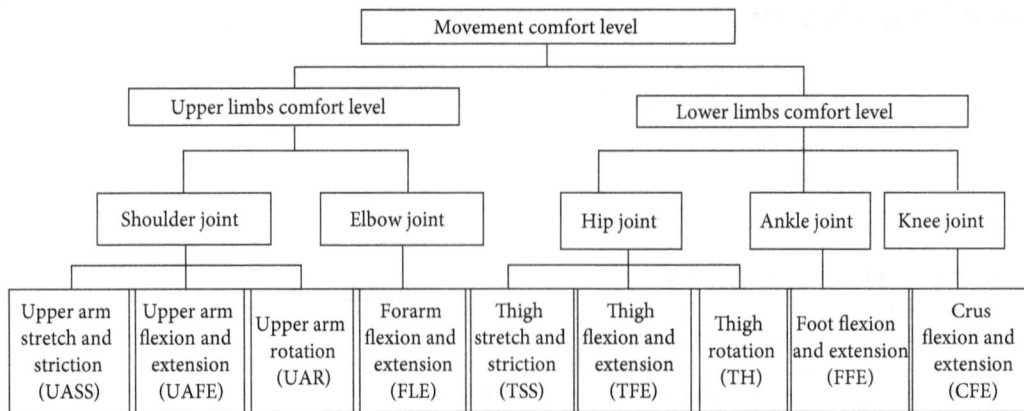

FIGURE 2: Joints that influence comprehensive comfort level.

To guarantee the accuracy, this research will choose 0–4 scoring method and analytic hierarchy process from various weight set quantized value methods. First, conduct subjective questionnaires among 11 firefighters and use 0–4 scoring method, in which firefighters grade the factors with their subjective judgment. If they are of identical importance, they are 2 points each; if one is more important than the other, then 3 points and 1 point graded; and if one is far more important, then it is graded 4 points and the other 0. After that, analytic hierarchy process is used with the evaluation results by experts.

In accordance with the factor set of fuzzy evaluation, five parts including shoulder, elbow, hip, ankle, and knee joints need to be compared. Since human body is symmetrical distribution, so whether the left or right joint will fit the experiment.

(1) The firefighters should be quite familiar with the procedure and requirement before the experiment.

(2) The firefighters should keep up-right and relaxed and move a single joint with other joints fixed. The duration is 3 minutes.

(3) Time interval is requisite between two items to prevent tiredness.

3.2. Weight Calculation. The joints are put in order by their importance after data processing. The higher it is graded, the higher it ranks and the more tired it becomes when human is in motion, which indicates its importance to human body. The sequencing results are listed in Table 1.

Kendall Coefficient of Concordance is used to judge the relevance of subjective judgment from the results. One has

$$W = \frac{S}{(1/12) K^2 (N^3 - N)}, \qquad (9)$$

where N is number of firefighters, K is number of firefighters who grade or standard number that grading depends on, and S is quadratic sum of dispersions between sum of scores R_i and averages of the sums.

Kendall coefficient of concordance calculation result $W =$ 0.8122 shows high concordance between the 11 subjective judgments according to "Kendall Coefficient of Concordance (W) Significance Critical Value Table." The importance order is B shoulder > D hip > C elbow > F knee > E ankle. Based on this, analytic hierarchy process is used to calculate weight. This method needs to collect the opinions of experts, and in this experiment, the results are shown in Table 2 after integration and comparison of the results in the set. The judgment matrix is established according to the D value of R_i. One has

$$E = \begin{bmatrix} 1 & e_{12} & \cdots & e_{1n} \\ e_{21} & 1 & \cdots & e_{2n} \\ \cdots & \cdots & \cdots & \cdots \\ e_{n1} & e_{n2} & \cdots & 1 \end{bmatrix}, \qquad (10)$$

$$w_i = \sqrt[n]{\prod_{j=1}^{n} e_{ij}}. \qquad (11)$$

The weight is obtained after unification. Similarly, the same method is used when calculating the weight of each joint, and the results are shown in Table 3.

4. Firefighting Operation Motion Moment Acquisition

4.1. Method Design. Software JACK is used to acquire the actual movement moment and joint motion angle. JACK is a mature, cross-platform environment for building, running, and integrating commercial-grade multiagent systems, which is built on a sound logical foundation. In JACK, agents are defined in terms of their beliefs, their desires, and their intentions. It is a set of long-term goals as well as an ever-evolving collection of thousands of lines of code and is used for human modeling and simulation in this research.

Through the angle simulation result obtained from the calculation of motion experiment by keyframe in JACK, the actual human body model moment is obtained [10]. The whole procedure is divided into three parts, data investigation of actual space, task simulation of virtual space, and quantitative output of simulation results. The implementation process is as Figure 3.

TABLE 1: Subjective evaluation results.

$N = 5$	Firefighters $K = 11$											
	1	2	3	4	5	6	7	8	9	10	11	Ri
Shoulder	5	4	5	5	3	4	5	4	5	3	4	47
Elbow	2	3	4	1	5	3	2	3	1	4	3	31
Hip	4	5	2	3	4	5	4	5	3	5	5	45
Ankle	1	1	1	2	1	1	3	1	2	2	1	16
Knee	3	2	3	4	2	2	1	2	4	1	2	26

TABLE 2: Factor importance judgment value.

x, y comparison	$fy(x)$	$fx(y)$	Illustration
x and y "same importance"	1	1	x, y contribute the same to some property
x "bit more important"	3	1/3	x contributes a bit more
x "bit much more important"	5	1/5	x contributes a bit much more
x "much more important"	7	1/7	x contributes much ore
x "absolutely more important"	9	1/9	x contributes absolutely more important
x, y between two judgments	2, 4, 6, 8	1/2, 1/4, 1/6, 1/8	Compromise between two judgments

4.2. Joint Movement Angle Acquisition by Capture Experiment. Movement capture system begins from image analysis method, which utilizes humans or other objects similar to animation model to show the movements to be analyzed in a three-dimensional space, and then the computer captures the movement data and quantitatively analyzes the motion track by dealing with the motion sequence. This laboratory is equipped with eight cameras to capture movements, and the location distribution is shown in Figure 4. To guarantee a fine reflection effect, a piece of blue cloth is laid in the center of the room.

During the process of movement capture, two firefighters of identical height are selected to conduct the experiment based on the investigation on firefighter human dimension. The information of the experimenters and the equipment is listed in Table 4.

11 key points are marked on the body according to the establishment of firefighter mathematical model. Taking into consideration that the human body coordinate system should be established during the movement capture process and the shelter effect of movement to the markers, the markers are selected as in Figure 5.

Through previous task analysis on investigation, four typical movements are extracted among different firefighting motions. 14 training samples are covered in the four movements. The motion track coordinate is established based on motion capture, shown in Table 5.

According to the analysis above, the coordinate of human body markers and the trajectory diagram are obtained. The calculation of motion angle is divided into two types: projection angle and absolute angle. For a joint with one degree of freedom, absolute angle is suitable; and for a joint with more degrees of freedom, projection angle is used.

The condition in which people stand with two arms upright down is treated as the initial movement and the joints are regarded as vectors. According to the vector formula, $\vec{a} \cdot \vec{b} = |a||b| \cos \theta$, the movement range and angles are calculated. The joint angle takes the initial condition as benchmark, so the bending angle of single degree of freedom is $180° - \theta$.

Taking water hose on high, for example, the angles of joints in this kind of motion are calculated as in Figure 6. Since this kind of motion mainly relates to shoulder joint and elbow joint with separately one degree of freedom and three degrees, the calculation results are shown in Figures 7 and 8.

Similarly, the joint angles change in different motions can be calculated through motion trail. The angle not only provides foundation for the following motion simulation, but also the joint movement maximum moment can be calculated through NASA power model; besides, it can provide data foundation for firefighting product design, such as clothing and equipment.

4.3. Firefighter Typical Movement Simulation. The ergonomic software JACK is used for simulation. The firefighting operation motion is simulated through key frame, which is utilized to be adjusted manually to change the motion of the virtual human. Ten key frames are chosen to simulate the joint angle, in order to acquire the moments in different time. Taking the motion of the disengaging water hose on high, for example, the motion simulation is shown in Figure 9. The joint moment output is illustrated in Table 6.

5. Fuzzy Judgment Result Analysis

The comfort membership is calculated in Matlab according to fuzzy comfort evaluation method. Taking the disengaging water hose on high at one moment, for example, the

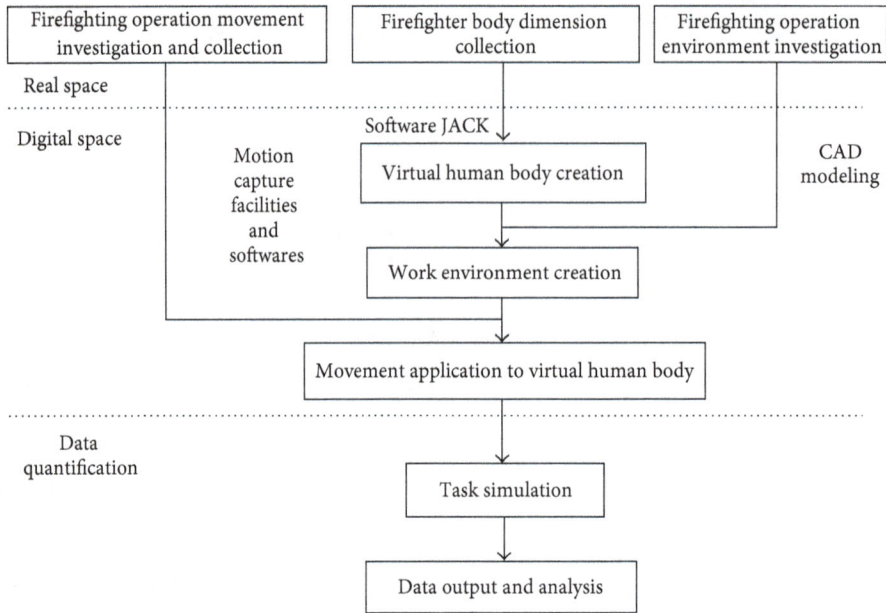

FIGURE 3: Research method frame diagram.

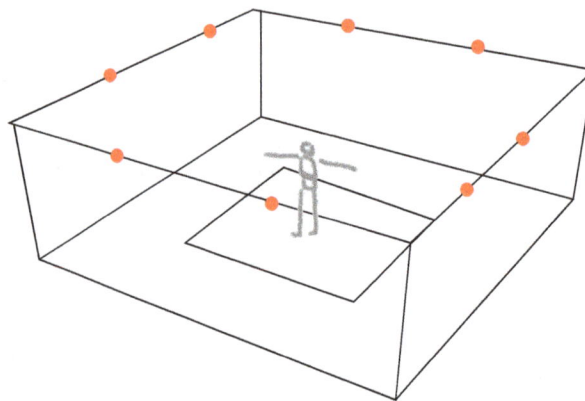

FIGURE 4: Experiment environment and cameras distribution.

FIGURE 5: Experimenter wearing effect and markers location.

TABLE 3: Weight statistical graph.

Weight	w_1 upper limbs 0.488		w_2 lower limbs 0.512		
	w_{11} shoulder 0.284	w_{12} elbow 0.204	w_{21} hip 0.2192	w_{22} ankle 0.145	w_{23} knee 0.147
Stretch and striction	0.463	—	0.379	—	—
Flexion and extension	0.396	1	0.423	1	1
Rotation	0.141	—	0.198	—	—

TABLE 4: Information of experimenter and experiment equipment.

	Experimenter	Gender	Height	Weight	Age
Information of experimenter	Experimenter 1	Male	170 cm	60 kg	22
	Experimenter 2	Male	173 cm	65 kg	22
	Equipment	Weight	Height (diameter)	Length	Pictures
Information of experiment equipment	Water hose	6.3 kg	0.12 m	20 m	Water hose
	Hydraulic giant	0.42 kg	0.10 m	0.25 m	
	Ladder	8 kg	0.45 m	2 m	Hydraulic giant
	Staircase	—	0.3 m	—	

TABLE 5: Typical motion capture experiment.

Ladder carrying training	Two people carry the ladder, one in the front and the other on the back. The front one lifts the ladder with hands and the back one holds it on shoulder. The ladder in heeling condition. The movements involved are legs running, arms swinging (legs 1), trunk forerake (arm 1), eye level watching (eye 4), and hands holding the ladder (hand 1).	
Ladder climbing training	One who leans on the corner holds the ladder, and the experimenter climbs it. The movements involved are lower limbs climbing, arms lifting, trunk bending, and hands holding the ladder.	
Water hose training on ground	The motion is divided into throwing the second hose, holding the hydraulic giant, and running. The motions are limbs running (legs 1), arms swinging, holding and clamping (arms 1/3/4), trunk forerake (trunk 1), and hands holding the hose (hand 1)	
Water hose training on high	The motion is divided into two hands alternatively stretching down the hose. The motions are holding and clamping with arms, trunk forerake and bending (trunk 1/3), and hands gripping and cambered open (hand 1/2).	
Rescue	Rescue includes backing, holding, shouldering, and lifting, of which this experiment chooses backing. The motions involved are lower limbs squatting and erecting (leg 4/2), arms holding (arm 3), trunk forerake (trunk 1), and hands gripping with cambered open shape (hand 1/2).	

TABLE 6: Output results of joint key frame simulation moment (N/m).

Time	Left elbow	Right elbow	Left shoulder X	Left shoulder Y	Left shoulder Z	Right shoulder Y	Right shoulder Y	Right shoulder Z
60	2.48	2.39	4.76	13.59	3.11	5.50	15.62	5.73
120	4.19	4.14	4.19	11.66	6.83	4.92	15.94	6.46
180	4.19	4.97	3.63	10.28	6.95	4.14	12.58	8.19
240	3.91	3.45	5.11	15.96	7.38	5.24	15.23	6.35
300	3.73	4.30	4.58	12.83	6.03	4.62	15.13	10.65
360	3.54	3.89	4.42	12.63	4.39	4.69	17.57	6.49
420	3.73	4.51	5.04	17.07	7.82	4.62	13.80	6.23
480	4.16	3.84	4.65	16.35	9.11	4.21	11.91	5.59
540	3.77	4.30	5.06	14.97	6.74	4.69	16.65	7.52
600	4.60	5.01	3.93	10.95	7.71	4.09	13.23	7.82

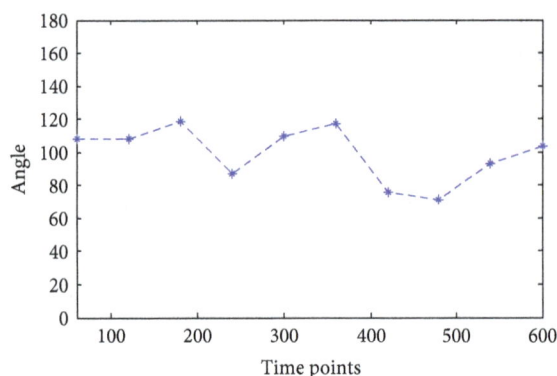

(a) Left elbow

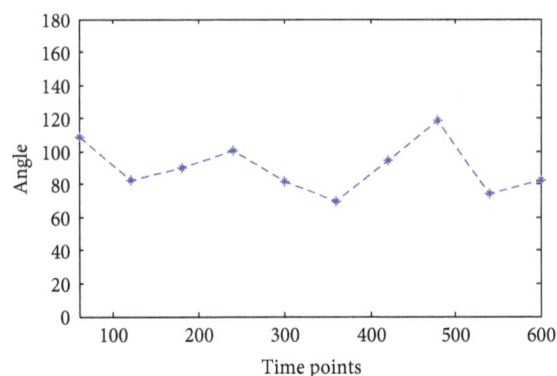

(b) Right elbow

FIGURE 6: Left and right elbow angles change with time.

first-order comprehensive fuzzy evaluation is conducted firstly. The factor set $U11 = \{U111, U112, U113\} = \{$Upper arm stretch and striction, upper arm flexion and extension, upper arm rotation$\}$, $U12 = \{U121\} = \{$Front arm flexion and extension$\}$ is calculated the membership $r_{ij} = (r_{i1}, r_{i2}, r_{i3})$ $(j = 1, 2, 3)$ of this motion is determined, and then the evaluation matrix is obtained.

Shoulder factor judgment matrix is as follows:

$$R_{1\text{Left}} = \begin{bmatrix} 0 & 0 & 1 \\ 0 & 0.755 & 0.245 \\ 0.231 & 0.769 & 0 \end{bmatrix},$$

$$R_{1\text{Right}} = \begin{bmatrix} 0 & 0 & 1 \\ 0.392 & 0.608 & 0 \\ 0.874 & 0.126 & 0 \end{bmatrix}. \tag{12}$$

Elbow factor judgment matrix is as follows:

$$R_{2\text{Left}} = \begin{bmatrix} 1 & 0 & 0 \end{bmatrix}, \qquad R_{2\text{Right}} = \begin{bmatrix} 0 & 0.717 & 0.283 \end{bmatrix}. \tag{13}$$

According to Equation $B = wR$:

$$B_{1\text{Left}} = \begin{bmatrix} 0.139 & 0.344 & 0.518 \end{bmatrix},$$

$$B_{1\text{Right}} = \begin{bmatrix} 0.278 & 0.259 & 0.463 \end{bmatrix},$$

$$B_{2\text{Left}} = \begin{bmatrix} 1 & 0 & 0 \end{bmatrix},$$

$$B_{2\text{Right}} = \begin{bmatrix} 0 & 0.717 & 0.283 \end{bmatrix}. \tag{14}$$

Second-order comprehensive fuzzy evaluation: taking B as the evaluation matrix R of the second-order comprehensive fuzzy evaluation, and through weight calculation $B = wR$, the comprehensive fuzzy evaluation is obtained as follows:

$$B_{\text{UpperLimbs}} = \begin{bmatrix} 0.330 & 0.325 & 0.345 \end{bmatrix}. \tag{15}$$

The comfort of some continuous motion is evaluated by the relationship of joint comfort change with time by Matlab. The membership of motion to comfort level is shown in Figures 10 and 11.

This motion involves two shoulders and two elbows and eight degrees of freedom, which forms the evaluation on upper limb movement. Through the previous weight calculation and proper unification, the comprehensive comfort level is evaluated.

From Figures 10, 11, and 12, the average of comprehensive comfort level in the latter half is higher than in

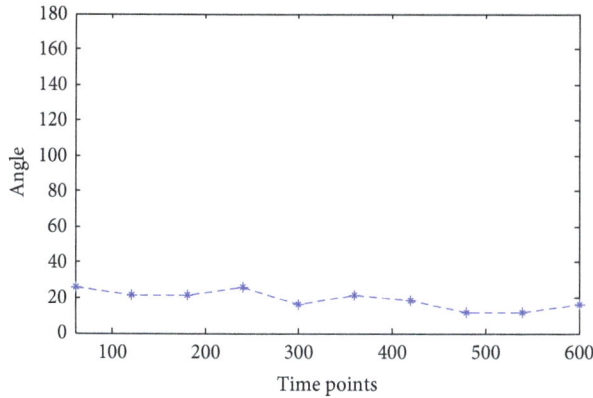

(a) X Stretch and striction

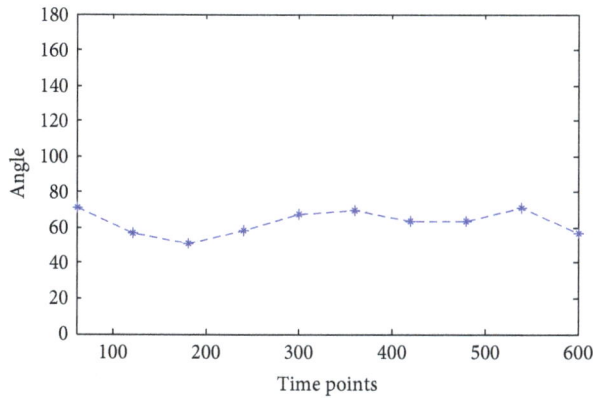

(b) Y Flexion and extension

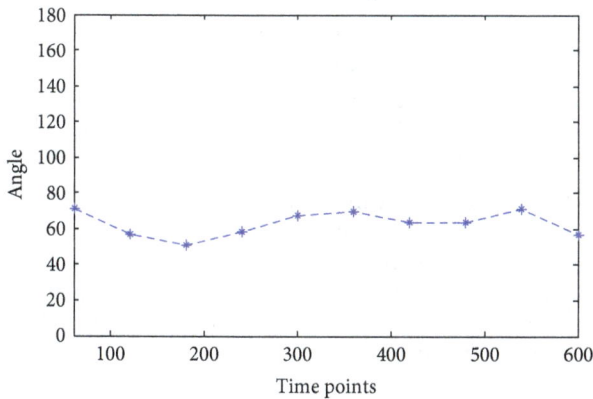

(c) Z rotation

FIGURE 7: Left shoulder angle in direction X, Y, and Z change with time.

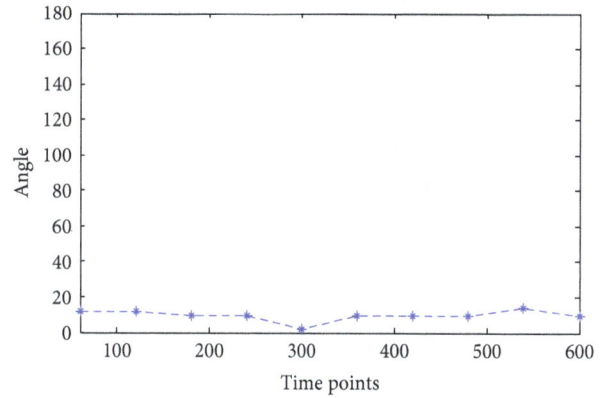

(a) X Stretch and striction

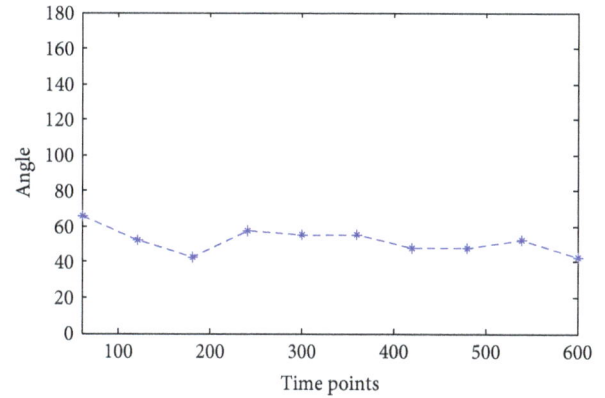

(b) Y Flexion and extension

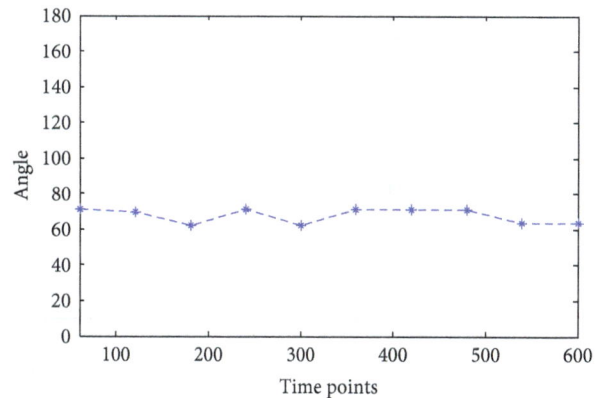

(c) Z rotation

FIGURE 8: Right shoulder angle in direction X, Y, and Z change with time.

the former half. Since this motion is of cyclicity, so the joint angle in the latter half fits comfort level effect better. Therefore, the latter half is always selected to conduct the experiment. According to this motion, the improvement in comfort means the improvement of work efficiency and safety. At the same time, when many experimenters take part, the best-comfortable motions will be taken as the model samples through comprehensive comfort evaluation.

6. Conclusion

This paper conducts ergonomic evaluation on firefighting operation motions with the method of fuzzy evaluation, which provides evaluation foundation on firefighting operation improvement and the establishment of standard motion database. The main contribution of the paper includes the following ways

(a) Motion simulation of disengaging water hose

(b) Joint moment output

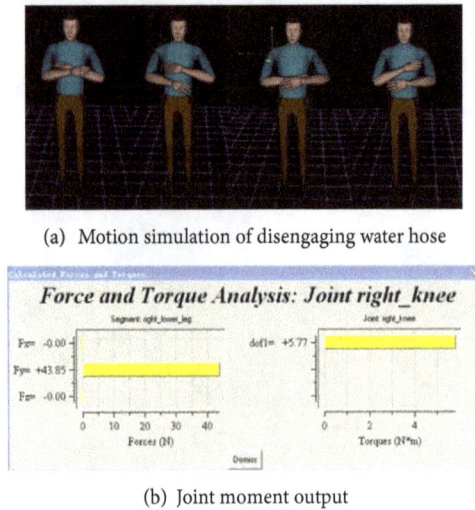

FIGURE 9: Motion simulation of disengaging water hose on high and moment result output.

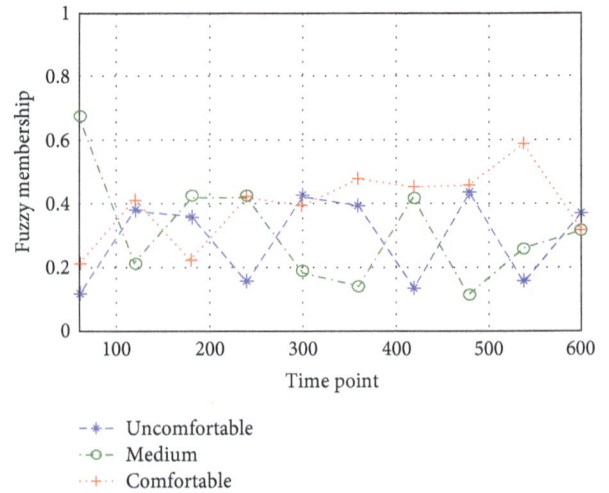

FIGURE 10: Left shoulder comfort change with time.

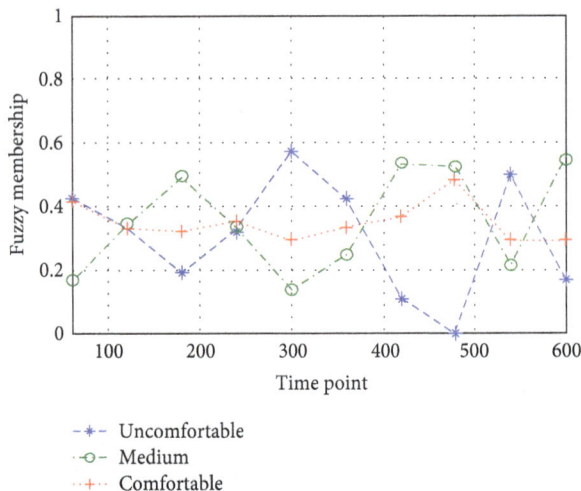

FIGURE 11: Right shoulder comfort change with time.

FIGURE 12: Upper limbs comfort change with time.

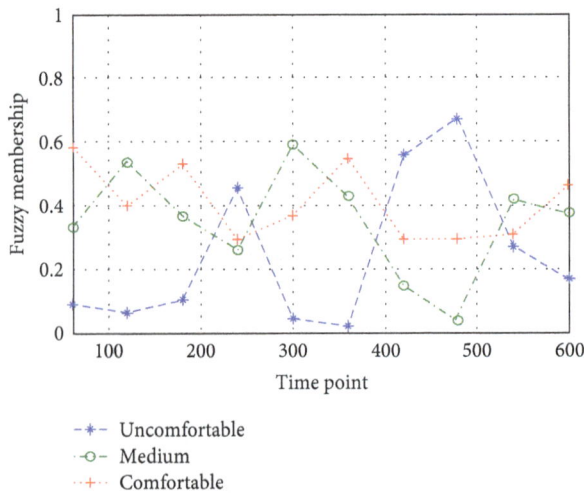

(1) The comprehensive fuzzy evaluation model of firefighting operation motion comfort level is established, which then is applied in the comfort evaluation.

(2) Joint weight determination through investigation experiments.

(3) The joint angles in every degree of freedom of typical motions are calculated in Matlab through the firefighter motion trails acquired by the motion capture equipment, which provide foundation to design firefighting equipment and clothing.

(4) The joint motion moment is acquired through motion simulation of disengaging water hose in JACK, and then the comprehensive fuzzy evaluation of comfort is conducted. The membership of joints to comfort level is obtained too.

This paper used the method of objective skeleton motion ergonomic evaluation, thus the results are objective and reliable. The research ideas will be applied in the future work to complete the database of motion evaluation, and the evaluation results will be dealt with statistically to build standard evaluation system.

References

[1] L. Wang, W. Hu, and T. Tan, "Recent developments in human motion analysis," *Pattern Recognition*, vol. 36, no. 3, pp. 585–601, 2003.

[2] A. Jiao, S. Yang, and L. Yuan, "The model for evaluation of fire training work effect applied by BP networks," *Fire Science and Technology*, vol. 24, no. 3, pp. 336–339, 2005.

[3] J. Chen and J. Li, "Designing of evaluation methods for fire simulation training system based on SAQ+B," *Journal of Chinese People's Armed Police Force Academy*, vol. 24, no. 2, pp. 33–35, 2008.

[4] S. J. Park, C. B. Kim, C. J. Kim, and J. W. Lee, "Comfortable driving postures for Koreans," *International Journal of Industrial Ergonomics*, vol. 26, no. 4, pp. 489–497, 2000.

[5] Y. Chen and G. Liu, "Comprehensive fuzzy evaluation for generalized product quality based on entropy weight," *Journal of Northeastern University (Natural Science)*, no. 2, pp. 241–244, 2010.

[6] B. Hongzhe and Z. Damin, *Computer Simulation for Aviation Man-Machining Engineering*, Electric Industry Press, Beijing, China, 2010.

[7] J. C. Bezdek, R. J. Hathaway, M. J. Sabin, and W. T. Tucker, "Convergence and theory for fuzzy C-means clustering: counter-examples and repairs," *IEEE Transactions on Systems, Man and Cybernetics*, vol. 17, no. 5, pp. 873–877, 1987.

[8] T. Padmaa and P. Balasubramanieb, "A fuzzy analytic hierarchy processing decision support system to analyze occupational menace forecasting the spawning of shoulder and neck pain," *Expert Systems with Applications*, vol. 38, no. 12, pp. 15303–15309, 2011.

[9] M. P. De Looze, L. F. M. Kuijt-Evers, and J. Van Dieën, "Sitting comfort and discomfort and the relationships with objective measures," *Ergonomics*, vol. 46, no. 10, pp. 985–997, 2003.

[10] S. Qin, L. Yang, P. Zhang, and Y. Li, "A new data visualisation methodology for evaluating product design with digital human models integrated with scanned body and captured motion," in *Proceedings of the 6th International Conference on Fuzzy Systems and Knowledge Discovery (FSKD '09)*, vol. 7, pp. 235–239, August 2009.

8

Parametric Optimization of Nd:YAG Laser Beam Machining Process Using Artificial Bee Colony Algorithm

Rajarshi Mukherjee, Debkalpa Goswami, and Shankar Chakraborty

Department of Production Engineering, Jadavpur University, Kolkata, West Bengal 700 032, India

Correspondence should be addressed to Shankar Chakraborty; s_chakraborty00@yahoo.co.in

Academic Editor: Josefa Mula

Nd:YAG laser beam machining (LBM) process has a great potential to manufacture intricate shaped microproducts with its unique characteristics. In practical applications, such as drilling, grooving, cutting, or scribing, the optimal combination of Nd:YAG LBM process parameters needs to be sought out to provide the desired machining performance. Several mathematical techniques, like Taguchi method, desirability function, grey relational analysis, and genetic algorithm, have already been applied for parametric optimization of Nd:YAG LBM processes, but in most of the cases, suboptimal or near optimal solutions have been reached. This paper focuses on the application of artificial bee colony (ABC) algorithm to determine the optimal Nd:YAG LBM process parameters while considering both single and multiobjective optimization of the responses. A comparative study with other population-based algorithms, like genetic algorithm, particle swarm optimization, and ant colony optimization algorithm, proves the global applicability and acceptability of ABC algorithm for parametric optimization. In this algorithm, exchange of information amongst the onlooker bees minimizes the search iteration for the global optimal and avoids generation of suboptimal solutions. The results of two sample paired t-tests also demonstrate its superiority over the other optimization algorithms.

1. Introduction

Increasing demand for advanced difficult-to-machine materials and availability of high-power lasers have stimulated interest among the researchers for the development of laser beam machining (LBM) processes [1]. The LBM, which is a thermal energy-based machining process, is now being widely applied to fulfill the present day requirements of high flexibility and productivity, noncontact processing, elimination of finishing operations, adaptability to automation, reduced processing cost, improved product quality, greater material utilization, processing of materials irrespective of electrical conductivity, minimum heat affected zone (HAZ), and green manufacturing. In this process, the material is removed by (a) melting, (b) vaporization, and (c) chemical degradation where the chemical bonds are broken causing the materials to degrade. When a high energy density laser beam is focused on a work surface, the thermal energy is absorbed which heats and transforms the work volume into a molten, vaporized, or chemically changed state that can

easily be removed by the flow of high pressure assist gas jet. This process also does not involve any mechanical cutting force and tool wear. Using LBM method, several material processing operations, such as laser microdrilling, cutting, microgrooving, microturning, marking, or scribing can be done [2, 3].

Among various types of lasers used for machining in industries, CO_2 and Nd:YAG lasers are the most established. Although CO_2 lasers have wide application in commercial sheet metal cutting operations, the benefits offered by Nd:YAG laser make it an interesting field of investigation. Experimental results show that Nd:YAG laser has some unique characteristics. Although the mean beam power is relatively low, the beam intensity can be relatively high due to smaller pulse duration and better focusing behavior. Smaller kerf width, microsize holes, narrower HAZ, and better cut edge kerf profile can be obtained in Nd:YAG LBM process. The smaller thermal load offered by Nd:YAG laser allows the machining of some brittle materials, such as SiC ceramics, which cannot be machined by CO_2 laser without crack damage.

As Nd:YAG LBM is a complex dynamic process with numerous parameters, like lamp current, pulse frequency, air pressure, pulse width, and cutting speed, so in order to maintain a high production rate and an acceptable level of quality for the machined parts, it is important to select the optimal combination of the process parameters, because these parameters directly affect the physical characteristics of the machined parts, as signified by kerf width, HAZ thickness, taper, and surface roughness. Experimental and theoretical studies show that the performance of Nd:YAG LBM process can be significantly improved by proper selection of the machining parameters [4]. For this purpose, the process engineers have to often rely on the manufacturer's data or handbook data. Hence, there is an ardent need for some sound optimization tools to determine the optimal machining parameters for Nd:YAG LBM process to have enhanced performance.

Mathew et al. [5] developed predictive models based on some important process parameters to determine the optimal process parameter ranges for pulsed Nd:YAG laser machining operation on carbon fibre reinforced plastic composites. Using response surface methodology (RSM), Kuar et al. [6] performed parametric analysis to determine the optimal setting of process parameters, like pulse frequency, pulse width, lamp current, and assist air pressure, for achieving minimum HAZ thickness and taper of microholes machined on zirconium oxide (ZrO_2) by pulsed Nd:YAG laser. Kuar et al. [7] studied the effects of several laser machining parameters on HAZ thickness and taper of the microdrilled holes on alumina-aluminium composites using RSM technique. Dhupal et al. [8] considered lamp current, pulse frequency, pulse width, assist air pressure, and cutting speed as the machining parameters during pulsed Nd:YAG laser micro-grooving operation and developed RSM-based equations to study the effects of those parameters on upper width, lower width, and depth of trapezoidal microgrooves. The optimal parametric combination was validated through experimentation and artificial-neural-network-(ANN-) based predictive model. Dubey and Yadava [9] presented a hybrid Taguchi method and RSM technique for simultaneous optimization of kerf width and material removal rate (MRR) for a laser beam cutting process. Dhupal et al. [10] investigated the effects of lamp current, pulse frequency, pulse width, assist air pressure, and cutting speed of workpiece on upper deviation, lower deviation, and depth characteristics of laser-turned microgrooves produced on cylindrical Al_2O_3 workpiece. Dubey and Yadava [11] simultaneously optimized kerf deviation and kerf width using Taguchi quality loss function during pulsed Nd:YAG laser beam cutting of aluminium alloy sheet. Dhupal et al. [12] developed RSM-based mathematical models and analyzed the machining characteristics of pulsed Nd:YAG laser during micro-grooving operation on aluminum titanate workpiece. Çaydaş and Hasçalık [13] presented a grey relational analysis-based approach for optimization of laser cutting process of St-37 steel with multiple performance characteristics. Dhupal et al. [14] selected lamp current, pulse frequency, pulse width, cutting speed, and assist gas pressure

as the major machining parameters for producing square micro-grooves on cylindrical surface. A predictive model for laser turning process parameters was developed using a feed-forward ANN technique, and an optimization problem was constructed based on RSM and then solved using genetic algorithm. Rao and Yadava [15] proposed a hybrid optimization approach for determining the optimal laser cutting process parameters to minimize kerf width, kerf taper, and kerf deviation together during pulsed Nd:YAG laser cutting of a thin sheet of nickel-based superalloy. Ciurana et al. [16] modeled the relationship between laser micromachining process parameters and quality characteristics using ANN and carried out multi-objective particle swarm optimization of the process parameters for minimum surface roughness and volume error. Based on RSM technique, Sivarao et al. [17] studied the effects of cutting speed, frequency, and duty cycle on surface roughness in the laser cutting process of mild steel. Doloi et al. [18] developed RSM-based mathematical models and analyzed the machining characteristics of pulsed Nd:YAG laser during micro-grooving operation on flat surface of aluminium titanate in order to optimize the parametric setting for achieving accurate taper angles of micro-grooves. Kuar et al. [19] performed RSM-based parametric analysis to investigate the change in the responses with the input parameters, such as pulse frequency, pulse width, lamp current, and assist air pressure, for achieving minimum height of the recast layer and maximum depth of the microgroove. Sharma et al. [20] performed parametric optimization of the kerf quality characteristics (kerf width, kerf taper, and kerf deviation) during pulsed Nd:YAG laser cutting of nickel-based superalloy thin sheet. Biswas et al. [21] investigated the effects of different process parameters on hole circularity at exit and taper of the hole during Nd:YAG laser microdrilling on gamma-titanium aluminide. Kibria et al. [22] performed experimental analysis on Nd:YAG laser microturning of cylindrical-shaped ceramic materials to achieve the desired responses, that is, depth of cut and surface roughness while varying the laser micro-turning process parameters, such as lamp current, pulse frequency, and laser beam scanning speed. Biswas et al. [23] observed the effects of five parameters on circularity and taper of holes in pulsed Nd:YAG laser microdrilling process and concluded that the circularity of the drilled hole at entry, exit, and taper were the important attributes influencing the quality of the hole. Biswas et al. [24] investigated the effects of lamp current, pulse frequency, pulse width, air pressure, and focal length of Nd:YAG laser micro-drilling process on hole circularity at entry and exit using RSM-based experimental results. Panda et al. [25] applied grey relational approach for determining the optimal process parameters to minimize HAZ and hole circularity and maximize MRR in pulsed Nd:YAG laser micro-drilling on high carbon steel. Sibalija et al. [26] presented a hybrid design strategy for determining the optimal laser drilling parameters in order to simultaneously meet all the requirements for seven quality characteristics of the holes produced during pulsed Nd:YAG laser drilling on a thin sheet of nickel-based superalloy.

Although the earlier researchers have applied different optimization techniques, like Taguchi method, grey relational analysis, desirability function, and genetic algorithm, for finding out the optimal process parameter values, in most of the cases, suboptimal or near optimal solutions have been reached In this paper, the application of artificial bee colony (ABC) algorithm is validated as an effective and efficient tool for parametric optimization of Nd:YAG LBM process. The optimization performance of ABC algorithm is also compared with that of other population-based algorithms, like genetic algorithm (GA), particle swarm optimization (PSO), and ant colony optimization (ACO) which proves the superiority of ABC algorithm.

2. Artificial Bee Colony Algorithm

Artificial bee colony algorithm is an evolutionary computational technique, developed by Karaboga et al. [27–30]. In this algorithm, the colony of artificial bees consists of three groups, that is, employed bees, onlookers, and scouts. The first half of the colony consists of the employed artificial bees and the second half includes the onlookers. For every food source, there is only one employed bee. Thus, the number of employed bees is equal to the number of food sources around the hive. The employed bee whose food source has been abandoned becomes a scout.

In this algorithm, the position of a food source represents a possible solution to the considered optimization problem and the nectar amount of the food source is proportional to the quality or fitness of the associated solution. The number of the employed bees or onlooker bees is equal to the number of solutions in the population. In the first step, the ABC algorithm generates randomly distributed predefined number of initial population, P, (position of the food sources) of SN populations, where $P \in$ SN. Each position of the food source, x_{ijk}, is three-dimensional with $i = 1, 2, \ldots, SN; j = 1, 2, \ldots, D; k = 1, 2, \ldots, V$, where D is the dimension of each variable and V is the number of variables in the objective function. After initialization, the population of the positions (solutions) is subjected to repeated cycles, $C = 1, 2, \ldots, MCN$ (maximum cycle number) of the search processes of the employed, onlooker, and scout bees.

An employed bee produces a modification on the position (solution) in its memory depending on the local information (visual information) and tests the nectar amount (fitness value) of the new food source (new solution). Provided that the nectar amount of the new source is higher than that of the previous one, the bee memorizes the new position and forgets the old one. Otherwise, it keeps the position of the previous source in its memory. When all the employed bees complete the search process, they share the nectar information of the food sources and their position information with the onlooker bees in the dance area. An onlooker bee evaluates the nectar information taken from all the employed bees and selects a food source with a probability related to its nectar amount. As in the case of an employed bee, the onlooker bee produces a modification on the position in its memory and checks the nectar amount of the candidate source. If its nectar amount is higher than that of the previous one, the onlooker bee memorizes the new position and forgets the old one.

An artificial onlooker bee selects a food source depending on the probability value associated with that food source, p_i, as given in the following equation:

$$p_i = \frac{\text{fit}_i}{\sum_{i=1}^{SN} \text{fit}_i}, \tag{1}$$

where fit_i is the fitness value of ith solution which is proportional to the nectar amount of the food source in ith position and SN is the number of food sources which is equal to the number of employed bees.

In order to produce a candidate food position from the old one in memory, the ABC algorithm adopts the following expression:

$$v_{ijk} = x_{ijk} + \varphi \left(x_{ijk} - x_{ljk} \right), \tag{2}$$

where v_{ijk} is the candidate food position. Although l is determined randomly, it has to be different from $i \cdot \varphi$ is a random number between -1 and 1. It controls the production of the neighborhood food sources around x_{ij} and represents the visual comparison of two food positions by a bee. From (2), it can be seen that as the difference between the parameters of x_{ijk} and x_{ljk} decreases, the perturbation on the position x_{ijk} gets decreased too. Thus, as the search process approaches to the optimal solution in the search space, the step length is adaptively reduced. If a parameter value produced by this operation exceeds its predetermined limit, the parameter can be set to an acceptable value. Here, the value of the parameter exceeding its limit is set to its limit value.

The food source of which the nectar is abandoned by the bees is replaced with a new food source by the scouts. In ABC algorithm, this is simulated by producing a random position and replacing it with the abandoned one. In this algorithm, providing that a position (solution) cannot be improved further through a predetermined number of cycles, then that food source is assumed to be abandoned. The value of the predetermined number of cycles is an important control parameter of ABC algorithm, which is known as "limit" for abandonment. Assume that the abandoned source is x_i and $j \in \{1, 2, 3, \ldots, D\}, k \in \{1, 2, 3, \ldots, V\}$, then the scout discovers a new food source to be replaced with x_i. This operation can be defined using the following equation:

$$x_{ijk} = x_{\min jk} + \text{rand}(0, 1) \left(x_{\max jk} - x_{\min jk} \right), \tag{3}$$

where $x_{\max jk}$ and $x_{\min jk}$ are the upper and lower bounds of kth variable, respectively. At each candidate source position, the value of v_{ijk} is searched out and evaluated by the artificial bees. Its performance is then compared with that of the old one. If the new food source has equal or better nectar amount than the old one, it is replaced with the old one in the memory. Otherwise, the old food source is retained in the memory. In other words, a greedy selection mechanism is employed as the selection process between the old and the candidate one. The main steps of ABC algorithm are given below [27].

(i) Initialize.

(ii) Repeat.

(a) Place the employed bees on the food sources in the memory.

(b) Place the onlooker bees on the food sources in the memory.

(c) Send the scouts to the search area for discovering new food sources.

(iii) Until (all the requirements are met).

The detailed pseudocode of ABC algorithm is presented as follows [31]:

(1) Initialize the population of solutions x_{ijk} ($i = 1, 2, \ldots, SN$; $j = 1, 2, \ldots, D$; $k = 1, 2, \ldots, V$).

(2) Evaluate the population.

(3) Cycle = 1.

(4) Repeat.

(5) Produce new solutions v_{ijk} for the employed bees and evaluate them.

(6) Apply the greedy selection process.

(7) Calculate the probability values for the solutions x_{ijk}.

(8) Produce the new solutions v_{ijk} for the onlookers from the solutions x_{ijk} selected depending on probability values and evaluate them.

(9) Apply the greedy selection process.

(10) Determine the abandoned solution for the scout, if exists, and replace it with a new randomly produced solution x_{ijk}.

(11) Memorize the best solution achieved so far.

(12) Cycle = cycle + 1.

(13) Until cycle = MCN.

3. Optimization of Nd:YAG LBM Processes

In order to validate the applicability and performance of ABC algorithm for parametric optimization of Nd:YAG LBM process, the experimental data and mathematical modeling of two LBM processes [6, 10] are analyzed here. For each of these processes, both the single and multi-objective optimizations of the responses are performed. For application of ABC algorithm, a computer code is developed in MATLAB 7.6 (R2008a) with the following control parameters: swarm size = 10, number of employed bees = 50% of the swarm size, number of onlookers = 50% of the swarm size, number of scouts per cycle = 1, number of cycles = 2000, and runtime = 2.

The role of various control parameters in ABC algorithm is also quite important, which mainly drive the operational aspect of this algorithm. For example, swarm size, number of employed bees, and number of onlookers directly influence selection of the initial starting point for this algorithm and control the number of bees participating in the search process (in this case, beginning of simultaneous search conditions initiated by each bee). Also, the number of onlookers directly

TABLE 1: Machining parameters with their levels.

Parameter	Levels				
	−2	−1	0	1	2
Lamp current (x_1) (amp)	17	19	21	23	25
Pulse frequency (x_2) (kHz)	1	2	3	4	5
Air pressure (x_3) (kg/cm^2)	0.6	1	1.4	1.8	2.2
Pulse width (x_4) (%)	2	6	10	14	18

influences how quickly the potential food sources are evaluated: higher number of onlooker bees means quick collection of information from the employed bees and thus selection or rejection of food sources will be faster. Similarly, number of scouts per cycle will quicken the search process for new food source in every cycle. Finally, the number of cycles represents how many times the algorithm will be run before termination and may prove to be useful where there are a large number of variables to be evaluated.

Changing the values of various control parameters in ABC algorithm may increase/decrease the number of iterations to reach the optimal solution, but there will not be any significant change in the optimal solution. In this paper, the control parameters are selected based on the nature of the mathematical model (second-order equations with four/five variables) and capacity of CPU used (1.83 GHz Core 2 DUO processor with 1 GB RAM): high end CPU may run with higher number of cycles and higher number of swarm size, but for CPU with limited resource, higher values of control parameters may cause CPU to freeze and RAM to overflow.

3.1. Example 1. Kuar et al. [6] performed laser beam microdrilling operation on zirconia (ZrO$_2$) ceramics of size 20 × 20 mm and 1 mm thick and studied the influences of four process parameters, that is, lamp current, pulse frequency, air pressure, and pulse width on HAZ thickness and taper of the drilled holes. Each of those four process parameters was set at five different levels, as given in Table 1.

To determine the multiparametric optimal combinations for pulsed Nd:YAG laser beam microdrilling process on ZrO$_2$ ceramics, experiments were carried out according to a central composite rotatable second-order design plan based on RSM technique and the following two equations were developed for HAZ thickness and taper:

$$
\begin{aligned}
Y_u \text{ (HAZ)} =\ & 0.3796 + 0.07888x_1 - 0.04120x_2 \\
& - 0.04301x_3 - 0.00570x_4 + 0.02146x_1^2 \\
& - 0.00957x_2^2 + 0.00266x_3^2 - 0.01234x_4^2 \\
& - 0.0228x_1x_2 - 0.00679x_1x_3 - 0.03158x_1x_4 \\
& + 0.01341x_2x_3 - 0.00983x_2x_4 - 0.00497x_3x_4,
\end{aligned}
$$

$$
\begin{aligned}
Y_u \text{ (Taper)} =\ & 0.07253 + 0.00912x_1 + 0.00887x_2 \\
& - 0.00606x_3 + 0.00449x_4 + 0.00153x_1^2 \\
& + 0.00225x_2^2 + 0.00233x_3^2 + 0.00399x_4^2 \\
& + 0.00431x_1x_2 - 0.00646x_1x_3 - 0.00519x_1x_4 \\
& - 0.00110x_2x_3 - 0.00023x_2x_4 - 0.07253x_3x_4.
\end{aligned}
$$

$$(4)$$

TABLE 2: Results for single objective optimization.

Optimization method	Response	Optimal value	Lamp current (amp)	Pulse frequency (kHz)	Air pressure (kg/cm²)	Pulse width (%)
Kuar et al. [6]	HAZ	0.0675	17	2	2	2
	Taper	0.0319	17	2	0.6	2
GA	HAZ	0.1066	19	1	2	2
	Taper	0.0843	23.86	2.29	1.38	13.92
PSO algorithm	HAZ	0.0604	18	1.25	2.12	2.4
	Taper	0.0458	20.23	4.10	1.81	11.95
ACO algorithm	HAZ	0.0324	17	1.5	2	2
	Taper	0.0377	18.04	4.47	1.73	14.18
ABC algorithm	HAZ	0.0174	17.0	4.8	2.1	2
	Taper	0.0202	18.2	1.25	0.6	2

TABLE 3: Single objective optimization performance.

Optimization method	Response	Optimal value	Mean	Standard deviation	Standard error
GA	HAZ	0.1066	0.1231	0.0102	0.0032
	Taper	0.0843	0.1056	0.0151	0.0047
PSO algorithm	HAZ	0.0604	0.0883	0.0212	0.0067
	Taper	0.0458	0.0662	0.0138	0.0043
ACO algorithm	HAZ	0.0324	0.0505	0.0129	0.0041
	Taper	0.0377	0.0507	0.0098	0.0031
ABC algorithm	HAZ	0.0174	0.0301	0.0094	0.0030
	Taper	0.0202	0.0346	0.0092	0.0029

3.1.1. Single Objective Optimization. In this case, the second-order RSM-based equations for the two responses are optimized separately. Here, both the responses are to be minimized with respect to the constraints set as $17 \leq x_1 \leq 25$, $1 \leq x_2 \leq 5$, $0.6 \leq x_3 \leq 2.2$, and $2 \leq x_4 \leq 18$. Kuar et al. [6] obtained the optimal settings of lamp current = 17 amp, pulse frequency = 2 kHz, air pressure = 2 kg/cm² and pulse width = 2%; lamp current = 17 amp, pulse frequency = 2 kHz, air pressure = 0.6 kg/cm², and pulse width = 2% for minimum values of HAZ thickness of 0.0675 mm and taper of 0.0319 mm, respectively. These optimal parametric settings are shown in Table 2. This table also shows the results when ABC algorithm is applied to optimize these two RSM-based equations with respect to the given constraints. It is observed that while employing ABC algorithm, the minimum value of HAZ thickness is drastically reduced from 0.0675 to 0.0174 mm and the minimum taper is also decreased from 0.0319 to 0.0202 mm. The optimal process settings are also changed. The optimization results for GA, PSO, and ACO algorithms are also given in Table 2, which proves the superiority of ABC algorithm over the others with respect to their optimization performance. Figure 1 shows the convergence diagram for all the considered optimization techniques for HAZ thickness. The termination criterion for each algorithm is set at 500 iterations; that is, after 500 iterations, the algorithm will be terminated and all the 500 solutions will be plotted on the convergence diagram. The best value is taken as the optimal solution of the objective function obtained by the algorithm. From Figure 1, it is

clear that ABC algorithm outperforms the other population-based algorithms while achieving the minimum value of HAZ thickness.

In order to study the optimization performance of ABC, ACO, PSO, and GA algorithms in details, the mean, standard deviation, and standard error of the obtained optimal values are computed, as given in Table 3. It is noted that the optimization performance of ABC algorithm is better than that of ACO, PSO, and GA with respect to the dispersion of the optimal solution values. The results of two sample paired t-tests, as exhibited in Table 4, show that the differences in optimization performance between ABC algorithm and other considered population-based algorithms are statistically significant at 5% significance level. It is also observed that the optimization performance of ABC algorithm is relatively more consistent than that of other algorithms. Table 5 compares the required computational (CPU) times for all the considered algorithms when run in an Intel Core 2 DUO, 1.83 GHz, 1 GB RAM CPU computer platform. It is interesting to note that although ABC algorithm has excellent optimization performance, its CPU time is not so very high compared to the other algorithms under consideration. Hence, it can be an effective optimization tool for finding out the best parametric combination of Nd:YAG LBM process for its enhanced machining performance.

Figure 2 shows the variations of HAZ thickness with respect to the four LBM process parameters. It is observed that with the increase in lamp current and pulse width, HAZ thickness increases, whereas it decreases with increasing

TABLE 4: Two sample paired t-tests between different algorithms.

Optimization methods	Response	Results of paired t-test
GA versus ABC algorithm	HAZ	95% CI for mean difference: (0.08886, 0.09616), t-test of mean difference = 0 (versus not = 0), t value = 50.97, P value = 0.000
	Taper	95% CI for mean difference: (0.06522, 0.07526), t-test of mean difference = 0 (versus not = 0), t value = 28.13, P value = 0.000
PSO algorithm versus ABC algorithm	HAZ	95% CI for mean difference: (0.05068, 0.06520), t-test of mean difference = 0 (versus not = 0), t value = 16.05, P value = 0.000
	Taper	95% CI for mean difference: (0.02703, 0.03450), t-test of mean difference = 0 (versus not = 0), t value = 16.55, P value = 0.000
ACO algorithm versus ABC algorithm	HAZ	95% CI for mean difference: (0.01633, 0.02440), t-test of mean difference = 0 (versus not = 0), t value = 10.13, P value = 0.000
	Taper	95% CI for mean difference: (0.01000, 0.01835), t-test of mean difference = 0 (versus not = 0), t value = 6.82, P value = 0.000

TABLE 5: CPU time for ABC, ACO, PSO, and GA algorithms.

Optimization method	Average CPU time (in sec)
ABC algorithm	13.8
ACO algorithm	13.6
PSO algorithm	14.1
GA	13.4

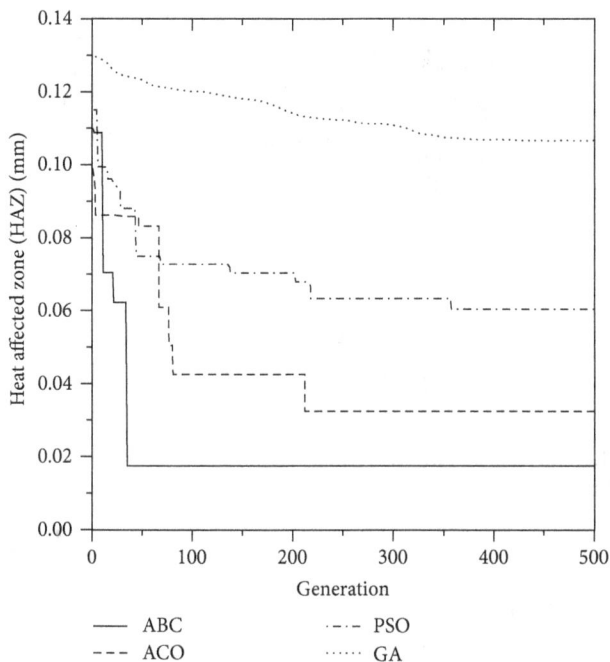

FIGURE 1: Convergence of ABC, ACO, PSO, and GA algorithms for HAZ thickness.

values of pulse frequency and assist air pressure. The energy of a laser beam is directly proportional to lamp current. High lamp current generates high thermal energy, which produces high HAZ thickness. At low pulse frequency, peak power of the laser beam is higher causing excessive material removal. But at higher pulse frequency as the peak power of laser beam is less, HAZ thickness is comparatively lower. It has also been found that the assist air pressure has an almost linear relationship with HAZ thickness [6]. Low assist air pressure is unable to remove the excess heat generated at the micro-drilling zone as well as being unable to assist the removal of ejected material. This phenomenon causes high HAZ thickness. However, at higher assist air pressure, the excess amount of heat can be rapidly removed which also helps in ejecting the molten material. As a result, low HAZ thickness can be observed. At lower pulse width, highly concentrated laser beam can easily penetrate into the material causing less HAZ thickness. Then HAZ thickness rapidly increases due to surface deposition of the molten material.

During pulsed Nd:YAG laser micro-drilling operation, minimization of taper of the micro-hole is highly required for maintaining quality and accuracy of the hole. Kuar et al. [6] observed that taper increases significantly with lamp current. High lamp current generates high thermal energy, and as a result, the top surface of work sample where the laser beam is focused gets melted and vaporized instantly, and large volume of material is removed from the top surface, which produces large taper. At very high pulse frequency, relatively large taper is observed, but at low pulse frequency, low taper is generated. At very low pulse frequency, the beam energy is slightly high but the time between two successive incident beams is more; therefore, material is removed only from the narrow focusing spot on the top surface of work sample. It has been observed that taper is significantly increased with the increase in assist air pressure [6]. Zirconia has a very low thermal conductivity and the higher assist air pressure cools the localized heating zone causing slower rate of material removal to penetrate up to the whole thickness of the work sample. As a result, laser beam energy for longer period causes large area material removal from the top surface of the hole, resulting in an increase in taper with increasing assist air pressure. At low pulse width, highly concentrated laser beam energy causes faster rate of penetration, and as a result less taper is formed. These same observations are also obtained in Figure 3 where the variations of taper with respect to four LBM process parameters are exhibited.

3.1.2. Multiobjective Optimization. In multi-objective optimization of Nd:YAG LBM process, instead of treating the

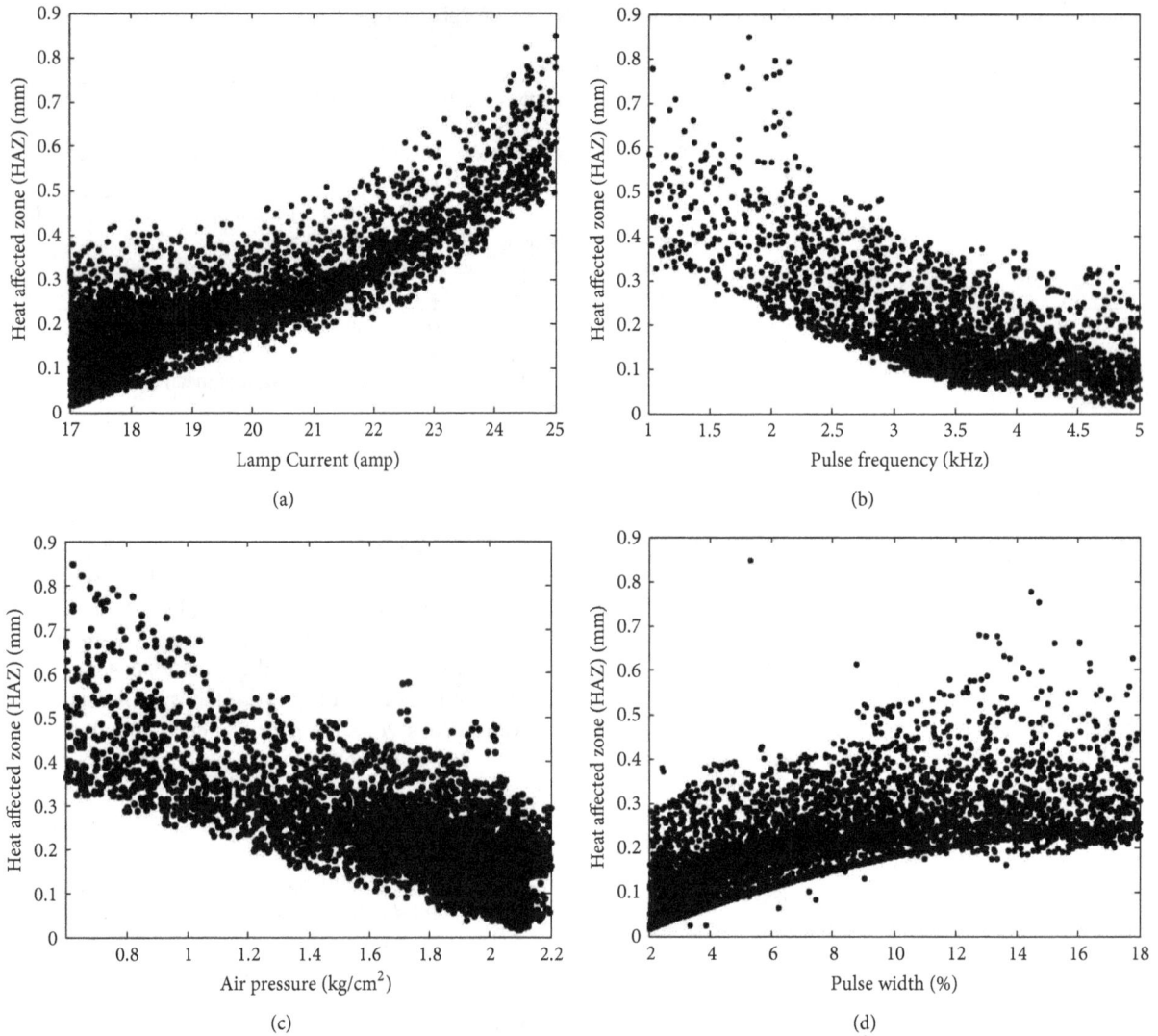

FIGURE 2: Variations of HAZ thickness with LBM process parameters.

two responses separately, both are simultaneously optimized, based on the following objective function [31]:

$$\text{Min}\,(Z_1) = \frac{w_1 Y_u\,(\text{HAZ})}{\text{HAZ}_{\text{min}}} + \frac{w_2 Y_u\,(\text{Taper})}{\text{Taper}_{\text{min}}}, \qquad (5)$$

where $Y_u(\text{HAZ})$ and $Y_u(\text{Taper})$ are the second-order RSM-based equations for HAZ thickness and taper, respectively; HAZ_{min} and $\text{Taper}_{\text{min}}$ are the minimum values of HAZ thickness and taper, respectively; w_1 and w_2 are the weights or priority values assigned to HAZ thickness and taper, respectively. These weights can be anything such that $w_1 + w_2 = 1$. Assignment of the weights (relative importance) to different responses is entirely based on the knowledge and experience of the concerned process engineers. Sometimes, analytic hierarchy process [32] is employed to determine these weight values. The HAZ_{min} and $\text{Taper}_{\text{min}}$ values are obtained from the single objective optimization results. Here, equal weights for both the responses, that is, $w_1 = w_2 = 0.5$, (case 1) are first considered and the results obtained

after solving this multi-objective optimization problem using ABC algorithm are given in Table 6. The constraints for this multi-objective optimization problem are the same as set for single objective optimization. The minimum HAZ thickness and taper values are obtained as 0.1019 mm and 0.0248 mm, respectively, which are quite better than those observed by Kuar et al. [6]. The optimal solution (Z_1) is 0.0634. Table 6 also shows the results of multi-objective optimization where two other weighting schemes to the responses (case 2: $w_1 = 0.9$ and $w_2 = 0.1$, and case 3: $w_1 = 0.1$ and $w_2 = 0.9$) are considered. In case 2, maximum weight is assigned to HAZ thickness and in case 3, minimization of taper is given more importance. In both these cases, the optimal process settings are changed. Table 7 gives a comparative analysis of the multi-objective optimization performance of ABC algorithm for all the three cases and it is important to note that the best performance is achieved when equal importance is given to the responses. Thus, based on these optimization results, it is always recommended to assign equal weights to all the responses.

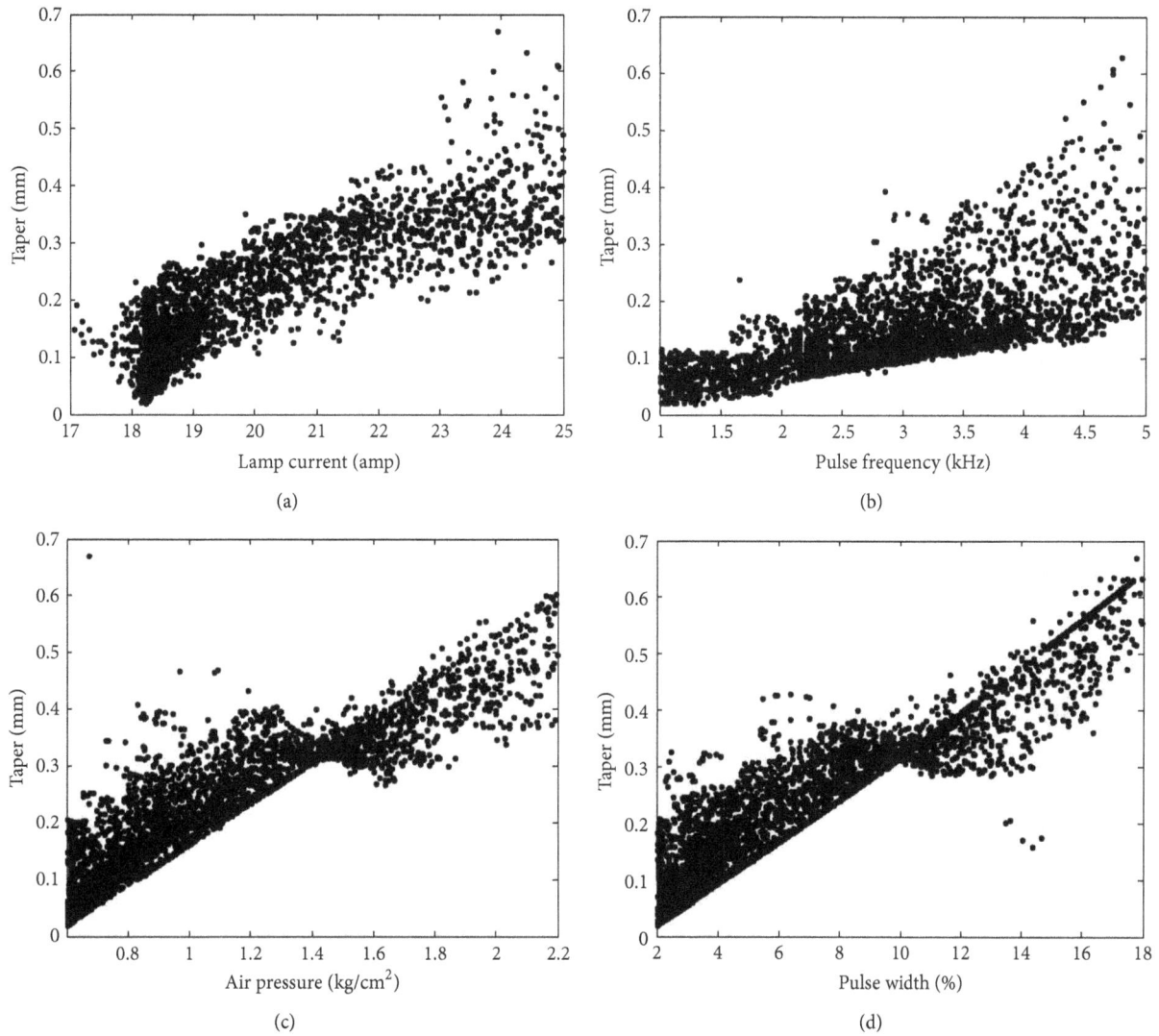

FIGURE 3: Variations of taper with various LBM process parameters.

TABLE 6: Results for multi-objective optimization.

Case	Response	Optimal value	Z_1	Lamp current (amp)	Pulse frequency (kHz)	Air pressure (kg/cm^2)	Pulse width (%)
Kuar et al. [6]	HAZ	0.1296	0.0848	17	1.62	1.04	2
	Taper	0.0400					
Case 1 ($w_1 = 0.5$ and $w_2 = 0.5$)	HAZ	0.1019	0.0634	17.18	1.5	1.33	2
	Taper	0.0248					
Case 2 ($w_1 = 0.9$ and $w_2 = 0.1$)	HAZ	0.0281	0.2007	17.2	1.5	2.16	2
	Taper	0.3733					
Case 3 ($w_1 = 0.1$ and $w_2 = 0.9$)	HAZ	0.3124	0.1891	18.66	1.34	1.20	7.95
	Taper	0.0329					

TABLE 7: Multi-objective optimization performance.

Case	Z_1	Mean	Standard deviation	Standard error
Case 1 ($w_1 = 0.5$ and $w_2 = 0.5$)	0.0634	0.0854	0.0151	0.0048
Case 2 ($w_1 = 0.9$ and $w_2 = 0.1$)	0.2007	0.2532	0.0413	0.0130
Case 3 ($w_1 = 0.1$ and $w_2 = 0.9$)	0.1891	0.2196	0.0309	0.0098

TABLE 8: LBM process parameters with their levels.

Parameter	Levels				
	−2	−1	0	1	2
Air pressure (x_1) (kg/cm^2)	0.3	0.8	1.3	1.8	2.3
Lamp current (x_2) (A)	13	16	19	22	25
Pulse frequency (x_3) (kHz)	1	2	3	4	5
Pulse width (x_4) (%)	2	4	6	8	10
Cutting speed (x_5) (rpm)	7	12	17	22	27

3.2. Example 2. Using an Nd:YAG laser-turning system, Dhupal et al. [10] performed micro-grooving operation on cylindrical Al$_2$O$_3$ workpiece (10 mm diameter and 40 mm length) and also investigated the effects of five process parameters (air pressure, lamp current, pulse frequency, pulse width, and cutting speed) on the upper deviation (Y_{uw}), lower deviation (Y_{lw}), and depth deviation (Y_d) of the machined micro-groove. Each of the five process parameters was set at five different levels, as shown in Table 8.

Dhupal et al. [10] conducted experiments based on a central composite rotatable second-order design plan and developed the following RSM-based equations for the considered three responses:

$$\begin{aligned}
Y_u(uw) = &-0.00376 - 0.01690x_1 - 0.00251x_2 \\
&- 0.00288x_3 + 0.00048x_4 + 0.00185x_5 \\
&+ 0.00678x_1^2 + 0.00232x_2^2 + 0.00276x_3^2 \\
&- 0.00012x_4^2 + 0.00207x_5^2 + 0.00004x_1x_2 \\
&- 0.00134x_1x_3 + 0.00188x_1x_4 - 0.00225x_1x_5 \\
&- 0.00149x_2x_3 - 0.00081x_2x_4 - 0.00052x_2x_5 \\
&+ 0.00114x_3x_4 - 0.00262x_3x_5 + 0.00120x_4x_5,
\end{aligned}$$

$$\begin{aligned}
Y_u(lw) = &\ 0.01857 - 0.01330x_1 - 0.00247x_2 \\
&- 0.00268x_3 + 0.00120x_4 - 0.00391x_5 \\
&+ 0.00299x_1^2 + 0.00224x_2^2 - 0.00137x_3^2 \\
&- 0.00122x_4^2 + 0.00051x_5^2 + 0.00235x_1x_2 \\
&- 0.00122x_1x_3 - 0.00168x_1x_4 + 0.00197x_1x_5 \\
&- 0.00197x_2x_3 - 0.00175x_2x_4 + 0.00166x_2x_5 \\
&- 0.0078x_3x_4 - 0.00211x_3x_5 + 0.00378x_4x_5,
\end{aligned}$$

$$\begin{aligned}
Y_u(d) = &\ 0.01265 - 0.02510x_1 - 0.00263x_2 \\
&+ 0.00451x_3 + 0.00479x_4 - 0.00229x_5 \\
&+ 0.00338x_1^2 + 0.00383x_2^2 + 0.00168x_3^2 \\
&+ 0.00157x_4^2 - 0.00112x_5^2 - 0.00214x_1x_2 \\
&- 0.00472x_1x_3 - 0.00264x_1x_4 + 0.00260x_1x_5 \\
&- 0.00035x_2x_3 - 0.00314x_2x_4 - 0.00365x_2x_5 \\
&- 0.00425x_3x_4 + 0.00006x_3x_5 + 0.00393x_4x_5.
\end{aligned}$$

$$(6)$$

3.2.1. Single Objective Optimization. The three above-mentioned RSM-based second-order equations for the responses are now optimized using ABC algorithm while treating the responses separately. The constraints are set as $0.3 \le x_1 \le 2.3$, $13 \le x_2 \le 25$, $1 \le x_3 \le 5$, $2 \le x_4 \le 10$, and 7

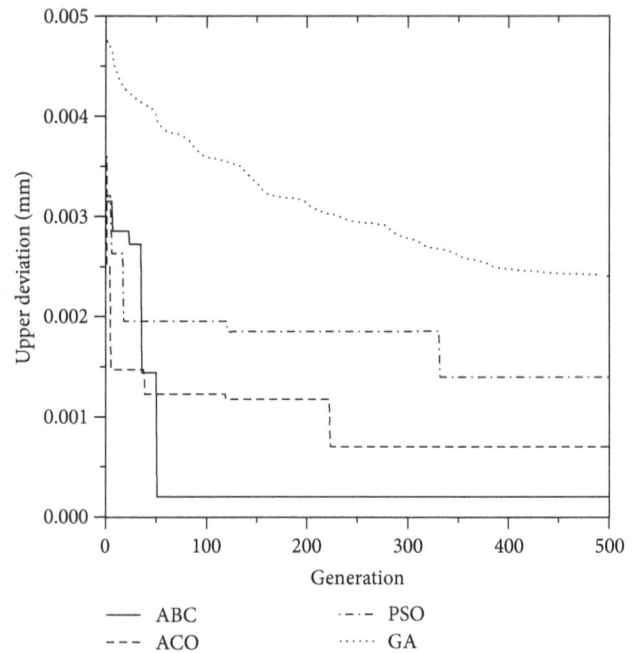

FIGURE 4: Convergence of ABC, ACO, PSO, and GA algorithms for upper deviation.

$\le x_5 \le 27$. The results of this single objective optimization of the responses are given in Table 9. In this case, all the three responses need to be minimized. It is observed from Table 9 that the minimum values for all the three responses are obtained when ABC algorithm is employed as the optimization tool. The performance of ABC algorithm is also better than the other population-based optimization methods, as shown in Table 9. Here, it is not possible to compare the results obtained using ABC algorithm with those of Dhupal et al. [10] as they did not consider the single objective optimization of Nd:YAG laser-turning process. The convergences of ABC, ACO, PSO, and GA algorithms for upper deviation of the machined micro-groove are shown in Figure 4. Table 10 compares the single objective optimization performance of the considered algorithms which again proves the superiority of ABC algorithm over the others.

The variations of upper deviation with respect to air pressure, lamp current, pulse frequency, pulse width, and cutting speed are exhibited in Figure 5. The dimensional upper deviation from the target is to be minimized. Dhupal et al. [10] observed that the upper deviation becomes lower with increasing values of lamp current and pulse frequency. As the lamp current increases, the laser beam energy increases and the top surface of the work material melts at a faster rate. High-energy laser beam produces low upper deviation because it removes material from the top surface and penetrates at a faster rate into the material to obtain the desired depth. It has also been observed that the change in upper deviation with pulse frequency is less compared to that of lamp current. The pulse width has moderate effect on upper deviation as compared with lamp current. At low pulse width, the upper deviation of the micro-groove approaches to zero,

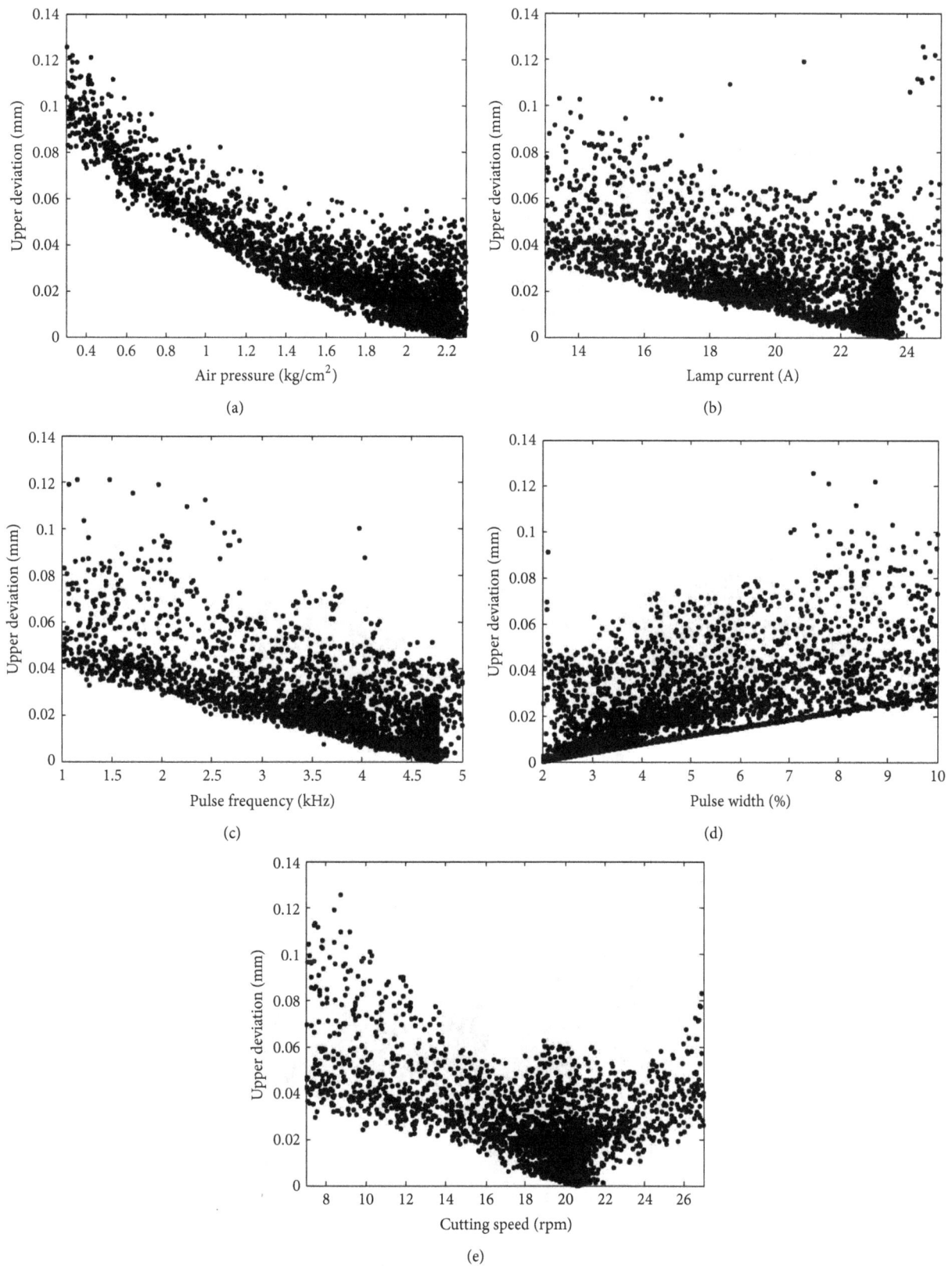

FIGURE 5: Variations of upper deviation with various LBM process parameters.

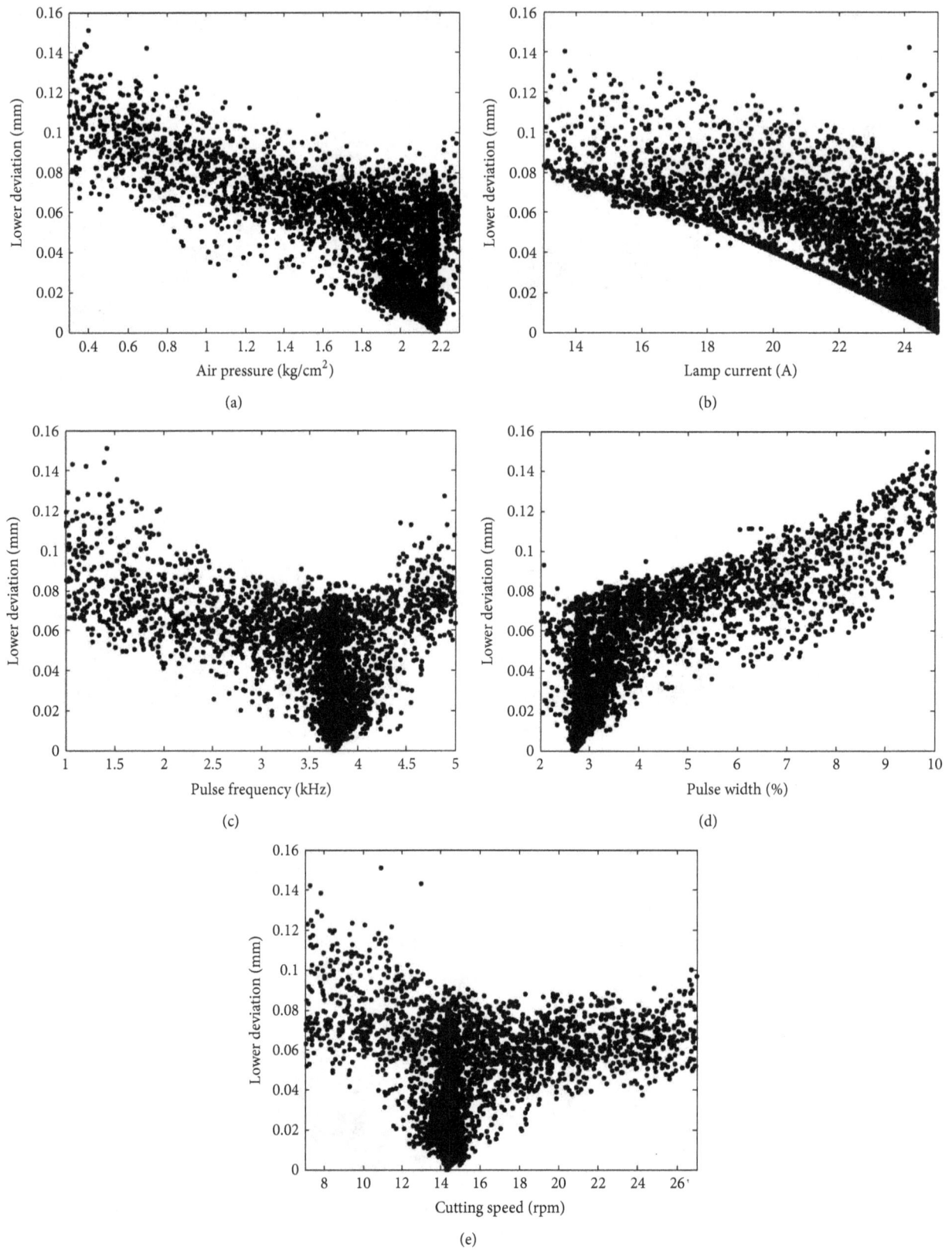

FIGURE 6: Variations of lower deviation with various LBM process parameters.

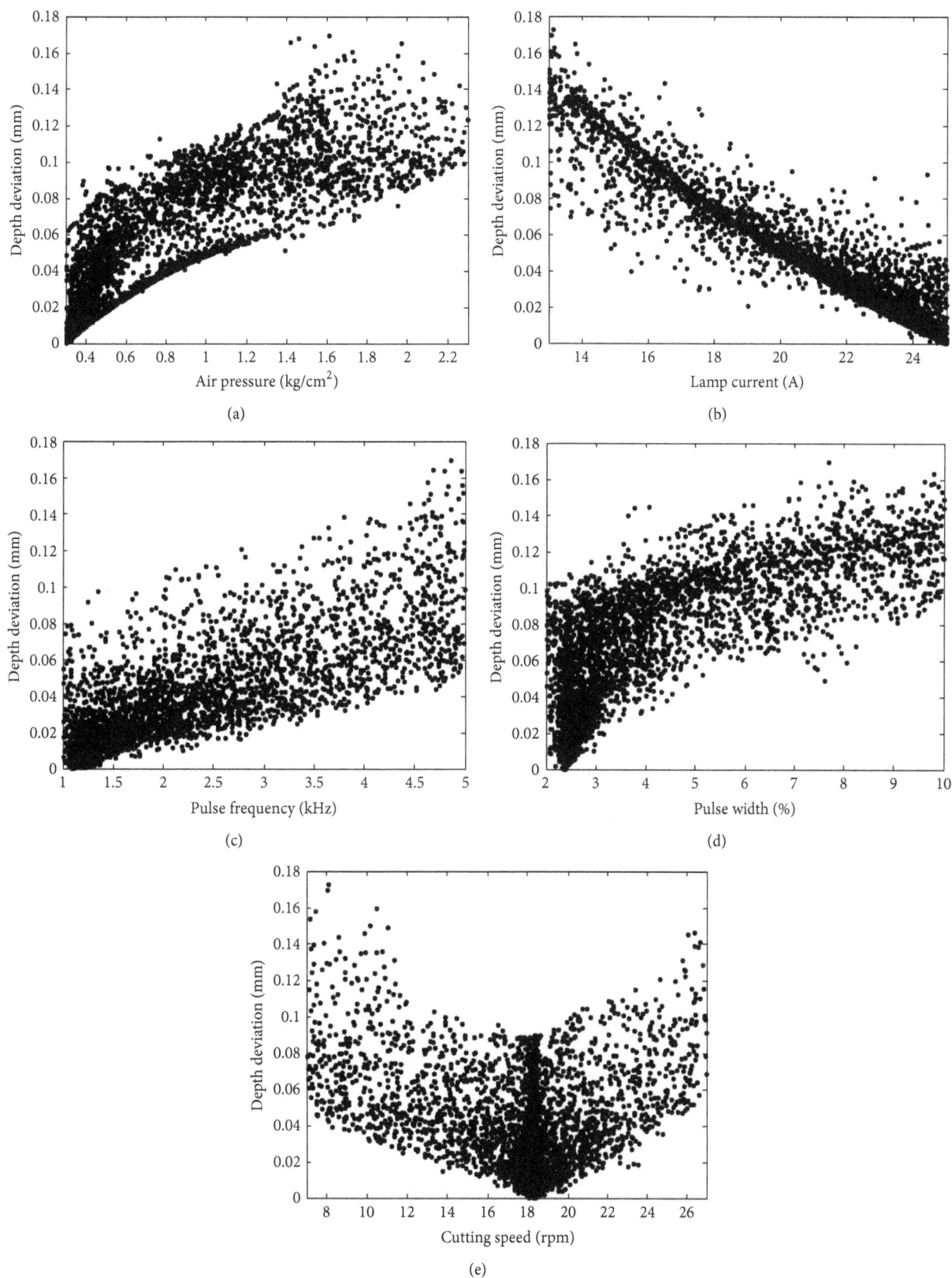

FIGURE 7: Variations of depth deviation with five LBM process parameters.

TABLE 9: Single objective optimization results.

Optimization method	Response	Optimal value	Air pressure (kg/cm²)	Lamp current (amp)	Pulse frequency (kHz)	Pulse width (%)	Cutting speed (rpm)
GA	Y_{uw}	0.0024	1.73	18.25	1.39	3.62	21.48
	Y_{lw}	0.0038	2.22	16.89	4.57	5.15	15.28
	Y_d	0.0037	1.68	23.50	1.75	7.60	17.25
PSO algorithm	Y_{uw}	0.0014	1.73	14.85	2.78	9.35	17.79
	Y_{lw}	0.0024	2.18	21.61	4.17	5.04	17.21
	Y_d	0.0022	1.93	17.20	4.00	8.40	24.50
ACO algorithm	Y_{uw}	0.0007	1.70	19.51	1.74	8.11	19.45
	Y_{lw}	0.0008	2.22	21.91	4.27	4.64	25.19
	Y_d	0.0006	1.85	20.20	2.67	9.26	23.00
ABC algorithm	Y_{uw}	0.0002	2.22	23.56	4.74	2.00	20.54
	Y_{lw}	0.0003	2.20	25.00	3.75	2.72	14.25
	Y_d	0.0003	0.30	25.00	1.10	2.35	18.30

TABLE 10: Single objective optimization performance.

Optimization method	Response	Optimal value	Mean	Standard deviation	Standard error
GA	Y_{uw}	0.0024	0.0033	0.00067	0.00021
	Y_{lw}	0.0038	0.0049	0.00083	0.00026
	Y_d	0.0037	0.0049	0.00099	0.00031
PSO algorithm	Y_{uw}	0.0014	0.0025	0.00073	0.00023
	Y_{lw}	0.0024	0.0032	0.00060	0.00019
	Y_d	0.0022	0.0029	0.00057	0.00018
ACO algorithm	Y_{uw}	0.0007	0.0014	0.00049	0.00016
	Y_{lw}	0.0008	0.0017	0.00059	0.00019
	Y_d	0.0006	0.0015	0.00071	0.00022
ABC algorithm	Y_{uw}	0.0002	0.0008	0.00046	0.00014
	Y_{lw}	0.0003	0.0010	0.00048	0.00015
	Y_d	0.0003	0.0009	0.00041	0.00013

and the desired value of upper deviation can be achieved while performing laser turning operation at lower pulse width. Then, the upper deviation increases with increase in pulse width. Higher air pressure has been recommended for higher dimensional accuracy in the form of upper deviation. It has been found that at low cutting speed of the workpiece, the upper deviation becomes less. At low cutting speed, the material of the workpiece absorbs sufficient amount of heat energy for longer time, and as a result, the material is removed from the upper surface to obtain the required upper deviation. At high cutting speed, the ablation rate of material from the upper surface is higher, and as a result, the upper deviation becomes high. These findings of Dhupal et al. [10] exactly match with those observed in Figure 5.

The variations of lower deviation of the micro-groove with respect to five Nd:YAG laser-turning process parameters are exhibited in Figure 6. It is observed that with the increase in the values of air pressure and lamp current, the lower deviation decreases, and it increases with the increasing value of pulse width. The effects of pulse frequency and cutting speed on lower deviation are almost nonlinear. On the other hand, Figure 7 shows the effects of five machining parameters on depth deviation of the microgroove. Depth deviation almost linearly increases with the gradual increment in the values of air pressure, pulse frequency, and pulse width,

whereas minimum depth deviation is achieved at higher values of lamp current. Cutting speed has a nonlinear effect on depth deviation. Dhupal et al. [10] extensively studied and analyzed the influences of the five process parameters on lower deviation and depth deviation of the machined microgroove.

3.2.2. Multiobjective Optimization. The same optimization problem is now solved using ABC algorithm while giving equal weights to all the three responses. For this multi-objective optimization problem, the following objective function is developed and solved with respect to the constraints as imposed in the case of single objective optimization as

$$\text{Min}\,(Z_2) = \frac{0.33Y_u\,(uw)}{uw_{min}} + \frac{0.33Y_u\,(lw)}{lw_{min}} + \frac{0.33Y_u\,(d)}{d_{min}}, \quad (7)$$

where $Y_u(uw)$, $Y_u(lw)$, and $Y_u(d)$ are the second-order RSM-based equations for upper deviation, lower deviation, and depth deviation respectively; uw_{min}, lw_{min}, and d_{min} are the minimum values of upper deviation, lower deviation and depth deviation, respectively. These minimum values are obtained from the results of single objective optimization. The multi-objective optimization results are shown in Table 11. Dhupal et al. [10] applied desirability function

TABLE 11: Multi-objective optimization results.

Optimization method	Response	Optimal value	Air pressure (kg/cm^2)	Lamp current (amp)	Pulse frequency (kHz)	Pulse width (%)	Cutting speed (rpm)
Dhupal et al. [10]	Y_{uw}	−0.0001	0.93	22.51	1.48	2.39	10.43
	Y_{lw}	−0.0002					
	Y_d	−0.0009					
ABC algorithm	Y_{uw}	0.0002	1.64	18.87	3.21	9.80	7.67
	Y_{lw}	0.0006					
	Y_d	0.0003					

approach to optimize the multiple responses of Nd:YAG laser-turning process for generation of micro-groove and achieved the optimal values of the responses as negatives which are infeasible to obtain. Applying ABC algorithm, it is observed that a combination of air pressure = 1.64 kg/cm^2, lamp current = 18.87 amp, pulse frequency = 3.21 kHz, pulse width = 9.80%, and cutting speed = 7.67 rpm would simultaneously optimize all the three responses of the LBM process. The optimal value of the objective function (Z_2) is determined as 0.000365.

The ABC algorithm is based on the foraging behavior of the honey bee colonies. The model consists of three essential components, that is, employed and unemployed foraging bees, and food sources. It also defines two leading modes of behavior which are necessary for self-organizing and collective intelligence, that is, recruitment of foragers to rich food sources resulting in positive feedback and abandonment of poor sources by foragers causing negative feedback.

In ABC algorithm, a colony of artificial forager bees (agents) search for rich artificial food sources (good solutions for a given problem). To apply ABC algorithm, the considered optimization problem is first converted to the problem of finding the best parameter vector which minimizes the given objective function. Then, the artificial bees randomly discover a population of initial solution vectors and then iteratively improve them by employing the strategy of moving towards better solutions by means of a neighborhood search mechanism while abandoning poor solutions.

The most innovative feature of ABC algorithm is the concept of exchange of information amongst the onlooker bees to find out a better food source which minimizes the search iteration for the global optimal and avoids candidate solutions which are sub-optimal. The same point is observed in Figure 1 (convergence of ABC, ACO, PSO and GA algorithms for HAZ thickness) and Figure 4 (convergence of ABC, ACO, PSO, and GA algorithms for upper deviation) where it is evident that due to its superior searching methodology, ABC algorithm reaches the convergent solution much earlier than ACO, PSO, and GA.

4. Conclusions

In this paper, the parametric optimization problems for two Nd:YAG laser beam machining processes are solved applying ABC algorithm. For both the cases, the results of single as well as multi-objective optimization of the LBM process

are derived. It is observed that the optimal values of the responses derived by ABC algorithm are far better than those obtained by the past researchers. The comparison of the performance of ABC algorithm with other population-based algorithms proves its superiority and applicability as an effective optimization tool. The optimal response values obtained using ABC algorithm have minimum dispersion and are close to the target solutions. Although ABC algorithm gives excellent results, its CPU time is quite comparable with that of the other optimization algorithms. The results of two sample paired t-tests also demonstrate its superiority over the other considered algorithms. It is also observed that for multi-objective optimization, it is always preferable to assign equal importance to all the considered responses. The derived parametric combinations for Nd:YAG LBM process would now help the process engineers to set the operating levels of various process parameters at their optimal values to have enhanced machining performance. This algorithm may also be effectively applied for parametric optimization of other machining processes.

References

[1] V. K. Jain, *Advanced Machining Processes*, Allied Publishers Pvt. Limited, New Delhi, India, 2005.

[2] J. Meijer, "Laser beam machining (LBM), state of the art and new opportunities," *Journal of Materials Processing Technology*, vol. 149, no. 1–3, pp. 2–17, 2004.

[3] A. K. Dubey and V. Yadava, "Laser beam machining: a review," *International Journal of Machine Tools and Manufacture*, vol. 48, no. 6, pp. 609–628, 2008.

[4] A. K. Dubey and V. Yadava, "Experimental study of Nd:YAG laser beam machining: an overview," *Journal of Materials Processing Technology*, vol. 195, no. 1–3, pp. 15–26, 2008.

[5] J. Mathew, G. L. Goswami, N. Ramakrishnan, and N. K. Naik, "Parametric studies on pulsed Nd:YAG laser cutting of carbon fibre reinforced plastic composites," *Journal of Materials Processing Technology*, vol. 89-90, pp. 198–203, 1999.

[6] A. S. Kuar, B. Doloi, and B. Bhattacharyya, "Modelling and analysis of pulsed Nd:YAG laser machining characteristics during micro-drilling of zirconia (ZrO$_2$)," *International Journal of Machine Tools and Manufacture*, vol. 46, no. 12-13, pp. 1301–1310, 2006.

[7] A. S. Kuar, G. Paul, and S. Mitra, "Nd:YAG laser micromachining of alumina-aluminium interpenetrating phase composite using response surface methodology," *International Journal of*

Machining and Machinability of Materials, vol. 1, pp. 432–444, 2006.

[8] D. Dhupal, B. Doloi, and B. Bhattacharyya, "Optimization of process parameters of Nd:YAG laser microgrooving of Al_2TiO_5 ceramic material by response surface methodology and artificial neural network algorithm," *Proceedings of the Institution of Mechanical Engineers B*, vol. 221, no. 8, pp. 1341–1351, 2007.

[9] A. K. Dubey and V. Yadava, "Multi-objective optimisation of laser beam cutting process," *Optics and Laser Technology*, vol. 40, no. 3, pp. 562–570, 2008.

[10] D. Dhupal, B. Doloi, and B. Bhattacharyya, "Pulsed Nd:YAG laser turning of micro-groove on aluminum oxide ceramic (Al_2O_3)," *International Journal of Machine Tools and Manufacture*, vol. 48, no. 2, pp. 236–248, 2008.

[11] A. K. Dubey and V. Yadava, "Optimization of kerf quality during pulsed laser cutting of aluminium alloy sheet," *Journal of Materials Processing Technology*, vol. 204, no. 1–3, pp. 412–418, 2008.

[12] D. Dhupal, B. Doloi, and B. Bhattacharyya, "Parametric analysis and optimization of Nd:YAG laser micro-grooving of aluminum titanate (Al_2TiO_5) ceramics," *International Journal of Advanced Manufacturing Technology*, vol. 36, no. 9–10, pp. 883–893, 2008.

[13] U. Çaydaş and A. Hasçalık, "Use of the grey relational analysis to determine optimum laser cutting parameters with multi-performance characteristics," *Optics & Laser Technology*, vol. 40, pp. 987–994, 2008.

[14] D. Dhupal, B. Doloi, and B. Bhattacharyya, "Modeling and optimization on Nd:YAG laser turned micro-grooving of cylindrical ceramic material," *Optics and Lasers in Engineering*, vol. 47, no. 9, pp. 917–925, 2009.

[15] R. Rao and V. Yadava, "Multi-objective optimization of Nd:YAG laser cutting of thin superalloy sheet using grey relational analysis with entropy measurement," *Optics and Laser Technology*, vol. 41, no. 8, pp. 922–930, 2009.

[16] J. Ciurana, G. Arias, and T. Ozel, "Neural network modeling and particle swarm optimization (PSO) of process parameters in pulsed laser micromachining of hardened AISI H13 steel," *Materials and Manufacturing Processes*, vol. 24, no. 3, pp. 358–368, 2009.

[17] Sivarao, T. J. S. Anand, and A. Shukor, "RSM based modeling for surface roughness prediction in laser machining," *International Journal of Engineering & Technology*, vol. 10, pp. 32–37, 2010.

[18] B. Doloi, D. Dhupal, and B. Bhattacharyya, "Modelling and analysis on machining characteristics during pulsed Nd:YAG laser microgrooving of aluminium titanate (Al_2TiO_5)," *International Journal of Manufacturing Technology and Management*, vol. 21, no. 1-2, pp. 30–41, 2010.

[19] A. S. Kuar, S. K. Dhara, and S. Mitra, "Multi-response optimisation of Nd:YAG laser micro-machining of die steel using response surface methodology," *International Journal of Manufacturing Technology and Management*, vol. 21, no. 1-2, pp. 17–29, 2010.

[20] A. Sharma, V. Yadava, and R. Rao, "Optimization of kerf quality characteristics during Nd: YAG laser cutting of nickel based superalloy sheet for straight and curved cut profiles," *Optics and Lasers in Engineering*, vol. 48, no. 9, pp. 915–925, 2010.

[21] R. Biswas, A. S. Kuar, S. Sarkar, and S. Mitra, "A parametric study of pulsed Nd:YAG laser micro-drilling of gamma-titanium aluminide," *Optics and Laser Technology*, vol. 42, no. 1, pp. 23–31, 2010.

[22] G. Kibria, B. Doloi, and B. Bhattacharyya, "Experimental analysis on Nd:YAG laser micro-turning of alumina ceramic," *International Journal of Advanced Manufacturing Technology*, vol. 50, no. 5–8, pp. 643–650, 2010.

[23] R. Biswas, A. S. Kuar, S. K. Biswas, and S. Mitra, "Effects of process parameters on hole circularity and taper in pulsed Nd:YAG laser microdrilling of Tin-Al_2O_3 composites," *Materials and Manufacturing Processes*, vol. 25, no. 6, pp. 503–514, 2010.

[24] R. Biswas, A. S. Kuar, S. K. Biswas, and S. Mitra, "Characterization of hole circularity in pulsed Nd:YAG laser micro-drilling of TiN-Al_2O_3 composites," *International Journal of Advanced Manufacturing Technology*, vol. 51, no. 9–12, pp. 983–994, 2010.

[25] S. Panda, D. Mishra, and B. B. Biswal, "Determination of optimum parameters with multi-performance characteristics in laser drilling—a grey relational analysis approach," *International Journal of Advanced Manufacturing Technology*, vol. 54, no. 9–12, pp. 957–967, 2011.

[26] T. V. Sibalija, S. Z. Petronic, V. D. Majstorovic, R. Prokic-Cvetkovic, and A. Milosavljevic, "Multi-response design of Nd:YAG laser drilling of Ni-based superalloy sheets using Taguchi's quality loss function, multivariate statistical methods and artificial intelligence," *International Journal of Advanced Manufacturing Technology*, vol. 54, no. 5–8, pp. 537–552, 2011.

[27] D. Karaboga and B. Basturk, "Artificial bee colony (ABC) optimization algorithm for solving constrained optimization problems," in *Foundations of Fuzzy Logic and Soft Computing*, vol. 4529, pp. 789–798, Springer, Berlin, Germany, 2007.

[28] D. Karaboga and B. Basturk, "A powerful and efficient algorithm for numerical function optimization: artificial bee colony (ABC) algorithm," *Journal of Global Optimization*, vol. 39, no. 3, pp. 459–471, 2007.

[29] D. Karaboga and B. Basturk, "On the performance of artificial bee colony (ABC) algorithm," *Applied Soft Computing*, vol. 8, no. 1, pp. 687–697, 2008.

[30] D. Karaboga and B. Akay, "A comparative study of artificial bee colony algorithm," *Applied Mathematics and Computation*, vol. 214, no. 1, pp. 108–132, 2009.

[31] S. Samanta and S. Chakraborty, "Parametric optimization of some non-traditional machining processes using artificial bee colony algorithm," *Engineering Applications of Artificial Intelligence*, vol. 24, no. 6, pp. 946–957, 2011.

[32] T. L. Saaty, *The Analytic Hierarchy Process*, McGraw-Hill, New York, NY, USA, 1980.

9

Integrative Approach to the Plant Commissioning Process

Kris Lawry,[1] and Dirk John Pons[2]

[1] Department of Chemical and Process Engineering, University of Canterbury, Private Bag 4800, Christchurch 8020, New Zealand
[2] Department of Mechanical Engineering, University of Canterbury, Private Bag 4800, Christchurch 8020, New Zealand

Correspondence should be addressed to Dirk John Pons; dirk.pons@canterbury.ac.nz

Academic Editor: Xueqing Zhang

Commissioning is essential in plant-modification projects, yet tends to be ad hoc. The issue is not so much ignorance as lack of systematic approaches. This paper presents a structured model wherein commissioning is systematically integrated with risk management, project management, and production engineering. Three strategies for commissioning emerge, identified as direct, advanced, and parallel. Direct commissioning is the traditional approach of stopping the plant to insert the new unit. Advanced commissioning is the commissioning of the new unit prior to installation. Parallel commissioning is the commissioning of the new unit in its operating position, while the old unit is still operational. Results are reported for two plant case studies, showing that advanced and parallel commissioning can significantly reduce risk. The model presents a novel and more structured way of thinking about commissioning, allowing for a more critical examination of how to approach a particular project.

1. Introduction

1.1. Background. Plant modifications are an ongoing process throughout the life of any process plant. Reasons for modification include efforts to improve reliability, production capacity, quality, or productivity. Seamless incorporation is the key concern associated with the installation of any new equipment in an operating plant due to the high cost of process downtime. Several steps can be taken to minimise the risk associated with the installation of new equipment such as hazard and operability studies, project management, development of redundancy plans, and commissioning of the new equipment.

Of these, commissioning is an essential activity in many plant-modification projects and has significant implications for project success. Yet paradoxically it tends to be approached in an ad hoc manner. It is often included in project plans, so it is not that people are ignorant of commissioning. Rather, the problem is that there is a lack of systematic approaches to commissioning, so it is frequently left to tradespeople and plant operators to manage in whatever way they see fit. This is an undesirable situation since it results in unpredictable outcomes. In some cases it can even cause serious problems. An extreme example would be the catastrophic failure of the Chernobyl nuclear power plant (1986), which was caused by operators attempting an ad hoc test of the efficacy of a modified emergency cooling system.

This paper presents a structured conceptual model for the commissioning process, and two cases studies showing application to operating plant.

2. Existing Models of Commissioning

2.1. Literature. Many authors have highlighted the value of commissioning from a range of different perspectives but they all agree that commissioning and the integration of a new project is critical to the success of any project [1–10]. However commissioning is poorly defined and is interpreted ambiguously [6, 11], which leads to inefficient utilisation within industry. In this paper "commissioning" is defined as the disciplined activity involving careful testing, calibration, and proving of all systems, software, and networks within the project boundary [5].

2.2. Current Models of Commissioning. Factors that are known to affect the commissioning process include the following.

Prevailing
assumption that
commissioning is
simply a routine set
of tasks
(limits options to
outcomes (a) and (b))

Existing
commission
strategies
(1)

(a) Ad hoc: action-
orientated
problem solving

(b) Template:
follow a checklist,
or method that
worked before, or
case study from a
similar application
elsewhere

(c) Methodological:
analysis of the
situational needs
and deliberate
selection of a
relevant method

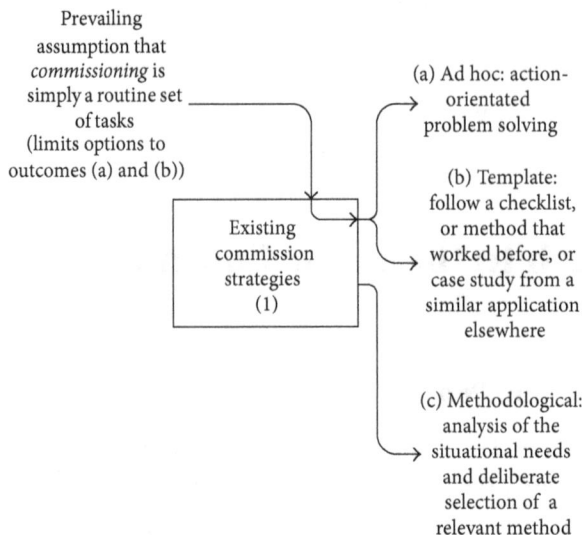

FIGURE 1: Three common strategies for commissioning, broadly reflecting the approaches described in the literature. If commissioning is perceived to be simply a routine set of tasks, which is a common assumption, then this tends to preclude any more thoughtful approach to the problem.

(i) Type of project. Thus situational variables are important; that is, the factors that resulted in a successful (or failed) commissioning outcome in one case are not necessarily transferrable to a different situation.

(ii) Who is in charge of the phase. Commissioning can be completed by a range of different groups depending on the project. It can be the equipment manufacturers, operation team, or a separate commissioning team depending on scale and requirements of the project [12]. The relationships between these people are also important (social dimension) [6, 13], hence also contractual obligations (see (iv) below).

(iii) Number/type of phases. Commissioning can also be broken down in several sections such as planning, precommissioning, testing, integration, monitoring, documenting, and handover depending on the level of complexity of the project. This requires careful project planning (see (iv) below).

(iv) Project planning and contractual sufficiency. It is widely recognised in the literature that commissioning requires deliberate planning, as opposed to ad hoc treatment. Thus it needs appropriate consideration in the work breakdown structure and project planning [14], allocation of resources, transferral of those costs into the initial contract [9, 15–18], and creation of specific operating procedures (especially important for safety-critical plant like boilers [19]). This corresponds to the "integration" tasks in the project management approach [6].

The commissioning process has been examined for a wide range of different projects [2–5, 8, 20, 21]. The predominate approach can be described as task specific; the literature tends to identify specific tasks that should be completed as part of commissioning. Thus the focus has been on completing multiple checks on a system to ensure it will operate as

expected. Thus there are many reports in the literature, too numerous to mention, about commissioning *experiences* in specific case studies. These are undoubtedly helpful, especially for lessons learned and application to comparable situations. They are also systematic, in a way, especially in the provision of templates and checklists to guide practitioners.

However there is a lack of holistic or integrative models. There is much less literature at the next higher level of abstraction, which is the commissioning process in general. At this level we are interested not so much in case-specific experiences but in the fundamental principles and the methodology. What exists at this level is primarily in the area of instrumentation and control; some examples are [5, 22, 23].

Thus the existing commissioning strategies in the literature can be categorised into three types, see Figure 1. These are (a) ad hoc, which is action-orientated problem solving; and (b) template, which involves using a checklist, or operating procedure that worked before or in another situation. Both (a) and (b) are premised on the assumption that commissioning is a routine set of tasks. The third strategy challenges that premise and calls for a deliberately thoughtful approach. Thus the third category in the literature is (c) methodological, which involves analysis of the situational needs and deliberate selection of the most relevant of several possible commissioning methods.

2.3. Issues and Problem Areas. A clear refrain in the literature is that commissioning (i) needs deliberate project management, but (ii) is too often not given the attention it deserves. One of the issues with commissioning, which contributes to problem (ii), is that the value thereof is hard to quantify. Justifying the value of commissioning may be completed using qualitative analysis similar to quantification of risk in a project [24]. This is based on the consequence and probability of the system failing to operate as anticipated. In other cases there is no attempt at justification at all, so the value is not appreciated.

Another issue is the tendency to underresource the commissioning in the project planning, which is issue (i) above. Underresourcing is due to several factors such as its omission in the project management. There is often a high level of variability as a result of the case-specific nature making it difficult to fit into the established planning structure. Existing project management frameworks, such as the PMBOK [9], are general approaches. While they acknowledge the commissioning stage they do not, and cannot reasonably be expected to, provide case-specific guidance on commissioning. They treat commissioning very lightly and rely on the practitioner to identify whether or not commissioning is an important part of the project. The literature suggests that practitioners too often fail to realise the importance and therefore fail to plan sufficiently. Alternatively, project managers may simply be overly optimistic about the risks associated with the installation of a new system. Whatever the reason, the result can be insufficient resources being allocated, with the consequence of poor completion. Incorporation of a broad conceptual model of commissioning into the project management practices would be the first logical step. Commissioning draws from several project knowledge areas such as integration, communication, and risk management. The logical approach is to incorporate into the project life cycle between the execution and closing phases [4, 6, 8].

2.4. Problem Definition. Current models of commissioning tend to be simplistic, and relevant only to specific areas. They are focused on the process and consequently tend to produce a somewhat prescriptive list of tasks that need to be performed. A higher-level reconceptualisation of the commissioning process, with the development of a more general theory, could be valuable.

The purpose of this work is to develop a more holistic and integrative theory of commissioning. The specific emphasis is on reducing process downtime, without compromising plant reliability. This is worth attempting as it has the potential to provide a general framework in which the other more process-specific models can be placed.

3. Approach

We start by reconceptualising commissioning in broad terms. We categorise the commissioning strategies according to the operational risk. This results in three categories: direct, parallel, and advanced. We then apply a system modelling method to embed these within the broader manufacturing context. Finally we apply the new framework to two case studies to demonstrate the applicability.

4. Results

4.1. Categorisation of Commissioning Projects

4.1.1. Starting Premise. We start with the premise that the value of commissioning is essentially one of systematic risk reduction, that is, used to minimise the risk associated with the installation of a new piece of equipment. More specifically the application of commissioning for the installation of new equipment into the process industry reduces the risk of equipment damage, environmental health and safety, and process downtime.

Thus commissioning is a strategy for treating risk [24]. This has the further important implication that the treatment, hence type of commissioning, can be aligned with the degree of technical risk that the organisation can accept. Thus we specifically link commissioning, as a treatment strategy, to the concept of "tolerable risk" within the risk management literature, and to the concept of strategic risk for the organisation as a whole [25]. This also has contractual implications in project-setup phase, where there is a need to differentiate between the commissioning risk elements and proportion them between the equipment manufacturer, project management organisation, and plant owner [26].

From this starting assumption we identify three categories of commissioning, as strategies in response to organisational risk-tolerance. These are direct commissioning, advanced commissioning, and parallel commissioning. Each has strengths and weaknesses. They can be deployed individually or together.

4.1.2. Direct Commissioning. Direct commissioning is the classical approach to commissioning where the new equipment is installed and the system must remain offline as commissioning is completed. Direct commissioning is the most straightforward approach as no additional equipment or simulation is required. The new equipment is installed into its operational position and the process cannot restart until the system has been commissioned and is running correctly. There is a high level of downtime in this process as the whole system cannot be operated until the new unit is electrically, mechanically, and operationally tested. There is also the risk of having to reinstall the old unit if there are significant complications at any phase of the commissioning process. Direct commissioning is often reserved for well-established unit operations such as new pumps and flow meters. Direct commissioning is most effective when it is used on well-established system and ones that are not a key requirement of the process.

4.1.3. Advanced Commissioning. Advanced commissioning is the process of operating the new unit in advance of installation and in isolation of the main process operation. Advanced commissioning requires the simulation of all proprietary systems that interact with the new unit. Simulation can be extremely complicated or simple depending on the level of interaction between the process and the new unit. (In this context "simulation" can refer to the artificial provision of physical inputs to the new machine or unit, smaller scale models, and mathematical modelling of the functional behaviour of the unit.) Advanced commissioning allows for the electrical, mechanical, and part of the operational testing to be completed. The full functionality of the unit cannot be proven as the system is being simulated by external means, which will always be an approximation of reality. Advanced process is extremely valuable for the development of new technology as it allows for the verification of novel processes at low risk. The most common type of advanced commissioning is the development of model systems which both simulate the operation of the system and the new unit. Advanced commissioning can also include computer simulation of new process which provides a cost effective method of developing concepts in the early stages of design. Advanced commissioning is valuable at proving conceptual designs of new technology. The main drawback of advanced commissioning is that the process is only simulated so there is still the potential that the unit can fail when installed into its operational environment.

4.1.4. Parallel Commissioning. Parallel commissioning is the testing of the new system in parallel to the operating system. Parallel commissioning is the most rigorous form of physical and operational commissioning. It allows for the new unit to be tested under full operational conditions, with low risk of significant process interruptions due to the added redundancy of the old system present in an operational capacity. However it also has the highest cost as it requires the duplicate hardware systems and additional structural space. The only risk associated with parallel commissioning is the integration between the two systems. Often there is some type of switching or merging component in these systems which may require minor process stoppage for installation. Parallel commissioning is often completed when it is critical that the process must not stop for any extended period of time. It often lends itself to processes with few interactions between new unit operation and the rest of the process. Parallel commissioning is seldom utilised due to the requirement of a process that can accommodate both the new and old unit.

4.2. Conceptual Model. Having identified three types of commissioning, we next seek to set those within a conceptual framework. This is worth doing as it has the potential to identify the situational variables relevant to each type of commissioning. This in turn can be used to further build a theoretical foundation, and provide guidance to practitioners.

4.2.1. Approach. The modelling method uses a structured, deductive process to decompose the process being analysed into multiple subactivities (functions) and for each deduce the initiating events, the controls that determine the extent of the outputs, the inputs required, the process mechanisms that are presumed to support the action, and the outputs. The model was then inductively reconciled with elements of the existing body of knowledge on this topic, and successively refined. The end result is a graphical model that describes the relationships between variables, thereby providing a synthesis of what is known and surmised about the topic. The model is expressed as a series of flowcharts using the integration definition zero (IDEF0) notation [27, 28]. With IDEF0 the object types are inputs, controls, outputs, and mechanisms (ICOM) and are distinguished by placement relative to the box, with inputs always entering on the left, controls above, outputs on the right, and mechanisms below.

4.2.2. Develop Production Capability (Prd-1). The broader context is that commissioning occurs as part of the development of production capability, and our model starts at this level. (This is already the second level into the model; the top level, which is not shown here, includes product design, operation of the plant, control of production flow, quality, distribution to market, packaging, health and safety, lean/JIT, among other activities. However the present paper focuses on the commissioning activities.) See Figure 2. Commissioning is included as element 5 and occurs towards the end of the plant-development process. Other important activities are the following.

(i) Determine manufacturing/production sequence.

(ii) Design of the production plant, which also includes the plant layout, material handling, plant control and automation, and (for manufacturing) the development of production tooling and flow control, for example, just-in-time (JIT). Analysis of technology risk (9) is another activity associated with the design phase.

(iii) Building the production system (4), and the associated project management activities.

(iv) Decommissioning the plant (7).

(v) Project management (8). We note the importance of project management methods in supporting many of the activities of commissioning. There are several models of project management that might be inserted here, including [9, 29], but these are not specific to commissioning and therefore not detailed further at this point.

We do not deal with these other activities here, but instead move the focus to the test and commission activities. Before doing so, we draw attention to some hollow arrows, which represent errors, in particular design and construction errors, at (2) and (4), respectively, and the possibility for unintended plant behaviours at (5): low productivity, safety issues, staff usability problems, product quality defects, and so forth. This point is important because the commissioning model that follows specifically seeks to address these risks.

4.2.3. Test and Commission Production System (Prd-1-5). The model for commissioning a new piece of plant equipment is shown in Figure 3 (Prd-1-5). The conventional commissioning process is included here, as are the new concepts for commissioning approach. One of the conventional activities is to verify instrumentation and control systems (1), which involves the systematic checking of installed hardware against plant schematics. The checks are progressively done for connectivity, cold operation, hot operation, and process control. We do not detail those processes here and instead refer the reader to source material [5] which has information that is useful to practitioners. The final objective of commissioning

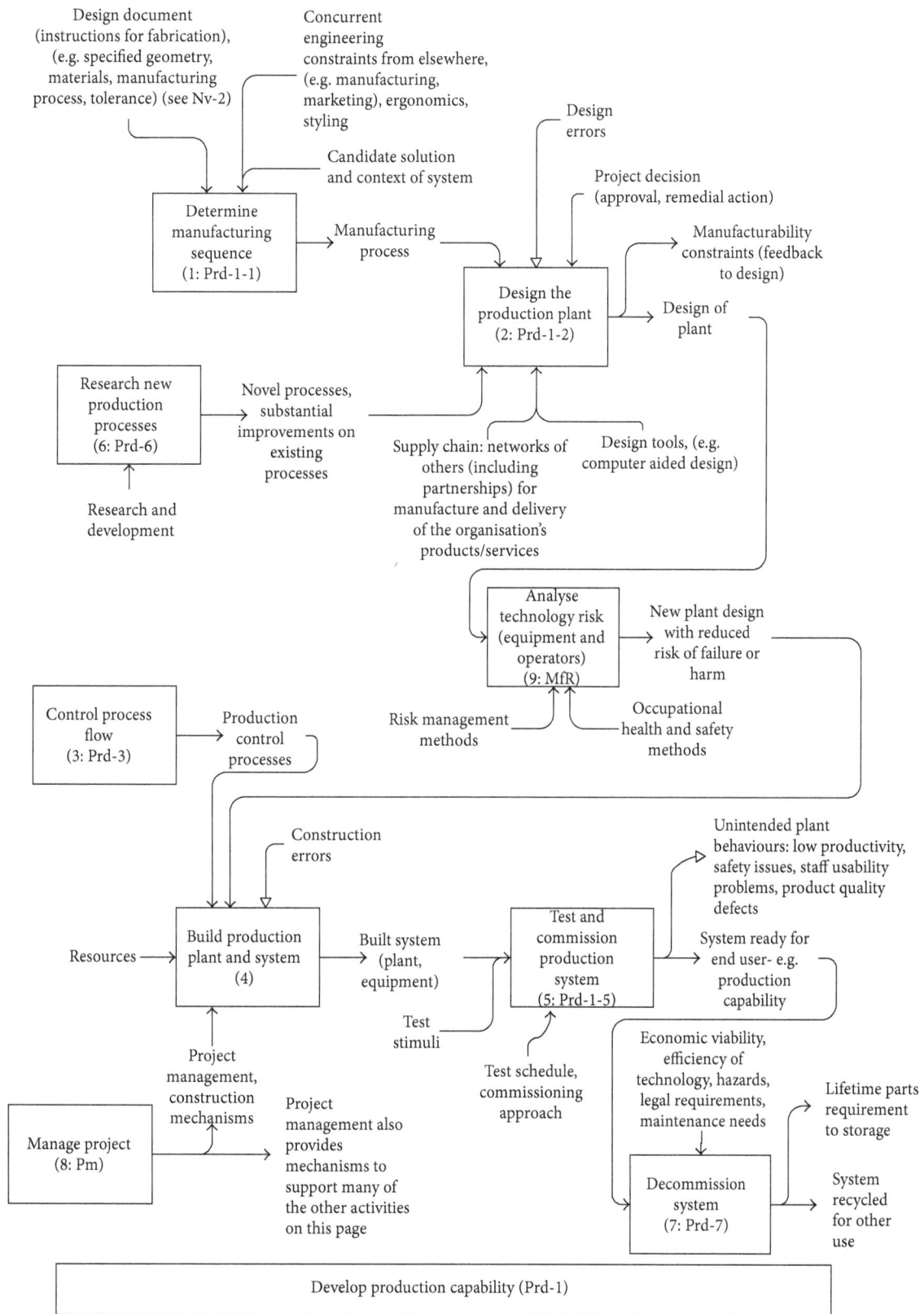

FIGURE 2: Model for the development of production capability.

Built hardware → **Verify instrumentation and control systems (1)** → New hardware with operational controls → **Physically set up machine in commissioning situation (3)** → New hardware ready to accept test loads

Systematic checking of installed hardware against plant schematics. Progressive checking for connectivity, cold operation, hot operation, process control (see "Practice note 09: commissioning capital plant")

Unintended plant behaviours: low productivity, safety issues, staff usability problems, product quality defects

System ready for end user, e.g., production capability

Perform tests on new machine (4: Prd-1-5-4) → Proven functionality in principle of new hardware → **New machine takes partial but real production load (5)** → Proven functionality of new hardware → **New machine takes a full production load (6)** → Plant operating as intended: meeting performance specifications

Contractual obligations fulfilled

Factors to consider:

(i) Plant usage, operational needs, desired continuity of production
(ii) Existing parallel flow that can keep plant operational
(iii) Degree to which new technology is well-established. Risks in the new technology
(iv) Feasibility of simulating plant behaviour
(v) Consequences if commissioning process encounters difficulty. Risk tolerance of organisation
(vi) Feasibility of the various commissioning approaches

Advanced commissioning: new machine initially operated in isolation with simulated inputs

Ongoing adjustments to increase productivity and quality (7) → Projects to further improvement machine, upgrades

Select commission approach (2: Prd-1-5-2)

Parallel commissioning: new machine operated in parallel with old and progressively takes load

The approaches can be combined

Direct commissioning: machine taken offline during changeover

Test and commission system (Prd-1-5)

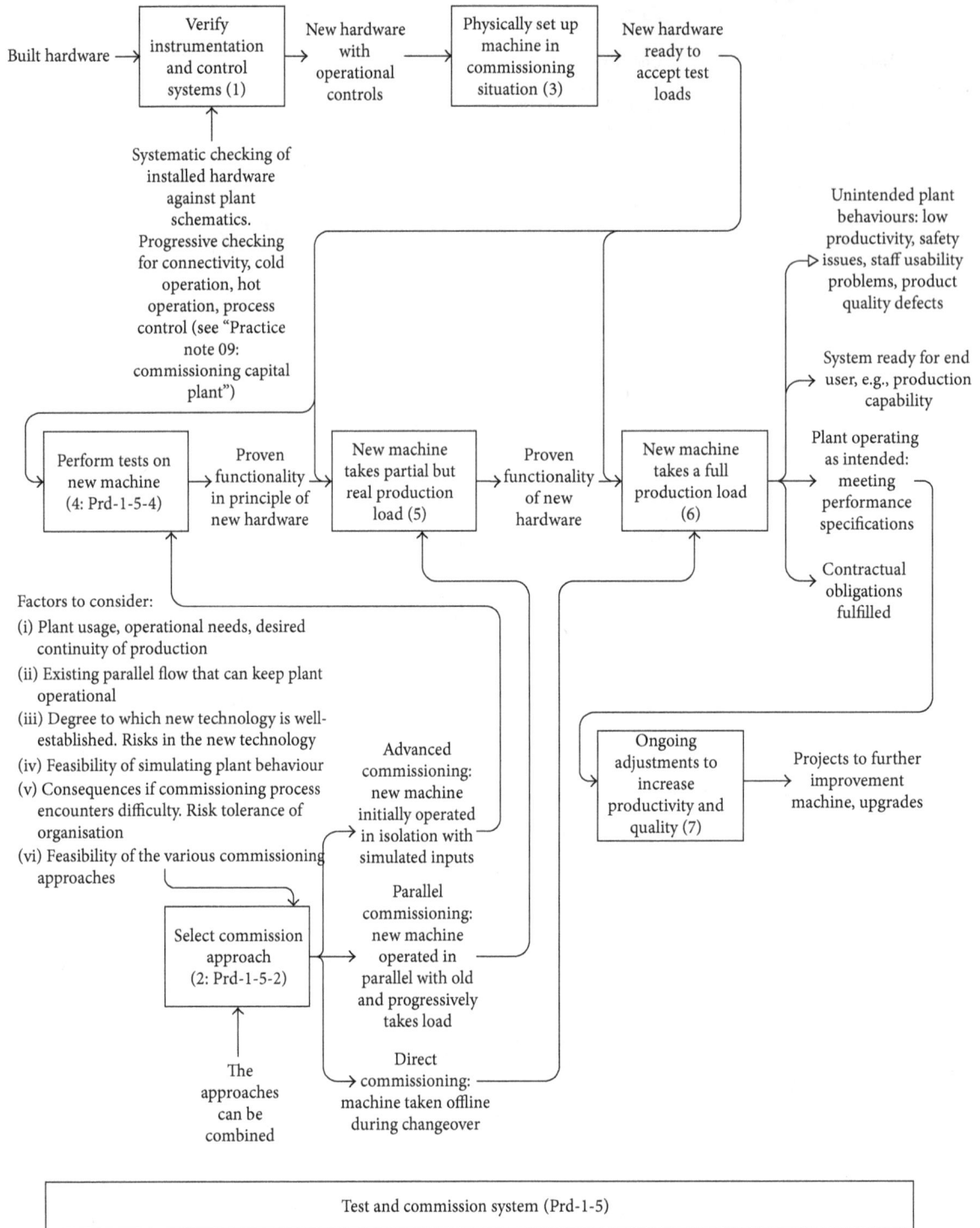

FIGURE 3: Model for the test and commissioning activities.

is also well known, to deliver an operational production system (6) that is ready for the client to use.

Where our model differs is the inclusion of a deliberate stage of deciding which of three commissioning approaches to use in the situation (2): direct, advanced, or parallel. We also note in passing that the quality and lean imperative for continuous improvement will generally mean that there

will be ongoing adjustments to increase productivity and quality (7) after the machine has been commissioned. Thus commissioning the machine and closing the contract with the client may be the end of the involvement of the machine builders, but are not the end of the life cycle for the machine itself. This again has contractual implications in the form of service and warranty support from the vendors, and

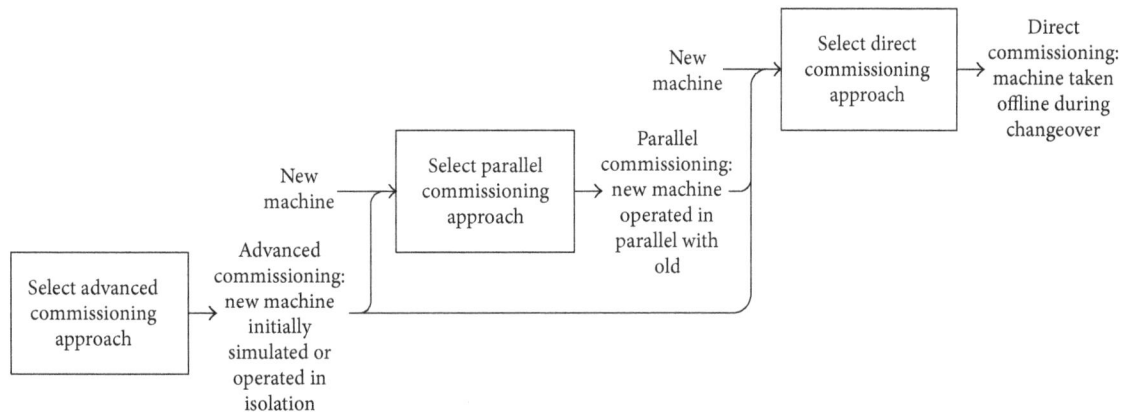

FIGURE 4: Relationship between the three commissioning approaches: advanced, parallel, and direct.

reliability centred maintenance by the plant operators. There is also the decommissioning to consider, which can be a project in itself. (In extreme cases, e.g., nuclear power plant, the cost of decommissioning is comparable to the initial construction cost. If there has been a catastrophic failure of the plant then the decommissioning cost can vastly exceed the construction cost.)

4.2.4. Select Commission Approach (Prd-1-5-2). The decision involves a choice of direct, advanced, or parallel commissioning. These are not mutually exclusive. Instead some of them may be done sequentially, as shown in Figure 4. For example, advanced commission may precede either of the other two.

The various factors relevant to this commissioning decision are anticipated and clustered in groups: ability to recover from a failed installation, assessed or perceived technology risk, desired operational continuity, and timing considerations. The detailed model and the factors within each cluster are shown in Figure 5. At this stage the model is primarily logical and qualitative and is intended as a debiasing tool and a guide to action rather than a decision algorithm. It is also a framework for further research in that it proposes subjective relationships of causality that can subsequently be tested and developed as appropriate. (It may even be that in certain areas it could be possible to develop a mathematical model to support the decision, particularly in well-defined areas. Specifically, the model incorporates risk assessment and it is not impossible that there could be well-defined situations where the variables can be determined with sufficient precision that a quantitative risk assessment coupled with (say) a Boolean consideration of the other factors might make for a sufficient mathematical model. However further research would be required to take it to this level of detail.)

Thus the model proposes that the following decision factors are relevant.

(i) *Advanced* commissioning is appropriate where technology risk is high, operational continuity is required, and timing constraints are tight.

(ii) *Parallel* commissioning is appropriate where operational continuity is required and timing constraints are tight.

(iii) *Direct* commissioning is appropriate where technology risk is low, operational continuity can be disrupted, and timing constraints are loose.

Finally, to complete this part of the conceptual framework, a model is provided for the testing activities of commissioning; see Figure 6.

5. Case Studies

Two cases studies were completed to determine the relevance of this commissioning model in the process industry. First was the development of a novel vertical screw system in the fertilizer industry which used the advanced approach to commissioning. Second was the installation of a ship unloader in the aluminium industry which used a parallel approach to commissioning.

5.1. Vertical Screw Project. Ballance Agri-Nutrients single superphosphate plant at Awarua (New Zealand) had recently designed and installed a new phosphate rock feed system. A new vertical screw system was developed to replace the old gravity feed vertical chute which was prone to blockages in the highly reactive and humid environment present in the reaction chamber. The vertical screw was designed to increase the reliability of the process by forcing the rock into the reaction chamber, hence reducing the number of blockages. An *advanced* approach to commissioning was completed for this new project due to the risk associated with the installation of a complex and untested unit into a critical position in the production line.

The advanced approach to commissioning allowed for rigorous testing to be completed in a controlled environment with low risk to production capacity. Commissioning was completed in two stages with the development of a model system and full scale commissioning of the new system before installation.

Factors to consider:
(i) Existing parallel flow that can keep
    plant operational
(ii) Permanence of the change
(iii) Ability to cover functionality with other
    plant
(iv) Ability to accept a temporary reduction
    in production

Factors to consider:
(i) Degree to which new technology
    is well established
(ii) Risks in the new technology

Tolerance levels for risk in this context
(may be implicit or explicit). See also "risk
attitude" (ISO31000).

Perceived size of risk load relative to
organisational capability to carry the
consequences, or personal self-efficacy to
find a solution if things don't work out

Ability to recover
from failed
installation (4)

Ease of
accommodating
a failed
installation

Assess technology
risk (5)

Low risk of
technical problems
(mild consequences)

Factors to consider:
(i) Plant usage, operational needs,
    desired continuity of production
(ii) Existing parallel flow that can keep
    plant operational
(iii) Service and warranty support from
    the vendors

Risk
management:
(ISO31000)
predict
consequence
and likelihood
of outcomes
(MfR-1-3)

Project
management:
estimate work,
resources, and time
required to complete
the commissioning,
including any time
required for further
tweaking to get plant
to target productivity/
yield

High risk (severe
consequences with
high likelihood) of
technical problems
at commissioning.
Note that multiple
approaches can be
combined

Desired
operational
continuity (6)

Low necessity for
operational
continuity

High necessity for
operational
continuity

Factors to consider:
(i) Availability of times of low
    demand on plant
(ii) Size of operational
    window compared to
    expected commissioning
    time

Timing
considerations (7)

Little or no timing
constraint

Timing is too tight
to accommodate
commissioning

Select direct
commissioning
approach (3)

Direct
commissioning:
machine taken
offline during
changeover

New
machine

Select parallel
commissioning
approach (2)

Parallel
commissioning:
new machine
operated in
parallel with
old

New
machine

Select advanced
commissioning
approach (1)

Advanced
commissioning:
new machine
initially simulated or
operated in isolation

Plant improvement cycle:
subsequent optimisation of plant
productivity or quality by testing,
modelling, and simulation

Select commission approach (Prd-1-5-2)

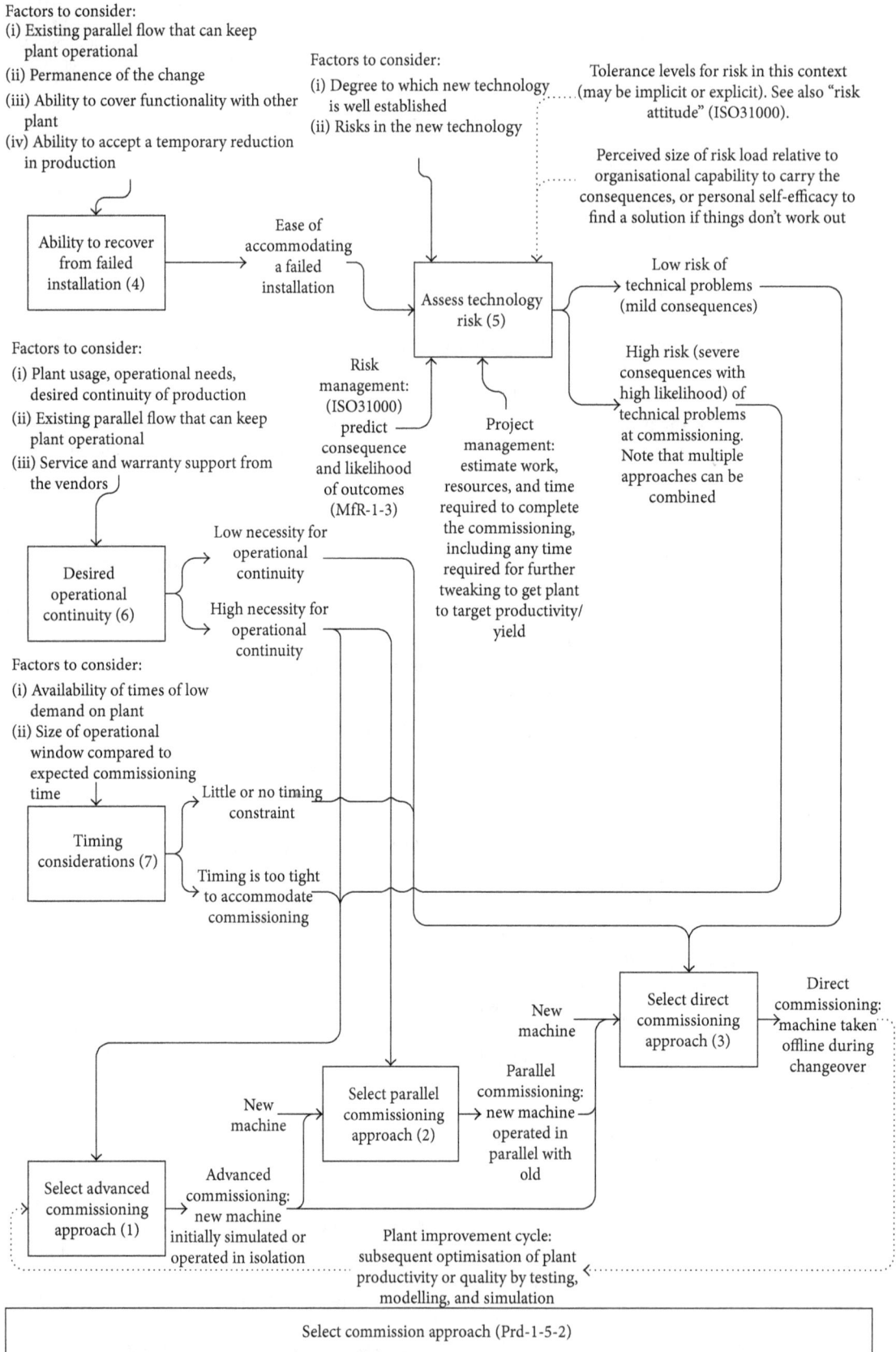

FIGURE 5: Factors relevant to the commissioning decision.

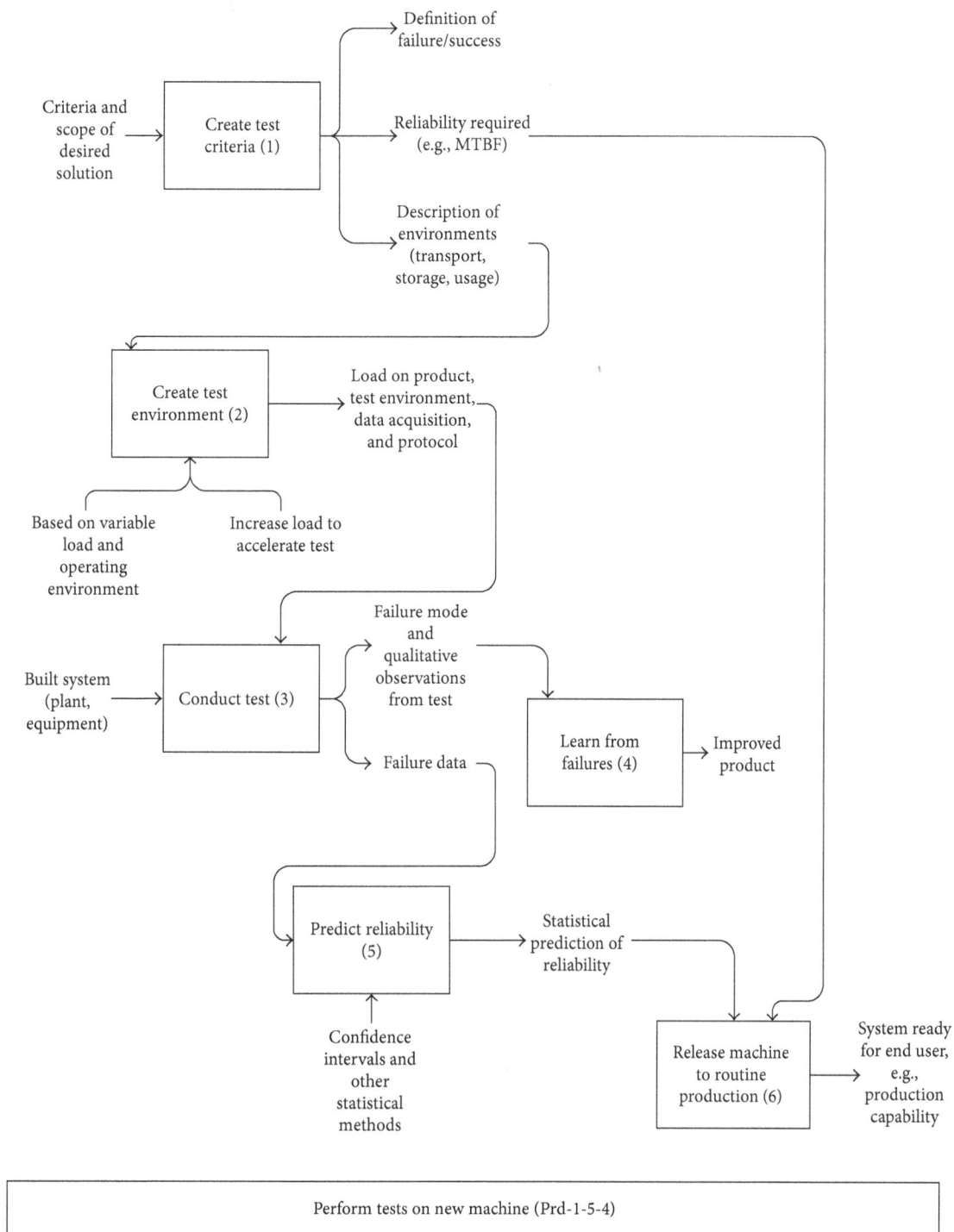

FIGURE 6: Model for the test activities.

Development of the model system allowed the basic concept of the system to be tested. The model was constructed out of crude materials and was tested under a range of conditions to determine the optimal operating parameters. The full scale vertical screw was designed and constructed based on the results obtained from the model system.

Commissioning was completed by operating the system under a set of conditions established to simulate normal operation. The simulation was completed by assembling the new system directly adjacent to its intended future position in the operational plant. It was wired into the system using all of the final wiring components but was not installed into the process. A feed hopper was fitted to the inlet of the screw to simulate the priority feed system and water and various other components were used to simulate the environment of the reaction chamber. The operation of the vertical screw

FIGURE 7: Advanced commissioning of vertical feed screw during plant operation. Image shows plant being fed with material via a temporary arrangement while being commissioned. (Photograph by K Lawry and used by permission of Ballance Agri-Nutrients.)

FIGURE 8: Parallel operation of old (left) and new (right) ship unloaders on the Tiwai wharf. (Photograph by K Lawry.)

during the advanced commissioning phase can be seen in Figure 7. The process of running the system under simulated operating conditions allowed the full commissioning of the mechanical and electrical systems. It also allowed for the partial commissioning of the operation capacity. During this process it was found that several components did not operate as expected. Changes were made to the system and were re-commissioned without any negative effect to the production capacity of the plant.

The process of commissioning prior to the installation of the vertical screw into the system was extremely successful. Full mechanical and electrical commissioning was completed as the full electrical system was used to drive the advanced commissioning process. The operational testing identified several design flaws in the vertical screw. It also reduced the uncertainty associated with the operation of this new technology.

Advanced commissioning has one significant drawback; it only reduced risk associated with the installation of the new vertical screw. The simulated process cannot represent the real process exactly. There are several factors such as continuous operation in the highly reactive environment that cannot be tested until the system is fully installed. Nonetheless the advanced approach was effective in eliminating latent defects and thereby reduced the overall risk.

5.2. Tiwai Point Ship Unloader. New Zealand Aluminium Smelters (NZAS) in Bluff installed a new ship unloader on the Tiwai wharf for the unloading of alumina and coke from incoming ships. The new unloader produced by Alesa Engineering Ltd. replaced the forty-year-old unloader that was installed when the smelter was first constructed in 1971. The new unloader has a significantly increased capacity capable of discharging at 1,000 tonnes per hour (TPH) of

alumina and 600 TPH of coke compared to the old unloader that was only capable of discharging 235 TPH of alumina and 250 TPH of coke [30]. The new ship unloader was installed with the aim of reducing the time required for a ship to spend unloading. Less time spent unloading will mean a more efficient use of shipping resources and the reduction of costs associated with slow turn around (demurrage).

Installation and commissioning was completed by Alesa under the guidance of specialist project engineering company Bechtel who work onsite at the Tiwai point aluminium smelter. This process had to be completed under tight constraints as it was critical that there were no process interruptions. The smelter is a continuous process which cannot be shut down or restarted without high associated cost.

It was decided to take a *parallel* approach to commissioning; the old unloader must remain in a fully operational state until the new system has been thoroughly commissioned and proven capable of carrying the full operational load.

Keeping the old ship unloader in an operational state significantly reduced the risk of supply interruptions, but introduced additional concerns relating to the integration of the two systems. Integration of both unloaders on the same wharf was completed by limiting the operation of the old unloader to the north half of the wharf, while the new unloader was installed and commissioned on the south half, as seen in Figure 8. Limiting the old unloader to half of the wharf increased the unloading time as the ship had to be manoeuvred around to allow access to all of the holds. This was taken as a minimal sacrifice to ensure consistent supply.

The main modification that was required for the integration of the new unloader into the existing infrastructure was the installation of a new conveyor system to replace the southern half of the existing wharf conveyor, clearly this has cost implications. This was completed before the new unloader was installed. Both the new and old conveyor systems operated as one continuous conveyor that serviced the new and old unloaders simultaneously. The upgrade of the conveyor acted to integrate the two unloaders into the overall process. The new conveyor and ship unloader were constructed and assembled off site. The units were then transported and lifted into place. The use of pre-constructed assembled units significantly reduced installation time, therefore allowing installation to be completed in the short window between scheduled shipping movements.

Several complications emerged during the commissioning, and these extended the project duration. However the parallel approach meant that these had no impact on the overall production capacity of the smelter. The reduction in unloading capacity caused by the limitation of the old unloader to the north half of the wharf was quickly offset by the high capacity of the new unloader even when it was operating at a reduced output. Parallel commission proved to be a successful method of commissioning as there were relatively minor additional costs and the risk of process downtime was completely mitigated. The old unloader was decommissioned once the new unit was fully operational [31].

6. Discussion

This paper makes several novel contributions. First, it provides a novel conceptual framework for the commissioning process. The model represents the decision making within the process, the broader context in which plant commissioning occurs, as well as making provision for the finer details. The novelty is creating a structured representation of the commission process, where models are otherwise sparse. Commissioning is generally an ad hoc process, and the value of this new framework is that it provides a structured theoretical foundation for this important activity.

A second contribution is the categorisation of commissioning into three main types: advanced, parallel, and direct. This exposes the variability of strategies within the commissioning process, so it becomes apparent that there is not merely one universal approach to commissioning. Achieving this adds choice to the project planning. It makes it clear that there are choices that practitioners can make, and stating these choices encourages a thoughtful consideration of the planning and resource implications thereof. This categorisation thus adds richness to the conceptual model and makes the decision points more explicit, without being prescriptive.

A third contribution is the development of a model for use by practitioners. The model captures and represents the proposed situational variables (contingency factors) involved in the process. This is valuable for informing the decision making of practitioners. The applicability of the model has been demonstrated by case studies.

A fourth contribution is the integration of commissioning into other management models. The model provides integration with the "risk management" and "project management" disciplines. This is valuable because it shows practitioners how commissioning may be approached in a more holistic manner. The commissioning model is also integrated into a wider model for the development of production plant, and thereby into "manufacturing engineering" including quality and lean manufacturing. Space does not permit full description of this integration, but the point is that the work shows that this integration is indeed feasible. The model is represented in IDEF0 notation, which is a production engineering notation, meaning that it is readily comprehendible in that context.

Overall what has been achieved is to replace the otherwise ad hoc process of commissioning with a systematic process complete with proposed decision factors and internal models of causality. There are implications for practitioners in the model, in the form of flowcharts identifying the critical success/failure factors for commissioning. Thus tentative recommendations can be made for the best commissioning approach for a given situation.

There are also implications for further research. The model is at least partly conjectural, and further research could be directed at establishing the validity of the proposed causal relationships. Another strand of research could be directed at further refinement of the model, and its extension deeper into specific cases, that is, further investigation of the situational variables.

7. Conclusion

Commissioning is extremely valuable to all projects but is poorly defined in the project management body of knowledge. The existing literature on commissioning is focussed on specific tasks, and holistic perspectives are lacking. This work has reconceptualised commissioning and shown that it is possible to identify three main types of commissioning (direct, parallel, and advanced) and construct a generalised conceptual framework around them. This approach to commissioning has been demonstrated by application to case studies.

The value of this work is that it presents a different and more structured way of thinking about commissioning. This allows for a more critical examination of how to complete the commission for a particular project, and ultimately the potential for a better commissioning outcome for practitioners. For theorists the benefit is that a generalised model has been developed, thus a foundation for future advancement of the subject. We have shown that the commissioning activities can be integrated into the risk management, project management, and production engineering bodies of knowledge.

Acknowledgments

The authors would like to thank Ballance Agri-Nutrients and Richard Sweney at New Zealand Aluminium Smelters for providing the information required for the cases studies. These cases studies provided a valuable insight into how commissioning is completed in industry and would not have been possible without the help from these organisations.

References

[1] R. Bernhardt, "Approaches for commissioning time reduction," *Industrial Robot*, vol. 24, no. 1, pp. 62–71, 1997.

[2] R. B. Brown, M. B. Rowe, H. Nguyen, and J. R. Spittler, "Time-constrained project delivery issues," *AACE International Transactions(PM. 09)*, pp. 1–7, 2001.

[3] P. Gikas, "Commissioning of the gigantic anaerobic sludge digesters at the wastewater treatment plant of Athens," *Environmental Technology*, vol. 29, no. 2, pp. 131–139, 2008.

[4] D. Horsley, Ed., *Process Plant Commissioning*, Institution of Chemical Engineers, Rugby, UK, 2nd edition, 1998.

[5] IPENZ, *Practice Note 09: Commissioning Capital Plant*, IPENZ, Wellington, New Zealand, 2007.

[6] J. Kirsilä, M. Hellström, and K. Wikström, "Integration as a project management concept: a study of the commissioning process in industrial deliveries," *International Journal of Project Management*, vol. 25, no. 7, pp. 714–721, 2007.

[7] NHS, *Project Management in a PCT Environment*, National Primary and Care Trust Development Programme, 2004.

[8] B. Peachey, R. Evitts, and G. Hill, "Project management for chemical engineers," *Education for Chemical Engineers*, vol. 2, no. 1, pp. 14–19, 2007.

[9] Project Management Institute (PMI), *A Guide to the Project Management Body of Knowledge (PMBOK Guide)*, Project Management Institute, Newtown Square, Pa, USA, 4th edition, 2008.

[10] P. V. Thomas, "Best practice for process plant modifications (fertilizer plants)," *Cost Engineering*, vol. 45, no. 5, pp. 19–29, 2003.

[11] V. S. Sohmen, "Capital project commissioning. Factors for success," in *Proceedings of the 36th Annual Transactions of the American Association of Cost Engineers (AACE' 92)*, Orlando, Fla, USA, June-July 1992.

[12] H. M. Guven and S. T. Spaeth, "Commissioning process and roles of pyers," in *Proceedings of the ASHRAE Winter Meeting*, la, New Orleans, La, USA, January 1994.

[13] S. K. Shome, "Integration of commissioning activities in project management in power sector," in *Proceedings of the Project Management in the Power Sector Seminar*, Ooty, India, November 1982.

[14] M. G. Tribe and R. R. Johnson, "Effective capital project commissioning," in *Proceedings of the 54th IEEE Pulp and Paper Industry Technical Conference (PPIC' 08)*, Piscataway, NJ, USA, June 2008.

[15] E. E. Choat, "Implementing the commissioning process," in *Proceedings of the Winter Meeting of ASHRAE Transactions, Part 1*, Chicago, Ill, USA, January 1993 1993.

[16] S. Doty, "Simplifying the commissioning process," *Energy Engineering*, vol. 104, no. 2, pp. 25–45, 2007.

[17] E. Schepers, "Commissioning chemical process plant," in *Proceedings of the 2nd National Chemical Engineering Conference*, pp. 60–69, Institution of Chemical Engineers, University of Queensland, Surfers Paradise, Australia, 1974.

[18] G. Shimmings, "Reflections on the causes of delays in commissioning automated materials handling projects," in *Proceedings of the 3rd International Conference on Automated Materials Handling*, Birmingham, UK, 1986.

[19] A. Levi and M. Stonell, "Project management and commissioning of industrial boiler plant," *Institution of Mechanical Engineers, Conference Publications*, pp. 55–67, 1979.

[20] E. Cagno, F. Caron, and M. Mancini, "Risk analysis in plant commissioning: the Multilevel Hazop," *Reliability Engineering and System Safety*, vol. 77, no. 3, pp. 309–323, 2002.

[21] V. Ramnath, "How you can precommission process plants systematically," *Hydrocarbon Processing*, vol. 90, no. 4, pp. 119–124, 2011.

[22] A. Rautenbach, "Site acceptance testing and commissioning of process control systems," *Elektron*, vol. 19, no. 5, pp. 40–44, 2002.

[23] G. Reid, "How to achieve successful startup and commissioning for instrumentation and controls project," in *Proceedings of the Advances in Instrumentation and Control Conference*, vol. 47, pp. 121–124, ISA Services, Houston, Tex, USA, October 1992 1992.

[24] ISO 31000, *Risk Management—Principles and Guidelines*, International Organization for Standardization, 2009.

[25] D. J. Pons, "Strategic risk management in manufacturing," *The Open Industrial and Manufacturing Engineering Journal*, vol. 3, pp. 13–29, 2010.

[26] J. Leitch, "Eliminating the risks to starting up your plant right the first time," *Hydrocarbon Processing*, vol. 85, no. 12, pp. 47–52, 2006.

[27] FIPS, "Integration definition for function modeling (IDEF0)," 1993, http://www.itl.nist.gov/fipspubs/idef02.doc.

[28] KBSI, "IDEF0 overview," 2000, http://www.idef.com/idef0.htm.

[29] D. J. Pons, "Ventures of co-ordinated effort," *International Journal of Project Organisation and Management*, vol. 4, no. 3, pp. 231–255, 2012.

[30] NZAS, "Unloader," in *Tiwai Pointer*, pp. 1–7, Newsletter of New Zealand Aluminium Smelters, 2011.

[31] NZAS, "A look at our new ship unloader," in *Tiwai Pointer*, pp. 1–7, Newsletter of New Zealand Aluminium Smelters, 2012.

10

A Price-Dependent Demand Model in the Single Period Inventory System with Price Adjustment

Kamran Forghani, Abolfazl Mirzazadeh, and Mehdi Rafiee

Department of Industrial Engineering, University of Kharazmi, Tehran, Iran

Correspondence should be addressed to Abolfazl Mirzazadeh; a.mirzazadeh@aut.ac.ir

Academic Editor: Jun Zhao

The previous efforts toward single period inventory problem with price-dependent demand only investigate the optimal order quantity to minimize the total inventory costs; however, there is no method in the literature to avoid unwanted costs due to the deviation between the actual demand and the previously estimated demand. To fill this gap, the present paper supposes that stochastic demand rate with normal distribution is sensitive to the selling price; this means that increasing the selling price would decrease the demand rate and vice versa. After monitoring the consumption trend within a section of the period, a new selling price is implemented to change the demand rate and reduce the shortage or salvage costs at the end of the period. Three functions were suggested to represent the demand rate as a function of selling price, and the numerical analysis was implemented to solve the proposed problem. Finally, an illustrative numerical example was solved for different configurations in order to show the advantages of the proposed model. The results revealed that there is a significant improvement in the system costs when price revision is considered.

1. Introduction

Inventory management is an important task in the business operations. The classical single period problem (SPP) deals with the purchasing inventory problem for single-period products, such as perishable or seasonal goods. The SPP has been popularly researched in the last decade, because of its extensive application in the inventory management of the products with short life cycles (e.g., fashion clothes and electronic products [1]). The authors made extensions to incorporate real world situations in SPP (for further information refer to Khouja [2]); however, there are two important problems in SPP; the first one is how to be in front of demand's uncertainty in the real dynamic global condition and the other one is how to deal with the demand pattern which is dependent to the selling price.

For the price-dependent demand background, Whitin [3] assumed that the expected demand is a function of price, and by using incremental analysis, he derived the necessary optimality condition. Whitin then provided closed-form expressions for the optimal price, which is used to find the optimal order quantity for a demand with a rectangular distribution. Mills [4] assumed demand to be a random variable with an expected value that is decreasing in price and with constant variance. Mills derived the necessary optimality conditions and provided further analysis for the case of demand with rectangular distribution. Karlin and Carr [5] replaced the additive relation presented by Mills [4] with the multiplicative relation for the price-dependent demand where the expected price is an exponential function of two positive constants with the additional restriction that random variable obeys a cumulative distribution function with a mean of 1. A. H. L. Lau and H. S. Lau [6] developed a price-dependent demand model, in which the demand's mean, standard deviation, skewness, and kurtosis are all functions of price. The proposed model was constructed based on the method of moments, which equates the moments of the distribution to their empirical averages in the data. Polatoglu [7] considered the simultaneous pricing and procurement decisions. Three kinds of models (including additive, multiplicative, and riskless models) were used to represent the random demand as a function of selling

TABLE 1: Consumption trend for the first 15 days.

Day	1	2	3	4	5	6	7	8	9	10	11	12	13	14	15
Demand	16	12	19	24	24	27	7	17	23	13	15	10	9	13	14

TABLE 2: Computational result without price revision.

Primary inventory (I_0)	400	500	600
Expected NPV without price revision (NPV_W)	1320.0	7058.4	5490.0

price. Polatoglu analyzed the SPP under general demand uncertainty to reveal the fundamental properties of the model independent of the demand pattern. Polatoglu and Sahin [8] studied a periodic inventory model with stochastic demand, which is dependent on the unit price. Abad and Jaggi [9] considered the problem of setting unit price and length of the credit period with price sensitive demand. A. H. L. Lau and H. S. Lau [10] studied the effect of five different price-dependent deterministic demand functions and showed that sometimes a slightly different demand function can lead to large change in the optimal solutions of the model. Banerjee and Sharma [11] considered a deterministic inventory model for a commodity with seasonal demand. In this model, the product under consideration has general price- and time-dependent seasonal demand rate. Simultaneous decisions were taken regarding the number of seasonal intervals before replenishment, the optimal order quantity, and the selling price, so as to maximize the net profit per unit time.

From the other side, to overcome the problem of difference between the real demand and estimated demand, that causes overstock or understock at the end of period, Bitran and Mondschein [12] characterized the optimal pricing policies as functions of time and inventory to present a continuous time model where the seller faces a stochastic arrival of customers with different valuations of the product. They showed that the optimal pricing policy is to successively discount the product during the season and promote a liquidation sale at the end of the planning horizon. Khouja [13] extended the SPP to the case in which demand is price dependent. Multiple discounts with prices under the control of the newsvendor were used to sell excess inventory. Two algorithms were developed to determine the optimal number of discounts under fixed discounting cost for a given order quantity and realization of demand. Also, the joint determination of the order quantity and initial price was analyzed. Şen and Zhang [14] considered the newsboy model with multiple demand classes, each with a different selling price. Optimal order quantity and demand values in each of the demand classes were determined by following a policy based on protecting the sales in the higher fare class and limiting the sales in the lower fare class. Feng and Xiao [15] investigated optimal times to switch between prices based on the remaining season and inventory; it was assumed that demand is price dependent with Poisson process. Petruzzi and Dada [16] presented a multiperiod newsvendor model to determine optimal stocking and pricing policies over time when a given market parameter of the demand process, though fixed, but initially is unknown. Chung et al. [17]

extended the classical SPP by allowing the retailer to make an in-season price adjustment after conducting a review and using the realized demand to obtain an accurate estimate of the remaining demand. In order to obtain optimal order quantity and adjusted price after the review time, they considered a zero salvage value, small variance after review time, deterministic function of price, and information before review time.

It is noticeable that pricing policies and price-dependent demand models have extensive applications in the other inventory fields; for instance, You and Chen [18] attempted to explore the inventory replenishment policy in EOQ setting by assuming that demand is sensitive to stock and selling price. In the proposed approach, optimal decision rules were surveyed without price change, with a single price change, and with two price changes. Also In the extension of this paper, Mo et al. [19] analyzed the optimal decisions for general value of n changes. Sinha and Sarmah [20] developed a single-vendor multibuyer discount pricing model under stochastic demand information. In this model, the vendor offers multiple pricing schedules to encourage the buyers to adopt the global optimal policy instead of their individual optimal ordering policy. Chen et al. [21] studied a coordination contract for a supplier-retailer channel with stochastic price-dependent demand. They supposed a two-stage optimization problem where in the first stage the supplier decides the amount of capacity reservation, and in the second stage after updating demand information, the retailer determines the order quantity and the retail price. Also, Chiu et al. [22] presented a policy for channel coordination in supply chains with both additive and multiplicative price-dependent demands. The proposed policy combines the use of wholesale price, channel rebate, and returns.

The previous efforts toward single period inventory problem with price-dependent demand only investigate the optimal order quantity to minimize the total inventory costs; however, there is no method in the literature to avoid unwanted costs due to the deviation between the actual demand and previously estimated demand. To fill this gap, the present paper supposes that, the stochastic demand rate with normal distribution is sensitive to the selling price; this means that increasing the selling price would decrease the demand rate (by affecting mean and variance of the demand rate distribution) and vice versa. After monitoring the consumption trend within a section of the period, a new selling price is implemented to change the demand rate and reduce the shortage or salvage costs at the end of the period. Three functions are suggested to represent the demand

TABLE 3: Computational results considering $I_0 = 400$.

Ratio function	α	β	p_E^*	NPV_E^*	p_A^*	NPV_A^*	$NPV_A(p_E^*)$	Imp. (%)
	—	2	110.6	8529	114.5	9442	9061	586.4%
Linear	—	1.8	103.9	7633	107.3	8404	7986	505.0%
	—	1.5	94.4	6311	96.8	6859	6439	387.8%
	6	2	110.6	8529	114.5	9442	9061	586.4%
Two-segment	5	1.8	103.9	7633	107.3	8404	7986	505.0%
	7	1.4	91.4	5877	93.4	6348	5935	349.6%
	1.2	1.9	92.1	5928	94.8	6516	6096	361.8%
Exponential	1.7	2.1	89.6	5594	91.8	6101	5674	329.9%
	1.2	0.8	100.8	7067	104.9	7933	7533	470.7%
	0	2	99.2	6865	102.7	7646	7267	450.6%

TABLE 4: Computational results considering $I_0 = 500$.

Ratio function	α	β	p_E^*	NPV_E^*	p_A^*	NPV_A^*	$NPV_A(p_E^*)$	Imp. (%)
	—	2	90.7	7165	92.3	8782	8761	24.1%
Linear	—	1.8	85.3	6795	87.9	8267	8137	15.3%
	—	1.5	80.7	6528	83.5	7657	7275	3.1%
	6	2	90.7	7165	92.3	8782	8761	24.1%
Two-segment	5	1.8	85.3	6795	87.9	8267	8137	15.3%
	7	1.4	79.4	6678	82.4	7496	6628	−6.1%
	1.2	1.9	79.5	6530	81.9	7404	6852	−2.9%
Exponential	1.7	2.1	79.2	6557	81.4	7323	6691	−5.2%
	1.2	0.8	80.8	6530	84.1	7715	7301	3.4%
	0	2	80.8	6530	83.8	7700	7301	3.4%

rate as a function of selling price, and numerical analysis is implemented to solve the proposed problem. Finally, a numerical example is solved for various configurations, and sensitivity analysis is performed to show the efficiency of the proposed approach.

2. The Classical Newsvendor Model

Single period problem or newsvendor problem is one of the most well-known inventory models which attempts to find order quantity within stochastic demand in a single period so as to maximize the expected profit or minimize the expected costs (these terms are equal). The assumptions of this model are as follows.

(i) Demand is a stochastic variable.

(ii) Unsold items are sold by a discount (salvaged) or disposed at the end of the period.

(iii) Shortage is allowed.

(iv) Only one purchase order is allowed and the ordered lot is placed at the beginning of the period.

For the classical newsvendor problem, the optimum order quantity (Q^*) can be easily obtained through the following equation [2]:

$$Q^* = F^{-1}\left(\frac{p+s-c}{p+s-v}\right), \qquad (1)$$

where p is the unit selling price, c is the unit procurement cost, v is the unit salvage value, s is the unit shortage cost, and $F^{-1}(\cdot)$ denotes the inverse of the cumulative distribution function of demand.

3. The Newsvendor Model with Inventory Revision

The previous efforts toward single period problem with price-dependent demand only investigate the optimal order quantity to minimize the total inventory costs; however, there is no method in the literature to avoid unwanted costs due to the deviation between the actual demand and previously estimated demand. To fill this gap, the present paper supposes that demand rate is sensitive to the selling price; this means that increasing the selling price would decrease the demand rate and vice versa. After monitoring the consumption trend within a section of the period, a new selling price is implemented to change the demand rate and reduce the shortage or salvage costs at the end of the period. For illustration, in Figure 1, demand rate within the interval $[0, t_0]$ shows that it is most likely to have some remaining stock at the end of the period; hence, we should decrease the selling price to increase the demand rate and consequently avoid the salvage cost. From the other side, in Figure 2, demand rate within the interval $[0, t_0]$ shows that it is most likely to have some unsatisfied demand at the end of the period, and therefore we should increase the selling price in

TABLE 5: Computational results considering $I_0 = 600$.

Ratio function	α	β	p_E^*	NPV_E^*	p_A^*	NPV_A^*	$\mathrm{NPV}_A(p_E^*)$	Imp. (%)
	—	2	89.9	4174	89.9	5827	5827	6.1%
Linear	—	1.8	81.9	3885	81.9	5507	5507	0.3%
	—	1.5	71.0	4455	72.5	6077	6027	9.8%
	6	2	74.7	7572	76.1	8450	7726	40.7%
Two-segment	5	1.8	73.6	7134	75.4	8142	7362	34.1%
	7	1.4	75.4	7873	76.8	8675	7874	43.4%
	1.2	1.9	71.5	6294	73.7	7423	6791	23.7%
Exponential	1.7	2.1	72.6	6824	74.6	7822	7037	28.2%
	1.2	0.8	68.8	4978	71.8	6404	5930	8.0%
	0	2	69.1	4831	71.8	6319	5985	9.0%

order to decrease demand rate and consequently reduce the shortage cost.

The assumptions in this work are as follows.

(i) The demand rate is a stochastic normal variable.

(ii) The parameters of the demand rate distribution are unknown before the beginning of the period.

(iii) The demand rate is sensitive to the selling price.

(iv) Mean and variance of the demand rate are a deterministic functions of the selling price.

(v) The changed demand rate is stable after price revision.

(vi) The order quantity is known and placed at the beginning of the period as the initial inventory level.

Note that, since the parameters of the demand rate distribution are unknown before the beginning of the period, we can use the expert's opinion to obtain initial inventory level, or the information of the similar commodities can be used to obtain it by using (1).

Also, the following notations are used in the mathematical model.

I_0: initial inventory level at the beginning of the period.

t_0: revision point.

p_0: selling price before price revision.

p: unit selling price after price revision.

c: unit procurement cost.

c_0: unit salvage (or dispose) value.

s: unit shortage cost.

D: stochastic demand rate after price revision, where $D \sim N(\mu_D, \sigma_D^2)$.

R_p: variation ratio in the demand rate when revising selling price.

p_A^*: optimum revised price based on the actual demand rate distribution.

p_E^*: optimum revised price based on the estimated demand rate within the interval $[0, t_0]$.

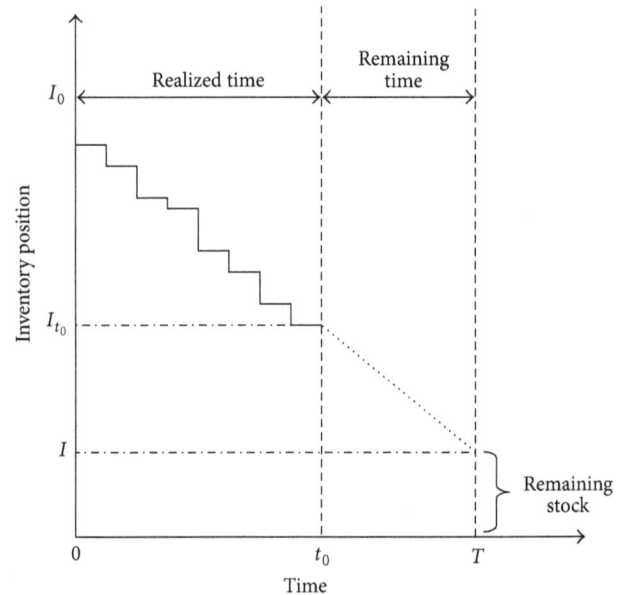

FIGURE 1: Consumption trend and predicted remaining stock at the end of the period.

NPV_A: optimum net present value based on the actual demand rate distribution.

NPV_E: optimum net present value based on the estimated demand rate.

As it was mentioned, the parameters of the demand rate distribution are unknown at the zero time. These parameters can be estimated based on the consumption trend. So $\overline{D}$ and S_D^2 are introduced as the mean and variance of demand rate before the revision point (t_0), respectively. These parameters are estimated as follows:

$$\overline{D} = \frac{(I_0 - I_{t_0})}{t_0}, \qquad S_D^2 = \frac{\sum_{i=1}^{t_0} (I_{i-1} - I_i - \mu_D)^2}{t_0 - 1}. \tag{2}$$

In (2), t_0 should be as large enough to appropriately estimate $\overline{D}$ and S_D^2; however, if we take t_0 too large, less time remains to adjust the inventory level. Therefore, we need to establish a compromise between the remaining time and the use of

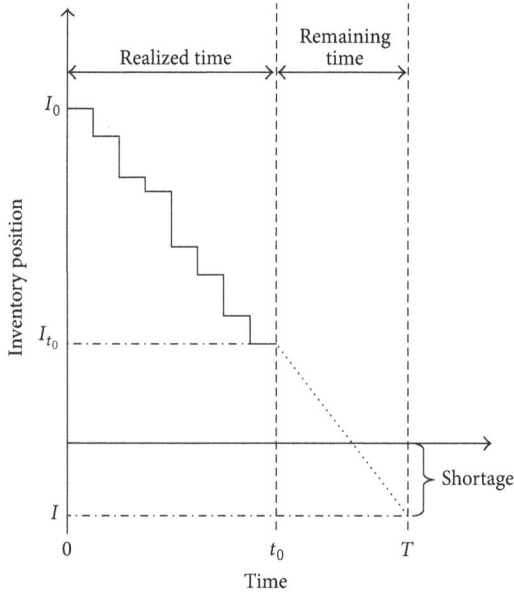

FIGURE 2: Consumption trend, predicting shortage at the end of the period.

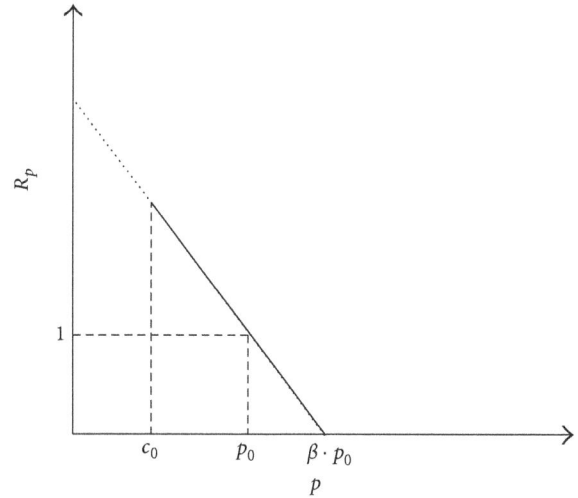

FIGURE 3: Linear ratio function.

the past information. For commodities, which have a selling period more than 30 days, using the information of 10–15 days may be sufficient to estimate the parameters of the demand rate distribution.

Now, in order to represent the demand rate after revision point (D) in terms of selling price p, we assume that $D \sim N(\mu_D, \sigma_D^2)$, where μ_D and σ_D^2 are the mean and variance of demand rate distribution after revision point, respectively. Since D is directly proportional to R_p, μ_D and σ_D^2 are estimated as follows:

$$\mu_D = \overline{D} \cdot R_p = \frac{\left(I_0 - I_{t_0}\right) R_p}{t_0},$$

$$\sigma_D^2 = S_D^2 \cdot R_p^2 = \frac{\sum_{i=1}^{t_0} \left(I_{i-1} - I_i - \mu_D\right)^2}{t_0 - 1} R_p^2. \quad (3)$$

In (3), R_p is inversely proportional to p, that is, increasing p would decrease R_p and consequently decrease D, and vice versa. Therefore, D is sensitive to p.

To represent the inverse relation between p and R_p, we present several ratio functions that satisfy the following requirements:

(i) $R_p = 1$, for all $p = p_0$.

(ii) $\partial R_p / \partial p \leq 0$.

(iii) $\lim_{p \to \infty} R_p = 0$.

(iv) $R_p = 1$, for all $p = p_0$.

Then, the first requirement indicates that the demand rate would not change if the selling price is the same before being offered, the second one represents that R_p and p have inverse relation, the third one implies that the demand rate would become almost zero if the selling price increases significantly,

and the last one represents that demand rate at any moment is a positive value. It is necessary to note that the revised price should be greater than c_0, because we are sure that the remaining stock can be sold without any trouble.

3.1. Linear Ratio Function.
In the linear ratio function, it is assumed that the relation between R_p and p is approximately linear (see Figure 3); therefore, the following function can be implemented to represent it.

$$R_p = \begin{cases} \dfrac{p - \beta \cdot p_0}{p_0 \left(1 - \beta\right)}, & p \leq \beta \cdot p_0, \\ 0, & p > \beta \cdot p_0. \end{cases} \quad (4)$$

In (4), β is the price elasticity; the higher the β, the more elasticity there is.

3.2. Two-Segment Ratio Function.
Since the linear ratio function is not more flexible, we present the two-segment ratio function. This function consists of two linear sections (see Figure 4). The inverse relation between R_p and p for the two-segment ratio function can be expressed as follows:

$$R_p = \begin{cases} \dfrac{1 - \alpha}{p_0 - c_0} \left(p - p_0\right) + 1, & \forall p < p_0, \\ \dfrac{p - \beta \cdot p_0}{p_0 \left(1 - \beta\right)}, & \forall p_0 \leq p \leq \beta \cdot p_0, \\ 0, & \forall p > \beta \cdot p_0. \end{cases} \quad (5)$$

In (5), α is the variation ratio in the demand rate, when it is assumed that $p = c_0$, and β is the price elasticity.

3.3. Exponential Ratio Function.
Another flexible function which corresponds to the inverse relation between p and R_p

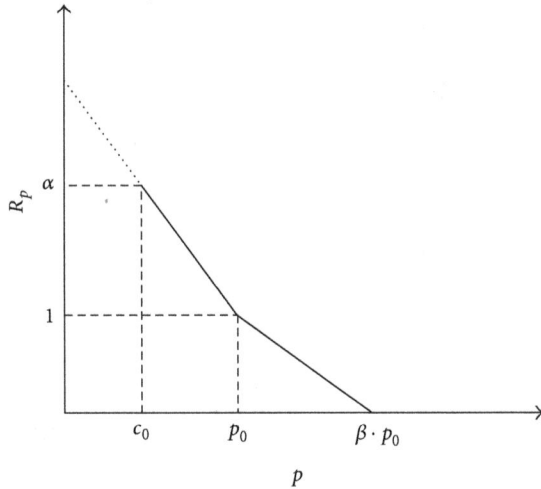

FIGURE 4: Two-segment ratio function.

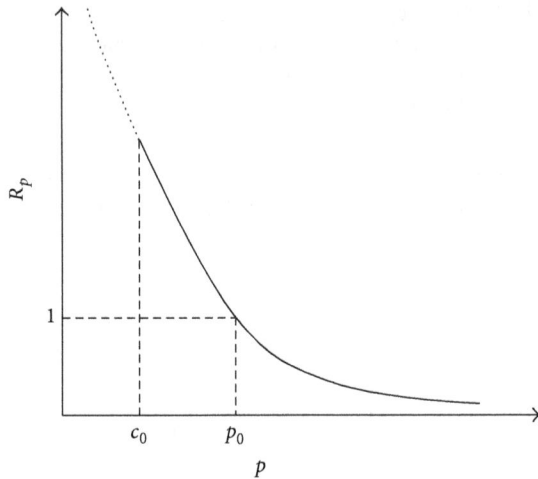

FIGURE 5: Exponential ratio function.

is the exponential ratio function (see Figure 5). This function can be expressed as follows:

$$R_p = \left(\frac{p_0}{p}\right)^\alpha \exp\left(\frac{\beta(p_0 - p)}{p_0}\right). \tag{6}$$

Various values for α and β in (6) can yield different ration functions.

The presented parameters in each ratio function, that is, α and β, can be obtained according to the expert's opinion or using statistical methods such as regression.

After selecting an appropriate ratio function for the problem, in the next step, in order to evaluate the net present value (NPV), we should specify the probability distribution of the inventory position (I) at the end of the period. As D is distributed normally, that is, $D \sim N(\mu_D, \sigma_D^2)$, and since

$I = I_{t_0} - D(T - t_0)$, I is also distributed normally with parameters μ_I and σ_I^2, that is, $D \sim N(\mu_I, \sigma_I^2)$, where

$$\mu_I = I_{t_0} - \frac{(I_0 - I_{t_0})R_p}{t_0}(T - t_0),$$

$$\sigma_I^2 = \frac{\sum_{i=1}^{t_0}(I_{i-1} - I_i - \mu_D)^2}{t_0 - 1}R_p^2(T - t_0)^2. \tag{7}$$

After specifying the distribution of I, we write NPV based on I to analyze optimum value of selling price.

$$\text{NPV} = \begin{cases} (I_{t_0} - I)p - I_{t_0}c + Ic_0, & \forall I \geq 0, \\ I_{t_0}p - I_{t_0}c + Is, & \forall I < 0. \end{cases} \tag{8}$$

In (8), if $I \geq 0$, then we have salvage cost; otherwise, we have shortage cost.

Now, we define μ_{NPV} as the expected value of NPV and obtain it as follows:

$$\mu_{\text{NPV}} = E[\text{NPV}] = \int_0^\infty \left(-I_{t_0}c + (I_{t_0} - I)p + Ic_0\right)f(I)\,d(I)$$

$$+ \int_{-\infty}^0 \left(-I_{t_0}c + I_{t_0}p + Is\right)f(I)\,d(I)$$

$$= -I_{t_0}c + (I_{t_0} - \mu_I)p + \mu_I c_0$$

$$+ \int_{-\infty}^0 I(s + p - c_0)f(I)\,d(I)$$

$$= -I_{t_0}c + (I_{t_0} - \mu_I)p + \mu_I c_0 + (s + p - c_0)$$

$$\times \int_{-\infty}^0 \frac{I}{\sqrt{2\pi\sigma_I^2}}\exp\left(-\frac{(I - \mu_I)^2}{2\sigma_I^2}\right)d(I). \tag{9}$$

After simplifying (9), we have

$$\mu_{\text{NPV}} = -I_{t_0}c + (I_{t_0} - \mu_I)p + \mu_I c_0 + (s + p - c_0)$$

$$\times \left(\mu_I \phi\left(\frac{-\mu_I}{\sqrt{\sigma_I^2}}\right) - \sqrt{\frac{\sigma_I^2}{2\pi}}\exp\left(-\frac{\mu_I^2}{2\sigma_I^2}\right)\right), \tag{10}$$

where $\phi(\cdot)$ is the cumulative distribution function of the standard normal distribution.

As it is obvious in (10), the term of $\phi(-\mu_I/\sqrt{\sigma_I^2})$ is dependent on p and can only be numerically obtained; hence, we cannot solve $d\mu_{\text{NPV}}/dp = 0$ to obtain the optimum revised price (p^*). Therefore, we implement the numerical analysis to obtain p^*. The pseudocodes of the proposed numerical analysis have been included in Pseudocodes 1 and 2.

4. An Illustrative Example

In this section, an illustrative numerical example has been included to show the efficiency of the proposed approach.

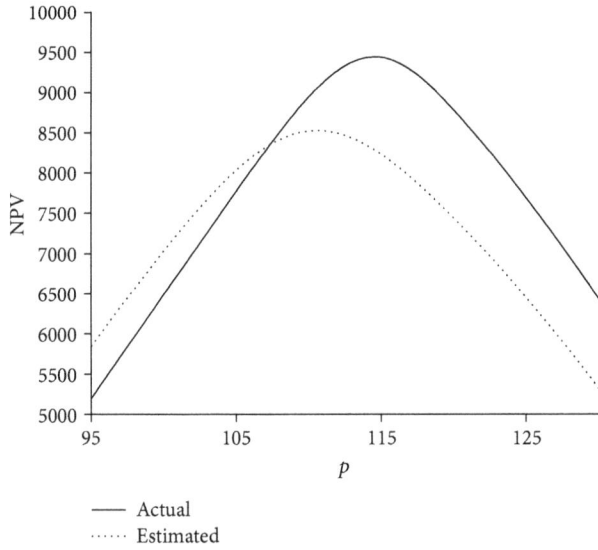

FIGURE 6: Expected NPVs with different p, considering linear ratio function, $I_0 = 400$ and $\beta = 2$.

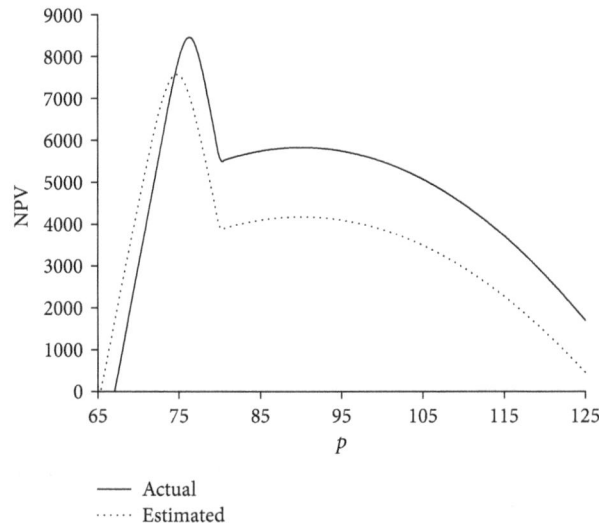

FIGURE 8: Expected NPVs with different p, considering exponential ratio function, $I_0 = 500$, $\alpha = 1.2$ and $\beta = 0.8$.

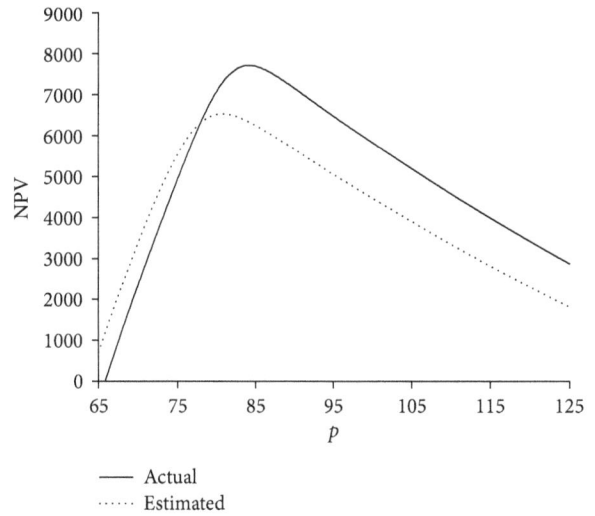

FIGURE 7: Expected NPVs with different p, considering two-segment ratio function, $I_0 = 600$, $\alpha = 6$ and $\beta = 2$.

The results of the proposed model (with price revision) are compared against that without price revision. The consumption trend for the first 15 days has been generated randomly according to the normal distribution with $\mu = 18$, $\sigma^2 = 25$ and included in Table 1. The other parameters are assumed as follows: $c = 50$, $c_0 = 20$, $s = 30$, $p_0 = 80$, $T = 30$, and $t_0 = 15$.

Table 2 shows the expected profit without price revision. This example is investigated for the three types of ratio functions and three levels of the initial inventory (I_0), including 400, 500, and 600, and the results are summarized in Tables 3, 4, and 5, respectively. In these tables, the columns of p_E^* and NPV_E^* indicate the optimum solutions based on the estimated demand rate within the interval $[0, t_0]$ (i.e., the proposed approach), the columns of p_A^* and NPV_A^* show the optimum solutions based on the actual demand rate, with the mean 18 and variance 25 (which are unknown for us), the column of $NPV_A(p_E^*)$ shows the expected NPV based on the actual demand rate for p_E^*, and finally the last column shows the relative gap between $NPV_A(p_E^*)$ and expected NPV without the price revision included in the last row of Table 2 (i.e., NPV_W); this column has been calculated as follows: $((NPV_A(p_E^*) - NPV_W)/NPV_W) \times 100$. Also, Figures 6, 7, and 8 display the plots of NPV_E and NPV_A against p for three sample cases. In these figures, the terms of "actual" and "estimated" refer to NPV_A and NPV_E, respectively.

The computational results indicated that in all the cases, except three cases in Table 4, the proposed approach gives better results in comparison with the state in which price revision is not considered. Also, for those cases in which the improvement percent is negative, the amount of loss in the expected NPV is insignificant. It can be observed that the optimum revised price for the estimated demand rate (p_E^*) has no significant difference with that for the actual demand rate (p_A^*); this implies that monitoring demand rate within the interval $[0, t_0]$ to estimate demand rate can provide an efficient approximation for the problem.

5. Sensitivity Analysis

In this section, we investigate the model's sensitivity with respect to the different parameters of the ratio functions. The relationship between NPV_E and different values of α and β, for the two-segment and exponential ratio functions, have been plotted in Figures 9 and 10, respectively. From Figures 9 and 10, we can observe that the two-segment ratio function

```
Function p(a, b)
    ε ← ⟨⟩  ▷ Initialize ε; where ε is the length of each step at each iteration
    Δ ← ⟨⟩  ▷ Initialize Δ; Δ > 2ε where Δ is the calculation accuracy
    p₁ = a
    p₂ = b
    repeat
                p = (p₁ + p₂)/2
                if μ_NPV(p) > μ_NPV(p + ε)  then
                        p₂ = p
                else
                        p₁ = p
                End if
    until  p₂ − p₁ < Δ              ▷ Check finishing criterion
    return  p
End Function
```

PSEUDOCODE 1: Pseudocode for obtaining the best revised price within the interval (a, b).

```
Function p*
    case ratio_function  of
        Linear:
                p* = p(c₀, βp₀)
        Two_Segment:
                p₁ = p(c₀, p₀)
                p₂ = p(c₀, βp₀)
                        if μ_NPV(p₁) > μ_NPV(p₂) then
                                p* = p₁
                        else
                                p* = p₂
                        End if
        Exponential:
                p* = p(c₀, ∞)
    End case
    return  p*
End Function
```

PSEUDOCODE 2: Pseudocode for obtaining the optimal revised price (p^*).

is more sensitive than the exponential ratio function. In the two-segment ratio function, α is more sensitive than β within the interval $[-40, 60)$; conversely, within the interval $[60, \infty)$, β is more sensitive than α. Also in the exponential ratio function, α is generally more sensitive than β.

6. Conclusion and Directions for Future Research

In this paper, we presented a price-dependent model for the single period inventory problem. Considering the fact that the distribution of the demand rate is unknown at the beginning of the period, we attempted to estimate the parameters of the demand rate distribution by investigating the consumption trend within a specific section of the period. Three functions were suggested to represent the demand rate as a function of selling price. We attempted to find the optimum revised price so as to maximize the expected NPV

at the end of the period. A numerical example was solved for different configurations, and the results were compared against the state in which price revision is not considered. The results revealed significant improvement when the price revision is considered. Furthermore, it was concluded that the optimum revised price for the estimated demand rate has no significant difference with that for the actual demand rate; this implies that monitoring the consumption trend within a section of the period to estimate the parameters of the demand rate distribution can provide an efficient approximation for the problem.

At the end, for the future researches, we suggest the following guidelines.

(i) Developing the proposed approach for dynamic price revision.

(ii) Using the other functions to represent the demand rate as a function of selling price, such as multisegment, combinational, and fuzzy functions.

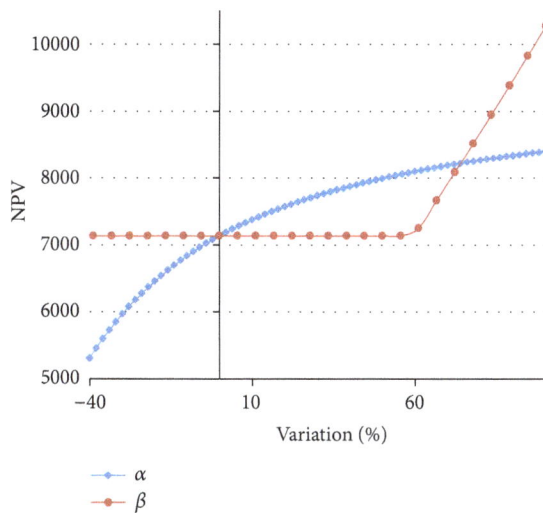

FIGURE 9: Sensitivity plot for two-segment ratio function with $I_0 = 680$ and initial parameters $\alpha = 5$ and $\beta = 1.8$.

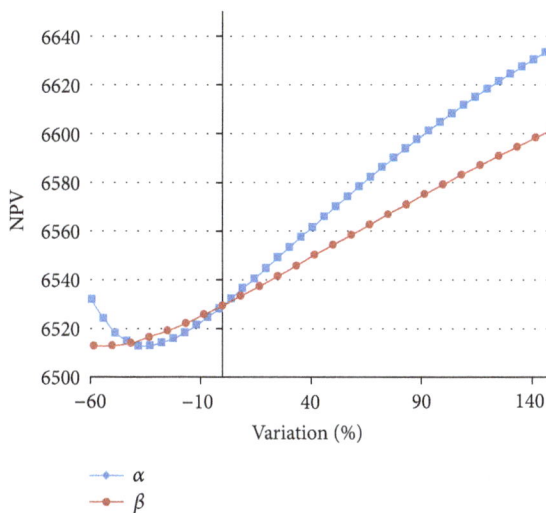

FIGURE 10: Sensitivity plot for exponential ratio function with $I_0 = 500$ and initial parameters $\alpha = 1.9$ and $\beta = 1.2$.

(iii) Developing a mathematical method in order to obtain revision point with more accuracy.

(iv) Considering time-dependent holding cost in the model.

(v) Using other distribution functions such as uniform distribution and Poisson distribution, as the distribution of the demand rate.

These suggestions remain critical issues for the future studies.

References

[1] J. L. Zhang, C. Y. Lee, and J. Chen, "Inventory control problem with freight cost and stochastic demand," *Operations Research Letters*, vol. 37, no. 6, pp. 443–446, 2009.

[2] M. Khouja, "The single-period (news-vendor) problem: literature review and suggestions for future research," *Omega*, vol. 27, no. 5, pp. 537–553, 1999.

[3] T. M. Whitin, "Inventory control and price theory," *Management Science*, vol. 2, pp. 61–68, 1955.

[4] E. S. Mills, "Uncertainty and price theory," *The Quarterly Journal of Economics*, vol. 73, pp. 116–130, 1959.

[5] S. Karlin and C. R. Carr, "Prices and optimal inventory policy," in *Studies in Applied Probability and Management Science*, K. J. Arrow and H. Scarf, Eds., pp. 159–172, Stanford University Press, Stanford, Calif, USA, 1962.

[6] A. H. L. Lau and H. S. Lau, "The newsboy problem with price-dependent demand distribution," *IIE Transactions*, vol. 20, no. 2, pp. 168–175, 1988.

[7] L. H. Polatoglu, "Optimal order quantity and pricing decisions in single-period inventory systems," *International Journal of Production Economics*, vol. 23, no. 1–3, pp. 175–185, 1991.

[8] H. Polatoglu and I. Sahin, "Optimal procurement policies under price-dependent demand," *International Journal of Production Economics*, vol. 65, no. 2, pp. 141–171, 2000.

[9] P. L. Abad and C. K. Jaggi, "A joint approach for setting unit price and the length of the credit period for a seller when end demand is price sensitive," *International Journal of Production Economics*, vol. 183, pp. 115–122, 2003.

[10] A. H. L. Lau and H. S. Lau, "Effects of a demand-curve's shape on the optimal solutions of a multi-echelon inventory/pricing model," *European Journal of Operational Research*, vol. 147, no. 3, pp. 530–548, 2003.

[11] S. Banerjee and A. Sharma, "Optimal procurement and pricing policies for inventory models with price and time dependent seasonal demand," *Mathematical and Computer Modelling*, vol. 51, no. 5-6, pp. 700–714, 2010.

[12] G. R. Bitran and S. V. Mondschein, "Periodic pricing of seasonal products in retailing," *Management Science*, vol. 43, no. 1, pp. 64–79, 1997.

[13] M. J. Khouja, "Optimal ordering, discounting, and pricing in the single-period problem," *International Journal of Production Economics*, vol. 65, no. 2, pp. 201–216, 2000.

[14] A. Şen and A. X. Zhang, "The newsboy problem with multiple demand classes," *IIE Transactions*, vol. 31, no. 5, pp. 431–444, 1999.

[15] Y. Feng and B. Xiao, "Optimal policies of yield management with multiple predetermined prices," *Operations Research*, vol. 48, no. 2, pp. 332–343, 2000.

[16] N. C. Petruzzi and M. Dada, "Dynamic pricing and inventory control with learning," *Naval Research Logistics*, vol. 49, no. 3, pp. 303–325, 2002.

[17] C. S. Chung, J. Flynn, and J. Zhu, "The newsvendor problem with an in-season price adjustment," *European Journal of Operational Research*, vol. 198, no. 1, pp. 148–156, 2009.

[18] P. S. You and T. C. Chen, "Dynamic pricing of seasonal goods with spot and forward purchase demands," *Computers and Mathematics with Applications*, vol. 54, no. 4, pp. 490–498, 2007.

[19] J. Mo, F. Mi, F. Zhou, and H. Pan, "A note on an EOQ model with stock and price sensitive demand," *Mathematical and Computer Modelling*, vol. 49, no. 9-10, pp. 2029–2036, 2009.

[20] S. Sinha and S. P. Sarmah, "Single-vendor multi-buyer discount pricing model under stochastic demand environment," *Computers and Industrial Engineering*, vol. 59, pp. 945–953, 2010.

[21] H. Chen, Y. Chen, C. H. Chiu, T. M. Choi, and S. Sethi, "Coordination mechanism for the supply chain with leadtime

consideration and price-dependent demand," *European Journal of Operational Research*, vol. 203, no. 1, pp. 70–80, 2010.

[22] C. H. Chiu, T. M. Choi, and C. S. Tang, "Price, rebate, and returns supply contracts for coordinating supply chains with price-dependent demands," *Production and Operations Management*, vol. 20, no. 1, pp. 81–91, 2011.

11

Optimizing Industrial Robots for Accurate High-Speed Applications

Hubert Gattringer, Roland Riepl, and Matthias Neubauer

Institute for Robotics, Johannes Kepler University, Altenbergerstraße 69, 4040 Linz, Austria

Correspondence should be addressed to Hubert Gattringer; hubert.gattringer@jku.at

Academic Editor: Aydin Nassehi

Today's standard robotic systems often do not meet the industry's demands for accurate high-speed robotic applications. Any machine, be it an existing or a new one, should be pushed to its limits to provide "optimal" efficiency. However, due to the high complexity of modern applications, a one-step overall optimization is not possible. Therefore, this contribution introduces a step-by-step sequence of multiple nonlinear optimizations. Included are optimal configurations for geometric calibration, best-exciting trajectories for parameter identification, model-based control, and time/energy optimal trajectory planning for continuous path and point-to-point trajectories. Each of these optimizations contributes to the improvement of the overall system. Existing optimization techniques are adapted and extended for use with a standard industrial robot scenario and combined with a comprehensive toolkit with discussions on the interplay between the separate components. Most importantly, all procedures are evaluated in practical experiments on a standard robot with industrial control hardware and the recorded measurements are presented, a step often missing in publications in this area.

1. Introduction

State-of-the-art robotic systems are equipped with highly sophisticated industrial hardware, capable of short sample times and offering high computational power. This increasing arithmetic performance may be used to improve positioning accuracy and dynamic accuracy and compute time/energy-optimal trajectories. The proper fundamentals are found in the underlying mathematical models.

This contribution focuses on giving an overview of various nonlinear optimizations. With each optimization, a certain aspect of an arbitrary robotic system is improved. Applying all of them in sequence will provide better results regarding positioning accuracy and dynamic accuracy and will reduce the cycle times. The basis for this optimization lies in the kinematic and dynamical modeling of the system. Therefore, Section 2 starts with the deviation of these models enhanced by a model-based control strategy. Special emphasis is laid on the implementation on an industrial system including time delays. In Section 3, the static position accuracy of the robot is improved by considering unavoidable tolerances in the robot kinematics. The main topic is obtaining optimal configurations for the identification process. The identification of the dynamic parameters, that are used for the model-based control, is described in Section 4. Also, the problem of optimal exciting trajectories is solved by nonlinear optimization techniques. Finally, Section 5 reviews time/energy optimal trajectory planning for continuous path trajectories and point-to-point trajectories.

Each section provides information on the used solvers for the various optimizations and also presents detailed results of the conducted experiments. Experiments are performed using an off-the-shelf robot, a Stäubli RX130L shown in Figure 3, with industrial control hardware and commercially available measurement hardware, to guarantee that all proposed methods are applicable for industry. For specifications of the articulated robot, please consult Table 1.

Previous works of the authors, which are also focused on industrial robotics, see, for example, [1], cover the basics like control and geometric path planning. Based on these works, a specific combination of algorithms is presented, each of which optimizes a specific aspect of the robotic system.

TABLE 1: Technical specifications.

Stäubli RX130L industrial robot	
Degrees of freedom	6
Maximum load capacity	10 kg
Reach at wrist	1660 mm
Weight	≈200 kg

2. System Modeling and Control

2.1. Kinematic Modeling.

The kinematic modeling can be divided in the direct kinematics problem and the inverse kinematics problem. The direct kinematics calculates the end-effector coordinates $\mathbf{z}_E^T = (\mathbf{r}_E^T \boldsymbol{\varphi}_E^T)$ as a function of the joint coordinates $\mathbf{q}$ and nominal geometric parameters $\mathbf{p}_n$ as

$$\mathbf{z}_E = \mathbf{f}_{DK}(\mathbf{q}, \mathbf{p}_n), \tag{1}$$

see Figure 3. Vector $\mathbf{r}_E$ indicates the end-effector position, which can be calculated by a sequential summation of the relative connecting vectors. For the orientation, a description in Cardan angles is used. A sequential multiplication of the relative rotation matrices between two successive joints delivers the rotation matrix $\mathbf{A}_{IE}$ between end-effector frame (E) and inertial frame (I) from which the Cardan angles can be extracted. In this paper, also the notation tool center point (TCP) is used for the end-effector. Inverse kinematics calculates the joint coordinates $\mathbf{q}$ as a function of end-effector coordinates $\mathbf{z}_E$ and $\mathbf{p}_n$:

$$\mathbf{q} = \mathbf{f}_{IK}(\mathbf{z}_E, \mathbf{p}_n). \tag{2}$$

Solutions for the inverse kinematics problem of such standard robotic systems can be found in many textbooks, for example, [2] or [3].

2.2. Dynamic Modeling.

The equations of motion for multibody systems, like a robot, can be calculated by several methods. The method with minimal effort is the Projection Equation, see [4]. Linear momenta $\mathbf{p} = m\mathbf{v}_c$ and angular momenta $\mathbf{L} = \mathbf{J}\boldsymbol{\omega}_c$ are projected into the minimal space (minimal velocities $\dot{\mathbf{q}}$) via the appropriate Jacobian matrices:

$$\sum_{i=1}^{N} \left[\left[\frac{\partial_R \mathbf{v}_c}{\partial \dot{\mathbf{q}}} \right]^T \left[\frac{\partial_R \boldsymbol{\omega}_c}{\partial \dot{\mathbf{q}}} \right]^T \right]_i \begin{pmatrix} {}_R\dot{\mathbf{P}} + {}_R\widetilde{\boldsymbol{\omega}}_{IR} \, {}_R\mathbf{P} - {}_R\mathbf{f}^e \\ {}_R\dot{\mathbf{L}} + {}_R\widetilde{\boldsymbol{\omega}}_{IR} \, {}_R\mathbf{L} - {}_R\mathbf{M}^e \end{pmatrix}_i = 0. \tag{3}$$

All the values like the translational velocity $\mathbf{v}_c$ or the rotational velocity of the center of gravity $\boldsymbol{\omega}_c$ can be inserted in arbitrary coordinate systems R. In contrast to $\boldsymbol{\omega}_c$, $\boldsymbol{\omega}_{IR}$ is the velocity of the used reference system. The matrix $\mathbf{J}$ is the inertia tensor, while $\widetilde{\boldsymbol{\omega}}\mathbf{p}$ characterizes the vector product $\boldsymbol{\omega} \times \mathbf{p}$. $\mathbf{f}^e$ and $\mathbf{M}^e$ are impressed forces and moments acting on the ith body. An evaluation of (3) yields the highly nonlinear equations of motion for the robot

$$\mathbf{M}(\mathbf{q})\ddot{\mathbf{q}} + \mathbf{g}(\mathbf{q}, \dot{\mathbf{q}}) = \mathbf{Q}_M. \tag{4}$$

They are composed of the minimal coordinates $\mathbf{q}$, the configuration dependent, symmetric and positive definite mass matrix $\mathbf{M}(\mathbf{q})$, the vector $\mathbf{g}(\mathbf{q}, \dot{\mathbf{q}})$ containing all nonlinear effects (like Coriolis, centrifugal, gravity, and friction forces), and the vector of the generalized actuating forces $\mathbf{Q}_M$.

Detailed dynamical modeling is an essential task because it is the basis for model-based control and optimal trajectory generation.

2.3. Control.

The hardware setup is composed of an industrial PC, communicating via Powerlink with six ACOPOS servo drives, which power the synchronous motors of the Stäubli robot. The servo drives use cascaded controllers for precise position control of the robot's joints. However, the articulated robot is a highly nonlinear system due to the serially connected arm/joint units. Typically, using linear joint controllers on a highly nonlinear mechanical system will not lead to sufficiently accurate and dynamic end-effector motion.

2.3.1. Model-Based Control.

This contribution suggests to effectively linearize the nonlinear mechanical system with a flatness-based feed-forward approach and use the servo drives' linear cascaded controllers to compensate model deviations and external disturbances. Thus, they are summarized for the sake of completeness. Introducing the state $\mathbf{x}^T = (\mathbf{q}^T \dot{\mathbf{q}}^T)$, the equations of motion, (4), in state space read

$$\frac{d}{dt}\begin{pmatrix} \mathbf{q} \\ \dot{\mathbf{q}} \end{pmatrix} = \begin{pmatrix} \dot{\mathbf{q}} \\ \mathbf{M}(\mathbf{q})^{-1}\left(-\mathbf{g}(\mathbf{q}, \dot{\mathbf{q}}) + \mathbf{Q}_M\right) \end{pmatrix} \tag{5}$$

with the input vector of motor torques $\mathbf{Q}_M$. With the flat output $\mathbf{y} = \mathbf{q}$ and its derivatives with respect to time,

$$\mathbf{y} = \mathbf{q},$$
$$\dot{\mathbf{y}} = \dot{\mathbf{q}}, \tag{6}$$
$$\ddot{\mathbf{y}} = \ddot{\mathbf{q}} = \mathbf{M}(\mathbf{q})^{-1}\left(-\mathbf{g}(\mathbf{q}, \dot{\mathbf{q}}) + \mathbf{Q}_M\right),$$

all state and input variables can be computed as a function of the flat output, see, for example, [5]. Consequently, the evolution of all system variables $\mathbf{q}$, $\dot{\mathbf{q}}$, and $\mathbf{Q}_M$ is determined by a sufficiently smooth trajectory $\mathbf{y}(t)$.

Thus, combining the superimposed feed-forward branch with the servo drives' internal controllers yields the control law for the motor torques $\mathbf{Q}_M$:

$$\mathbf{Q}_M = \underbrace{\mathbf{M}(\mathbf{q}_d)\ddot{\mathbf{q}}_d + \mathbf{g}(\mathbf{q}_d, \dot{\mathbf{q}}_d)}_{\text{feed-forward } \mathbf{Q}_{M,ff}} + \underbrace{\mathbf{K}_p(\mathbf{q}_d - \mathbf{q}) + \mathbf{K}_d(\dot{\mathbf{q}}_d - \dot{\mathbf{q}})}_{\text{feed-back}}. \tag{7}$$

Please note that the feed-forward term in (7) guides the system states along the desired trajectory, while the feedback control law ensures stability against disturbances and modeling errors with the positive diagonal matrices $\mathbf{K}_p$ and $\mathbf{K}_d$. The stability of the overall system, applying (7), is proven in [6]. Alternatively, the feed forward control law may also be computed by using the structural properties of the robot's subsystem representation, as described in [7].

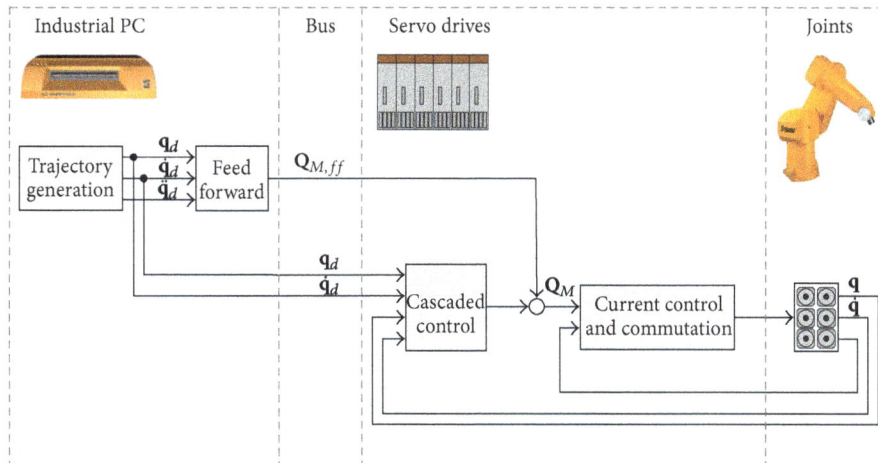

FIGURE 1: Block diagram of the control structure.

2.3.2. Implementation. In order to evaluate the quality of the suggested controls, the Stäubli RX130L robot is interfaced with the motion hardware. During this process, the robot's mechanics, AC motors, and resolvers are not changed or manipulated in any way.

The industrial PC is using $400\,\mu s$ as base sample time. Every sample step, the inverse kinematic problem is solved numerically to transform the desired end-effector movement into corresponding joint positions, velocities, and accelerations. With these reference joint variables ($\mathbf{q}_d$, $\dot{\mathbf{q}}_d$, and $\ddot{\mathbf{q}}_d$), the mathematical model is evaluated to compute the feed-forward motor torques as stated in (7). Then, for each joint, a separate Ethernet telegram is generated containing the reference positions and corresponding feed-forward motor torques.

The servo drives are parameterized to accept cyclic position inputs for the cascaded controller. To do so, the cyclically arriving network telegrams are received and the desired reference positions are passed to the set value inputs of the cascaded controllers. Also, the computed feed-forward torques are transformed to corresponding motor currents and added to the servo drives' current controllers set values. Figure 1 presents the corresponding block diagram.

2.3.3. Experimental Results. The quality of the implemented control strategy is validated by experiments with the Stäubli RX130L robot. Regarding industrial interests, the standardized trajectory found in the EN ISO NORM 9283, see [8], is used as reference trajectory. It consists of various elements: circles, straight lines, and squares in the typical workspace for an industrial application. The TCP velocity is chosen with $1\,\text{m/s}$ and its acceleration with $5\,\text{m/s}^2$.

Figure 2(a) depicts the the lag errors of the first three joints. The other axes show similar results. The positive influence of the feed-forward loop is clearly observable. To present insights on the TCP-accuracy during high dynamics, the TCP errors for the standard and the improved control law along the trajectory are presented in Figure 2(b). They are

measured using a high-precision laser tracker LTD600 from Leica at a sampling rate of $1\,\text{ms}$.

Obviously, the positive effects of the model-based feed-forward approach drastically reduce the TCP and joint errors. Mainly, they compensate static effects, for example, resulting from gravity, and dynamic effects, for example, arising from friction and high accelerations while the standard cascaded controller provides stiffness, stability, and robustness.

3. Static Position Accuracy

Industrial robotic systems usually offer a satisfying repeatability. However, the lack of sufficient accuracy of the end-effector is responsible for time-consuming adjustments while commissioning new tasks. The sources of these errors are nonmodeled manufacturing tolerances, assembly inaccuracies, and joint encoder offsets.

3.1. Kinematics with Error Parameters. The direct kinematics, which compute the end-effector position $\mathbf{r}_E$ and orientation $\boldsymbol{\varphi}_E$ as function of the joint angles $\mathbf{q}$ and nominal geometric parameters $\mathbf{p}_n$, can be extended by a set of error parameters $\mathbf{p}_e$ regarding lengths, joint offsets, and assembly inaccuracies. For example, the connection vector between points 2 and 3 reads

$$\mathbf{r}_{23} = \begin{pmatrix} l_2 \\ 0 \\ 0 \end{pmatrix} + \begin{pmatrix} p_{e2x} \\ p_{e2y} \\ p_{e2z} \end{pmatrix}, \qquad (8)$$

where l_2 is the nominal length of the connection (part of $\mathbf{p}_n$) and p_{e2x}, p_{e2y}, p_{e2z} are the position error values (part of $\mathbf{p}_e$). Thus, the robot's TCP position and orientation is

$$\mathbf{z}_{Ee} = \begin{pmatrix} \mathbf{r}_{Ee} \\ \boldsymbol{\varphi}_{Ee} \end{pmatrix} = \mathbf{f}_{DKe}\left(\mathbf{q}, \mathbf{p}_n, \mathbf{p}_e\right). \qquad (9)$$

For a standard six-axis industrial robot, as sketched in Figure 3, $\mathbf{p}_e$ contains $6 \cdot 3$ error parameters regarding length, 6 joint encoder offsets, $6 \cdot 2$ angles for misaligned coordinate

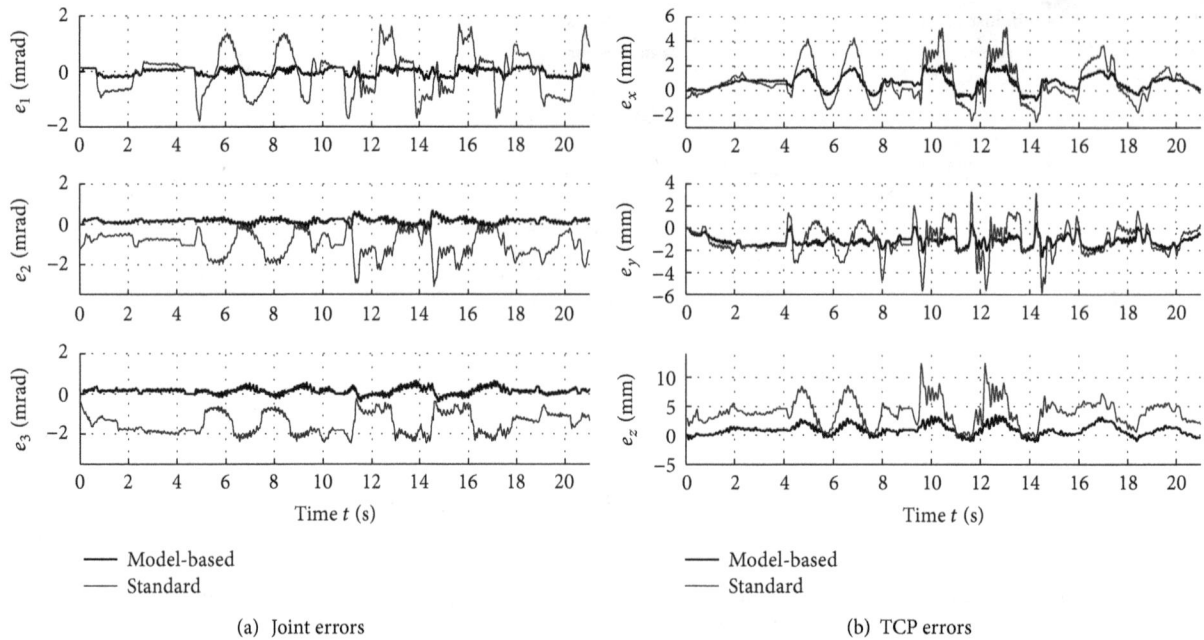

(a) Joint errors   (b) TCP errors

FIGURE 2: Measured errors with (model-based) and without feed-forward loop (standard).

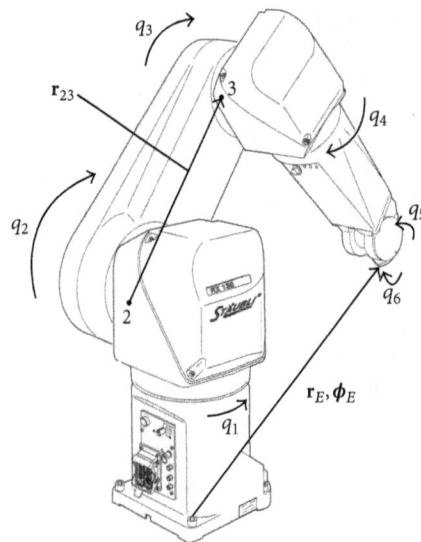

FIGURE 3: RX130L industrial robot.

frames between each link, and 3 tolerances for defining the inertial coordinate system. This results in a set of 39 parameters. To remove all linear dependencies, a QR-decomposition is applied, see Section 4.1 for details. Applying this method yields the final error parameter $\mathbf{p}_e \in \mathbb{R}^{30}$.

3.2. Geometric Calibration. The calibration process calculates valid values for the error parameter $\mathbf{p}_e$ that minimize the error

$\mathbf{e}$ between measured TCP-positions $\mathbf{z}_M$ and TCP-positions determined using the corresponding joint angles $\mathbf{q}$:

$$\mathbf{e} = \mathbf{z}_M - \mathbf{f}_{\mathrm{DK}e}\left(\mathbf{q}, \mathbf{p}_n, \mathbf{p}_e\right) = 0. \tag{10}$$

This highly nonlinear equation can be solved by an iterative method. Representing (10) as a Taylor-series around $\mathbf{p}_e$, neglecting higher order terms and demanding that the error converges towards zero, yields

$$\mathbf{e} = \mathbf{e}\left(\mathbf{z}_M, \mathbf{q}, \mathbf{p}_n, \mathbf{p}_e^{(n)}\right)$$

$$+ \underbrace{\left.\frac{\partial \mathbf{e}\left(\mathbf{z}_M, \mathbf{q}, \mathbf{p}_n, \mathbf{p}_e\right)}{\partial \mathbf{p}_e}\right|_{\mathbf{p}_e^{(n)}}}_{:= \widehat{\Theta}} \Delta\mathbf{p}_e = 0. \quad (11)$$

For m measurements, (11) becomes

$$\underbrace{\begin{pmatrix} \mathbf{e}\left(\mathbf{z}_{M1}, \mathbf{q}_1, \mathbf{p}_n, \mathbf{p}_e^{(n)}\right) \\ \vdots \\ \mathbf{e}\left(\mathbf{z}_{Mm}, \mathbf{q}_m, \mathbf{p}_n, \mathbf{p}_e^{(n)}\right) \end{pmatrix}}_{\mathbf{Q}} + \underbrace{\begin{pmatrix} \widehat{\Theta}\left(\mathbf{z}_{M1}, \mathbf{q}_1, \mathbf{p}_n, \mathbf{p}_e^{(n)}\right) \\ \vdots \\ \widehat{\Theta}\left(\mathbf{z}_{Mm}, \mathbf{q}_m, \mathbf{p}_n, \mathbf{p}_e^{(n)}\right) \end{pmatrix}}_{\Theta} \Delta\mathbf{p}_e$$

$$= 0, \quad (12)$$

which is an overdetermined equation system solvable by standard least-squares techniques. The minimum of the quadratic error is found with

$$\frac{\partial}{\partial \Delta\mathbf{p}_e}\left[\frac{(\mathbf{Q} + \Theta\Delta\mathbf{p}_e)^T(\mathbf{Q} + \Theta\Delta\mathbf{p}_e)}{2}\right]^T \quad (13)$$

$$= \Theta^T\Theta\Delta\mathbf{p}_e + \Theta^T\mathbf{Q} = 0,$$

yielding the update law for the error parameter vector

$$\Delta\mathbf{p}_e = -\left[\Theta^T\Theta\right]^{-1}\Theta^T\mathbf{Q} \text{ with } \mathbf{p}_e^{(n+1)} = \mathbf{p}_e^{(n)} + \Delta\mathbf{p}_e. \quad (14)$$

Initial values for $\mathbf{p}_e^{(0)}$ can be assumed to be zero because the error parameters regarding length, and so forth, are considerably smaller than the nominal geometric parameters. To finally obtain good position accuracy, the identified parameters $\mathbf{p}_e$ must be considered in the inverse kinematics algorithm. Any suitable numerical solution is viable and can handle the set of additional parameters.

3.3. Optimal Configurations for Calibration.
It is obvious that the geometric calibration only identifies parameters well that are excited accordingly. To overcome this problem, the whole work- or joint-space could be discretized for taking the necessary measurements. While this approach is viable for small systems, it is heavily time and cost consuming for systems with multiple degrees of freedom. Such complex systems are common in today's industrial applications. As a consequence, an optimal set of configurations, which guarantees well-excited error parameters, needs to be found.

3.3.1. Observation Index and Algorithm.
In order to ensure a well-conditioned problem formulation and also well-excited parameters $\mathbf{p}_e$, the minimum singular value of the covariance matrix $\Lambda = \Theta^T\Theta$ in (14) is used as observation index and optimization criterion and further described by $B = \sigma_m(\Lambda)$. As, for example, [9–11] show, there exist various choices for

valid observation indices with slightly different properties. The proposed algorithm, which has been presented in [12], allows the use of arbitrary observation indices.

Before introducing the optimization algorithm, two basic operations are defined:

(i) Add a configuration: adding a new configuration with the Jacobian $\widehat{\Theta}_j$ to the current information matrix Θ_i yields

$$\Theta_{i+1} = \begin{bmatrix} \Theta_i \\ \widehat{\Theta}_j \end{bmatrix}, \quad \text{and consequently } \Lambda_{i+1} = \Lambda_i + \widehat{\Theta}_j^T\widehat{\Theta}_j, \quad (15)$$

which directly influences the change of the chosen observation index

$$\Delta B = \sigma_m\left(\Lambda_{i+1}\right) - \sigma_m\left(\Lambda_i\right). \quad (16)$$

(ii) Remove a configuration: similar to the previous case, removing a configuration $\widehat{\Theta}_j$ from the current information matrix Θ_i is described by

$$\Theta_i = \begin{bmatrix} \Theta_{i+1} \\ \widehat{\Theta}_j \end{bmatrix} \text{ with } \Lambda_{i+1} = \Lambda_i - \widehat{\Theta}_j^T\widehat{\Theta}_j,$$

$$\Delta B = \sigma_m\left(\Lambda_{i+1}\right) - \sigma_m\left(\Lambda_i\right). \quad (17)$$

Additionally, the complete set of N measurable configurations is called the configuration-set Π. An arbitrary subset of it is denoted by the set ξ. These operations and definitions are the basics for the following optimization method.

(1) Initialization: a given number of n configurations is randomly selected from the initial configuration-set Π. This subset ξ_0 must lead to a nonsingular covariance-matrix Λ_0 to ensure a computation of the observation index $B(\Lambda_0)$. Should this not be the case for the chosen configurations, then the random selection process is simply repeated.

(2) Remove and replace: the configuration that leads to the minimal decrease of the observation index, $\Delta B(\Lambda_{i+1})$, is removed from the current configuration set ξ_i. Then, the resulting subset ξ_{i+1} is complemented by exactly that configuration of Π that maximizes $\Delta B(\Lambda_{i+2})$. For both operations, the correct element is found by an exhaustive search over the respective set. Note that due to the simple way to calculate ΔB (see ((15)–(17))), the computational effort for these searches is not as dramatic as it may sound. This remove-and-replace strategy effectively sorts out configurations that do not improve the overall conditioning of the covariance matrix and replaces them with ones that do. The step is repeated as long as one remove-and-replace operation still improves the observation index. Otherwise, the next step is executed.

(3) Add additional configurations: from the configuration-set Π exactly the one configuration (again,

FIGURE 4: Some exemplary optimal configurations.

found by exhaustive search) that, by itself, maximizes $\Delta B(\Lambda_{i+i})$, is added to the current subset ξ_i. This step is iterated until the observation index cannot be further increased by such an operation or until a user-specified maximum number of desired configurations is reached. Otherwise, execute the final step.

(4) Final remove and replace: analogously to step 2, each single configuration of the subset ξ_i is evaluated. If any remove and replace operation as described above still increases the observation index, this exchange is executed. Otherwise, the procedure is terminated.

The algorithm may be applied multiple times to prevent getting stuck at a local minimum. The solution with the maximum observation index is used for the experimental analysis. Nevertheless, finding these optimal configurations is time consuming and lasts about five minutes on a standard PC. Since these calculations are performed offline before any measurements are done, this is not a critical issue.

3.4. Experimental Results. To evaluate the performance of the proposed algorithm, detailed experimental results for the RX130L robot are presented. First, the robot's joint-space is discretized by 13 equally large intervals per degree of freedom. This results in a set of $13^6 \approx 4.8 \cdot 10^6$ configurations. In reality the coordinates of the end-effector are measured by the visual infrared tracker system 3D-Creator, manufactured by the Boulder Innovation Group. A probe emitting infrared light by 4 diodes is mounted on the end-effector of the robot. This light is recorded by three CCD cameras and the position and orientation in world coordinates are calculated by stereovision. Robot configurations that are not visible by this system are removed. The remaining set consists of approximately 4000 configurations. Then, the algorithm is used to compute 32 optimal configurations. As Figure 4 emphasizes, the optimal configurations differ strongly from each other to keep the condition number of the covariance matrix low, also leading to well-excited parameters. Compared to a linear scan of the workspace, the observation index is reduced by the factor 18.5 while keeping the number of necessary measurements very small.

For information on the position and orientation accuracy before (uncal., $\mathbf{p}_e = 0$) and after calibration (cal), see Table 2. The orientation errors are represented in Cardan angles.

4. System Identification

In this section, the identification of the dynamical parameters describing the robot behavior, see (4), is covered. This is clearly an essential task because it is the basis for model-based control and optimal trajectory generation.

4.1. Dynamic Parameter Identification. An important property of the equations of motion is the linearity with respect to dynamic parameters $\mathbf{p} \in \mathbb{R}^p$ characterizing the robot links and motors. By introducing the information matrix $\Theta \in \mathbb{R}^{n,p}$ (n degrees of freedom), (4) can be restructured to

$$\Theta \left(\mathbf{q}, \dot{\mathbf{q}}, \ddot{\mathbf{q}} \right) \mathbf{p} = \mathbf{Q}_M. \tag{18}$$

The parameter vector $\mathbf{p}$ is the composition of the the joint parameters $\mathbf{p}_{\mathrm{joint},i}$

$$\mathbf{p}_{\mathrm{joint},i}^T = \left(m \ \mathbf{r}_c^T \ \mathbf{J}^T \ d_c \ d_v \right)_i \tag{19}$$

containing the mass m_i, the center of mass $\mathbf{r}_{c,i}$, the elements of the inertia tensor $\mathbf{J}_i^T = (A \ B \ C \ D \ E)_i$, and Coulomb/viscous friction coefficients $d_{c,i}$ and $d_{v,i}$ for the ith joint-arm unit. Since the dynamical parameters are confidential manufacturer data, all these parameters have to be identified. Usually, the information matrix Θ and the parameter vector $\mathbf{p}$ contain linear dependencies, which have to be removed before the identification process. A *QR* decomposition and reformulation leads to

$$\widehat{\mathbf{Q}}^T \Theta = \widehat{\mathbf{R}}, \tag{20}$$

where $\widehat{\mathbf{Q}} \in \mathbb{R}^{n,p}$ is an orthogonal matrix and $\widehat{\mathbf{R}} \in \mathbb{R}^{p,p}$ is an upper triangular matrix. The nonidentifiable parameters are those whose corresponding elements on the diagonal of the matrix $\widehat{\mathbf{R}}$ are zero. In contrast, if $|\widehat{\mathbf{R}}_{ii}| > 0$, then the corresponding column in Θ is independent. Let the $p - s$ independent columns be collected in the matrix Θ_1 and the corresponding parameters be collected in $\mathbf{p}_1$. The dependent columns and parameters are represented by Θ_2 and $\mathbf{p}_2$, respectively, such that

$$\Theta \mathbf{p} = \begin{bmatrix} \Theta_1 & \Theta_2 \end{bmatrix} \begin{bmatrix} \mathbf{p}_1 \\ \mathbf{p}_2 \end{bmatrix} = \begin{bmatrix} \Theta_1 & \Theta_1 \kappa \end{bmatrix} \begin{bmatrix} \mathbf{p}_1 \\ \mathbf{p}_2 \end{bmatrix}, \tag{21}$$

is fulfilled, since Θ_2 is linearly dependent on Θ_1 by a matrix $\kappa \in \mathbb{R}^{p-s,s}$. In (21) κ can be shifted to the parameters leading to

$$\Theta \mathbf{p} = \Theta_1 \underbrace{(\mathbf{p}_1 + \kappa \mathbf{p}_2)}_{\mathbf{p}_b}, \tag{22}$$

with the base parameters $\mathbf{p}_b \in \mathbb{R}^{p-s}$. For the calculation of κ, (20) is again evaluated with separated matrices $[\Theta_1 \Theta_2]$ and $\widehat{\widehat{\mathbf{Q}}} = [\widehat{\mathbf{Q}}_1 \widehat{\mathbf{Q}}_2]$, respectively, yielding

$$\widehat{\widehat{\mathbf{Q}}}^T \begin{bmatrix} \Theta_1 & \Theta_2 \end{bmatrix} = \begin{bmatrix} \widehat{\widehat{\mathbf{R}}}_1 & \widehat{\widehat{\mathbf{R}}}_2 \end{bmatrix} = \begin{bmatrix} \widehat{\widehat{\mathbf{R}}}_{11} & \widehat{\widehat{\mathbf{R}}}_{12} \\ \widehat{\widehat{\mathbf{R}}}_{21} & \widehat{\widehat{\mathbf{R}}}_{22} \end{bmatrix} \tag{23}$$

and therefore

$$\widehat{\widehat{\mathbf{Q}}}^T \begin{bmatrix} \Theta_1 & \Theta_1 \kappa \end{bmatrix} = \begin{bmatrix} \widehat{\mathbf{R}}_1 & \widehat{\mathbf{R}}_1 \kappa \end{bmatrix} = \begin{bmatrix} \left(\dfrac{\widehat{\widehat{\mathbf{R}}}_{11}}{\widehat{\widehat{\mathbf{R}}}_{21}} \right) & \left(\dfrac{\widehat{\widehat{\mathbf{R}}}_{11}}{\widehat{\widehat{\mathbf{R}}}_{21}} \right) \kappa \end{bmatrix}. \tag{24}$$

Combining (23) and (24), κ can be obtained as

$$\kappa = \widehat{\widehat{\mathbf{R}}}_{11}^{-1} \widehat{\widehat{\mathbf{R}}}_{12}. \tag{25}$$

It contains geometric parameters like lengths, lengths squared, gear ratios, and so on, see [13] for details. The identification process can be performed by sequentially arranging $k \gg p - s$ measurements in (18) and (22) leading to

$$\mathbf{e} = \underbrace{\begin{bmatrix} \Theta_1 \left(\mathbf{q}_1, \dot{\mathbf{q}}_1, \ddot{\mathbf{q}}_1 \right) \\ \vdots \\ \Theta_1 \left(\mathbf{q}_k, \dot{\mathbf{q}}_k, \ddot{\mathbf{q}}_k \right) \end{bmatrix}}_{\Theta_f} \mathbf{p}_b - \underbrace{\begin{pmatrix} \mathbf{Q}_{M,1} \\ \vdots \\ \mathbf{Q}_{M,k} \end{pmatrix}}_{\mathbf{Q}_f} = 0. \tag{26}$$

The unknown base parameters $\mathbf{p}_b$ are calculated with the least-squares method leading to

$$\left\{ \frac{\partial}{\partial \mathbf{p}_b} \left[\frac{\mathbf{e}^T \mathbf{e}}{2} \right] \right\}^T = \Theta_f^T \Theta_f \mathbf{p} - \Theta_f^T \mathbf{Q}_f = 0, \tag{27}$$

and therefore the solution

$$\mathbf{p}_b = \left[\Theta_f^T \Theta_f \right]^{-1} \Theta_f^T \mathbf{Q} \tag{28}$$

is directly available since $\text{Rank}(\Theta_f) = p - s$.

4.2. Optimal Exciting Trajectories. Similar to the problem of finding optimal configurations for the geometric calibration, the dynamic parameter identification requires exciting trajectories. In contrast to the steady-state configurations for the calibration, the identification trajectories include an additional dimension, the continuous time. Again, the aim is to minimize the condition number (cond) of the covariance matrix $\Lambda_f = \Theta_f^T \Theta_f$ by choosing suitable trajectories. A promising approach is introduced with [14]. This section summarizes these ideas, presents solutions to the optimization problem, and verifies the results on the RX130L robot.

TABLE 2: Errors before (uncal., $\mathbf{p}_e = 0$) and after geometric calibration (cal.).

| | Position and orientation errors | | | |
| | Maximum | | Peak-to-peak | |
	uncal.	cal.	uncal.	cal.
e_x in mm	7.779	0.272	±6.171	±0.271
e_y in mm	19.026	0.310	±4.267	±0.303
e_z in mm	12.614	0.414	±1.402	±0.376
e_α in °	1.140	0.128	±0.272	±0.109
e_β in °	1.817	0.191	±0.335	±0.182
e_γ in °	1.160	0.131	±0.514	±0.116

4.2.1. Trajectory Description and Optimization. An adequate description for the identification trajectory for the ith joint is given with

$$q_i(t) = \sum_{l=1}^{N_i} \left\{ \frac{a_{i,l}}{\omega_f l} \sin\left(\omega_f l\, t \right) - \frac{b_{i,l}}{\omega_f l} \cos\left(\omega_f l\, t \right) \right\} + q_{i0}. \tag{29}$$

This Fourier-series is parameterized by the finite order N_i, the coefficients $a_{i,l}$ and $b_{i,l}$, the base angular frequency ω_f, and the joint offset q_{i0}. To guarantee a periodic excitation of the robot, ω_f is a common parameter for all joints. To speed up the optimization time, a discretization of (29) and derivatives for one period of the base frequency ($T_f = 2\pi/\omega_f$) is done. The signals are therefore evaluated at discrete time values $i\Delta T$, $i = 0 \cdots k$, $\Delta T = T_f/k$, where k is the number of discretization points. This provides the following optimization problem:

$$\begin{aligned} \underset{\mathbf{a},\mathbf{b}}{\text{Minimize}} \quad & \text{cond}\left(\Lambda_f \left(\mathbf{a}, \mathbf{b}, \omega_f, k \right) \right) \\ \text{subject to} \quad & \left| \mathbf{q}\left(\mathbf{a}, \mathbf{b}, \omega_f, k \right) \right| \leq \mathbf{q}_{\max} \\ & \left| \dot{\mathbf{q}}\left(\mathbf{a}, \mathbf{b}, \omega_f, k \right) \right| \leq \dot{\mathbf{q}}_{\max} \\ & \left| \ddot{\mathbf{q}}\left(\mathbf{a}, \mathbf{b}, \omega_f, k \right) \right| \leq \ddot{\mathbf{q}}_{\max} \\ & \left| \mathbf{Q}_M \left(\mathbf{a}, \mathbf{b}, \omega_f, k \right) \right| \leq \mathbf{Q}_{M,\max} \\ & \mathbf{r}_E \left(\mathbf{a}, \mathbf{b}, \omega_f, k \right) \subset \mathbb{R}^3_{\text{allowed}}. \end{aligned} \tag{30}$$

The constraints include the mechanical joint limits $\mathbf{q}_{\max}$, the permitted joint velocities $\dot{\mathbf{q}}_{\max}$, the expected maximum joint accelerations $\ddot{\mathbf{q}}_{\max}$, and a space of allowed end-effector positions $\mathbb{R}^3_{\text{allowed}}$, which is used to avoid collisions with the robot itself and its surroundings. Due to the flatness property of the system, see Section 2.3.1, also the motor torques can be calculated dependent on the joint coordinates and derivatives. In contrast to [15], this motor torques are also used in our formulation. The optimization problem is solved with the Matlab-Optimization-Toolbox ($f$mincon). The parameters order $N = 5$, angular frequency $\omega_f = 0.8\,\text{rad/s}$, and $k = 80$ turned out to work well to get adequate dynamical parameters, see Figure 5(b). For numerical scaling purposes, the columns of Θ_f should also be normalized. Solving the optimization problem yields the optimal joint trajectories.

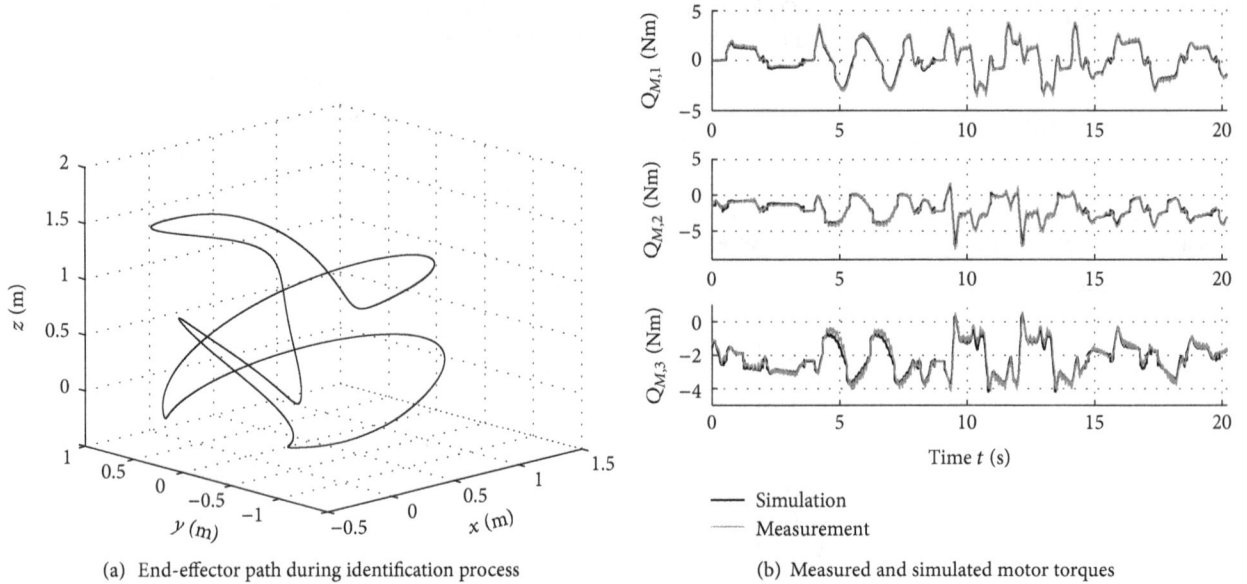

(a) End-effector path during identification process

(b) Measured and simulated motor torques

FIGURE 5: Identification trajectory and verification.

4.3. Experimental Results. The measurements for the overdetermined equations system (28) are taken to estimate the dynamical base parameters $\mathbf{p}_b$. Figure 5(a) presents a plot of the end-effector path. Along the path, velocities up to 5 m/s and accelerations up to 12 m/s^2 are achieved. To verify the quality of the identified parameters, the real robot's motor torques are recorded and compared to the simulated ones on a reference trajectory which is defined in the EN ISO NORM 9283, see [8]. Figure 5(b) emphasizes, by presenting the measured and simulated motor torques, the good parameter identification for the RX130L robot. Joints four, five, and six show an equivalent quality.

5. Optimal Trajectory Generation

There are two types of trajectory generation problems that have to be considered. The first one is the continuous path (CP) trajectory and the second one is the point-to-point (PtP) trajectory planning problem. For the first one, the geometric path in world coordinates is defined. Typical maneuvers are straight lines, circles, and splines. For the PtP planning, only the coordinates for the start-point $\mathbf{z}_{E,0}$ and the endpoint $\mathbf{z}_{E,\text{end}}$ of the trajectory are given. These points can directly be transformed to joint coordinates $\mathbf{q}_0, \mathbf{q}_{\text{end}}$, and the motion planning can be performed in this coordinates. For both cases, physical constraints like motor speeds and torques have to be taken into account. Additionally, the computed trajectories have to result in continuous velocities, accelerations, and jerks. This helps that the gear elasticities are not overly excited, leading to a soft, vibration-less endeffector movement. Again, the equations of motion, derived in Section 2.2 and identified in Section 4, provide the basis for this approach.

5.1. Continuous Path Trajectory. CP applications appear in many robotic applications like arc welding, deburring, and laser cutting. Due to the nonlinear kinematic interrelation between Cartesian space and the robot's joint space, high motor speeds and accelerations may occur, especially near kinematic singularities. The CP motion planning can be divided into two subproblems, namely, a geometric path planner, describing a path $\mathbf{z}_E(\sigma)$

$$\mathbf{z}_E = \begin{pmatrix} \mathbf{r}_E(\sigma) \\ \boldsymbol{\varphi}_E(\sigma) \end{pmatrix} = \mathbf{z}_E(\sigma) \in \mathbb{R}^6 \qquad (31)$$

as a function of a (scalar) trajectory parameter $\sigma \in [\sigma_0, \sigma_{\text{end}}]$ and a dynamic path planner that solves the problem of finding an optimized time behavior $\sigma(t)$, see also [16]. For example, a straight line in space is characterized by (39). The vector $\mathbf{r}_E$ contains the end-effector position while $\boldsymbol{\varphi}_E$ describes the evolution of the orientation in Cardan angles, both are functions of σ. The trajectory parameter σ itself is a function of time $\sigma = \sigma(t)$ and the corresponding trajectory velocity and acceleration are

$$\dot{\sigma} = \frac{d\sigma}{dt}, \qquad \ddot{\sigma} = \frac{d\dot{\sigma}}{dt}. \qquad (32)$$

The trajectory parameter and its derivatives (trajectory speed and acceleration) represent the trajectory and its evolution in a demonstrative and easily interpretable manner. Figure 6 shows an example for a straight line, beginning at σ_0 and ending at σ_{end}. With six degrees of freedom and six world coordinates, the analytical solution of the inverse kinematic problem, see for example, [3], for the RX130 robot is obtained

$$\mathbf{q} = \mathbf{f}(\mathbf{z}_E(\sigma)), \qquad (33)$$

which can be differentiated with respect to time, leading to

$$\dot{\mathbf{q}} = \frac{d}{dt}\mathbf{f}\left(\mathbf{z}_E(\sigma)\right) = \dot{\mathbf{f}}(\sigma, \dot{\sigma})$$

$$\ddot{\mathbf{q}} = \frac{d^2}{dt^2}\mathbf{f}\left(\mathbf{z}_E(\sigma)\right) = \ddot{\mathbf{f}}(\sigma, \dot{\sigma}, \ddot{\sigma}).$$

(34)

Inserting the kinematic relations (33) and (34) into the equations of motion (4) yields the following description of the robot's dynamics:

$$\mathbf{Q}_M = \mathbf{M}(\sigma)\,\ddot{\mathbf{q}}(\sigma, \dot{\sigma}, \ddot{\sigma}) + \mathbf{g}(\sigma, \dot{\sigma}).$$

(35)

This representation allows to compute the motor torques as a function of the trajectory position, velocity, and acceleration. For additional details, see [17]. Please note that the inverse kinematics does not necessarily have to be computed analytically. Also numerical calculations of the inverse kinematics can be used.

5.2. Optimization Strategy. To find an optimal evolution along a desired trajectory, the objective reads

$$J = \min \int_0^{t_e} \left(k_1 + k_2\,\mathbf{Q}_M^T\mathbf{Q}_M\right) dt,$$

(36)

where the motor torques $\mathbf{Q}_M$ are calculated in (35). The design parameters k_1 and k_2 are manipulated to weight between time and/or energy optimality. However, the upper limit of the integral in (36) is unknown in advance, merely a result of optimization.

Thus, a transformation with $\dot{\sigma} = d\sigma/dt \Rightarrow dt = (1/\dot{\sigma})d\sigma$ leads to an objective function with constant integration limits and σ as integration variable. Equation (38) summarizes the result. The objective function is also subject to a set of differential equations, describing the evolution of time along the trajectory parameter σ. The derivatives with respect to σ are used to compute all trajectory variables, which are found in the constraints. For instance, the trajectory acceleration is computed with

$$\ddot{\sigma} = \frac{d\dot{\sigma}}{dt} = \frac{d\dot{\sigma}}{d\sigma}\dot{\sigma} = \dot{\sigma}'\dot{\sigma},$$

(37)

where a prime indicates the differential operator $(\cdot)' = d(\cdot)/d\sigma$. The input u in the set of differential equations needs to be computed by a numerical solver in order to minimize the objective function while satisfying all constraints.

Finalizing the following complete description:

$$\underset{\sigma}{\text{Minimize}} \quad \int_{\sigma_0}^{\sigma_{\text{end}}} \frac{1}{\dot{\sigma}}\left(k_1 + k_2\,\mathbf{Q}_M^T\mathbf{Q}_M\right) d\sigma$$

$$\text{subject to} \quad \frac{d}{d\sigma}\begin{pmatrix} t \\ \dot{\sigma} \\ \dot{\sigma}' \end{pmatrix} = \begin{pmatrix} \frac{1}{\dot{\sigma}} \\ \dot{\sigma}' \\ u \end{pmatrix}$$

(38)

$$|\mathbf{Q}_M| \le \mathbf{Q}_{M,\text{max}}$$

$$|\dot{\mathbf{q}}| \le \dot{\mathbf{q}}_{\text{max}}$$

$$|\ddot{\mathbf{q}}| \le \ddot{\mathbf{q}}_{\text{max}}$$

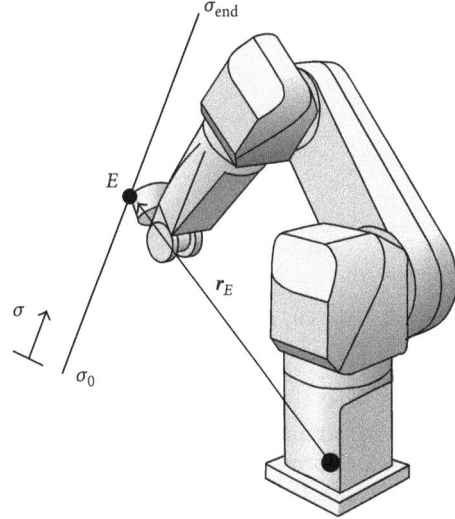

FIGURE 6: A sample trajectory.

the motor torques, motor velocities, and accelerations are constrained. To compute an optimal solution u for the optimization problem stated in (38), the MUSCOD-II software package is utilized. Based on [18], it is developed by the Interdisciplinary Center for Scientific Computing at the University of Heidelberg. Using the direct multiple-shooting method, see [19, 20], the original continuous optimal control problem is reformulated as a nonlinear programming (NLP) problem which is then solved by an iterative solution procedure, a specially tailored sequential quadratic programming (SQP) algorithm. For further details on the used software package, see [21].

5.3. Experimental Results—CP Trajectory. For a straight tool center movement with constant end-effector orientation, the trajectory is given by

$$\mathbf{z}_E = \begin{pmatrix} x_0 + \left(x_{\text{end}} - x_0\right)\sigma \\ y_0 + \left(y_{\text{end}} - y_0\right)\sigma \\ z_0 + \left(z_{\text{end}} - z_0\right)\sigma \\ 0 \end{pmatrix}, \quad \sigma \in [0, 1],$$

(39)

where index 0 indicates the start-point and index end the end-point, respectively. For a grid of 200 shooting intervals, the time optimal case $k_1 = 1$, $k_2 = 0$ is computed. Minimizing energy would lead to not only smoother, but also longer trajectories for σ and is therefore not considered in this example. The resulting trajectory parameters $t(\sigma), \dot{\sigma}$, and $\dot{\sigma}'$ are retransformed into the time domain to compute the corresponding reference joint angles $\mathbf{q}_d(t)$ and the constrained feed-forward torques $\mathbf{Q}_{M,ff}(t)$. The corresponding velocity profile is depicted in Figure 7. The velocity-drop occurs due to passing closely to a singular configuration.

The experiment with the Stäubli robot is conducted analogously to the previous section. The reference angles and torques are transferred to the servo drives where the occurring torques and motor velocities are measured for verification. Figure 8 shows the torques and velocities of the

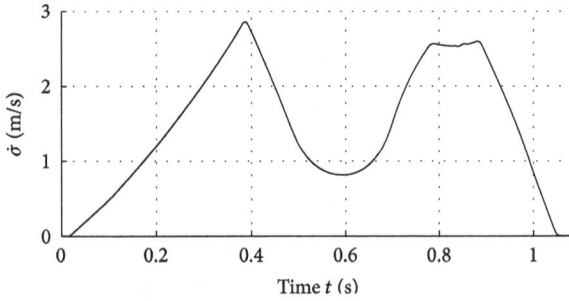

FIGURE 7: Velocity profile CP trajectory.

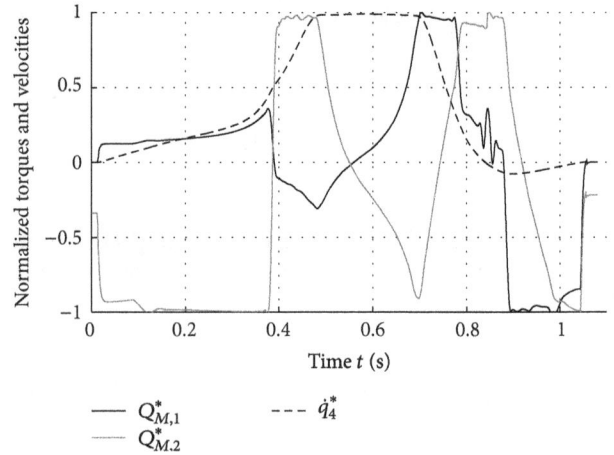

$$
\begin{array}{ll}
\text{———} \quad Q^*_{M,1} & \text{- - -} \quad \dot{q}^*_4 \\
\text{———} \quad Q^*_{M,2} &
\end{array}
$$

FIGURE 8: Active constraints recorded during experiment.

axes that actively constrain the trajectory. For readability reasons, the signals are normalized to their maximum values. The figure also shows that at least one constraint is always active, which is a necessary condition for the time-optimal solution.

5.4. Point-to-Point Trajectory. As an extension to the previous section, the PtP optimization is not limited to geometrically defined paths. Consequently, the one-dimensional variation of the trajectory parameter becomes a more complex two-point boundary value problem of the joint states with the additional task of minimizing time and/or energy, summarized in (40). Analogously to (38), the available joint velocities and torques are limited. Additionally, the joint limits $\mathbf{q}_{max}$ need to be considered and the desired initial and final configurations $\mathbf{q}_0$, $\dot{\mathbf{q}}_0$, $\mathbf{q}_e$, and $\dot{\mathbf{q}}_e$ are introduced.

A compromise between time and energy optimality is adjusted with the parameters k_1 and k_2. Also note that the highly nonlinear differential equations of motion are included in the set of constraints.

5.5. Experimental Results—PtP Trajectory. The optimization formulation for the point-to-point trajectory

$$
\begin{aligned}
&\underset{\mathbf{q},t_e}{\text{Minimize}} \quad \int_0^{t_e} \left(k_1 + k_2 \mathbf{Q}_M^T \mathbf{Q}_M \right) dt \\
&\text{subject to} \quad |\mathbf{Q}_M| \le \mathbf{Q}_{M,\max} \\
&\qquad\qquad\quad |\dot{\mathbf{Q}}_M| \le \dot{\mathbf{Q}}_{M,\max} \\
&\qquad\qquad\quad |\mathbf{q}| \le \mathbf{q}_{\max} \\
&\qquad\qquad\quad |\dot{\mathbf{q}}| \le \dot{\mathbf{q}}_{\max} \\
&\qquad\qquad\quad \frac{d}{dt}\begin{pmatrix} \mathbf{q} \\ \dot{\mathbf{q}} \end{pmatrix} = \begin{pmatrix} \dot{\mathbf{q}} \\ \mathbf{M}^{-1}(\mathbf{q})\left(-\mathbf{g}(\mathbf{q},\dot{\mathbf{q}}) + \mathbf{Q}_M\right) \end{pmatrix} \\
&\qquad\qquad\quad \mathbf{q}(0) = \mathbf{q}_0 \\
&\qquad\qquad\quad \dot{\mathbf{q}}(0) = \dot{\mathbf{q}}_0 \\
&\qquad\qquad\quad \mathbf{q}(t_e) = \mathbf{q}_e \\
&\qquad\qquad\quad \dot{\mathbf{q}}(t_e) = \dot{\mathbf{q}}_e
\end{aligned}
\tag{40}
$$

is solved with MUSCOD-II setting $k_1 = 1$ and $k_2 = 0$ (time optimality). While the one-dimensional variation problem of the previous section is computed in one second, the PtP trajectory takes about three minutes to be calculated with a standard workstation. After solving the optimization problem, the system states and torque inputs for the two-point boundary problem are known and may be utilized as reference values. The initial conditions $\mathbf{q}_0$, $\dot{\mathbf{q}}_0$, $\mathbf{q}_e$, and $\dot{\mathbf{q}}_e$ are chosen equals to the start- and endpoint of the previous section's straight-line trajectory. This allows a comparison between both methods. The overall time of the trajectory was reduced from $t_e = 1.04\,\text{s}$ to $t_e = 0.67\,\text{s}$. Figures 9(a) and 9(b) show the optimized torques and velocities. Again, the most interesting signals are selected for reasons of readability, all normalized to their maximum values. Experiments using these results on the RX130L robot verify validity of the optimized trajectory. The maximum end-effector velocity and acceleration that are reached along this trajectory are 5.2 m/s and 48.6 m/s², respectively. To avoid high jerks in joint space, which would induce vibrations, the torque rates are limited to $\dot{\mathbf{Q}}_{M,\max}$.

6. Conclusions

This work summarizes various optimization strategies for robotic systems to improve the overall performance. The individual results of each section not only verify the chosen approach but also offer values for expected accuracies of modern medium scale robotic systems. The suggested ideas are also applicable to smaller or larger scale robots with an arbitrary amount of degrees of freedom.

It is important to note that the quality of the control as well as the trajectory optimization is only as good as the underlying set of (kinematic or dynamical) parameters. As a consequence, the optimal configurations for calibration and the optimal-exciting trajectories for the identification process provide the basis for control and trajectory generations. All proposed methods can be automated by designing according to algorithms and used during the commissioning of robot applications.

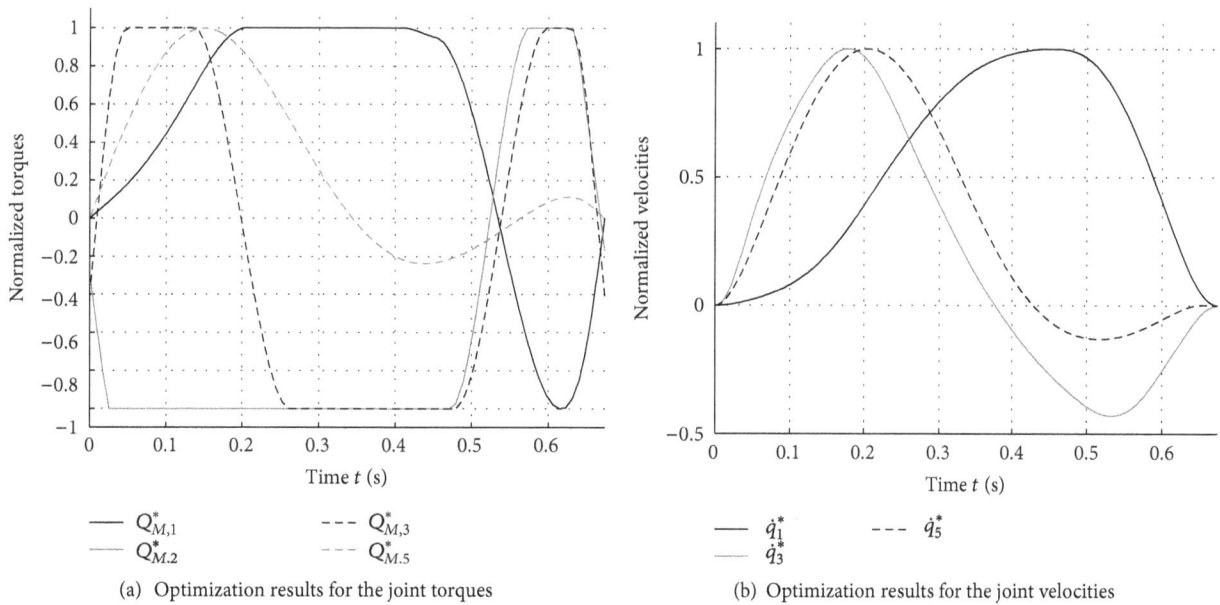

(a) Optimization results for the joint torques

(b) Optimization results for the joint velocities

FIGURE 9: Results PtP trajectories.

Acknowledgments

Support of the current work within the framework of the Austrian Center of Competence in Mechatronics (ACCM) is gratefully acknowledged. The support of the Interdisciplinary Center for Scientific Computing at the University of Heidelberg is acknowledged, especially to Moritz Diehl and Andreas Poschka who kindly gave support. One of the basic ideas of this contribution is to review and present details using industrial, not laboratory, equipment. However, the authors do not have a direct relation to any mentioned hardware and software products. This information is consciously included, enabling the reader to review technical specifications for further comparisons.

References

[1] R. Riepl and H. Gattringer, "Accurate high-speed robotics [Dissertation]," Johannes Kepler University, B&R Industrial Ethernet Award, 2009.

[2] W. Khalil and E. Dombre, *Modeling, Identification and Control of Robots*, Kogan Page Science, London, UK, 2004.

[3] B. Siciliano, L. Sciavicco, L. Villani, and G. Oriolo, *Robotics: Modelling, Planning and Control*, Advanced Textbooks in Control and Signal Processing Series, Springer, London, UK, 2009, Edited by A. J. J. Grimble.

[4] H. Bremer, *Elastic Multibody Dynamics: A Direct Ritz Approach*, Springer, New York, NY, USA, 2008.

[5] A. Isidori, *Nonlinear Control Systems*, Springer, New York, NY, USA, 1985.

[6] A. Kugi, "Introduction to tracking control of finite- and infinite-dimensional systems," Stability, Identification and Control in Nonlinear Structural Dynamics (SICON), 2008.

[7] H. Gattringer and H. Bremer, "A penalty shooting walking machine," in *IUTAM Symposium on Vibration Control of Nonlinear Mechanisms and Structures*, H. Ulbrich and W. G. Günthner, Eds., Springer, 2005.

[8] ISO NORM 9283, "Manipulating industrial robots—performance and criteria," Norm, EN ISO, 9283, 1998.

[9] C. H. Menq, J. H. Borm, and J. Z. Lai, "Identification and observability measure of a basis set of error parameters in robot calibration," *Journal of Mechanisms, Transmissions, and Automation in Design*, vol. 111, no. 4, pp. 513–518, 1989.

[10] M. R. Driels and U. S. Pathre, "Significance of observation strategy on the design of robot calibration experiments," *Journal of Robotic Systems*, vol. 7, no. 2, pp. 197–223, 1990.

[11] Y. Sun and J. M. Hollerbach, "Observability index selection for robot calibration," in *Proceedings of the IEEE International Conference on Robotics and Automation (ICRA '08)*, pp. 831–836, May 2008.

[12] Y. Sun and J. M. Hollerbach, "Active robot calibration algorithm," in *Proceedings of the IEEE International Conference on Robotics and Automation (ICRA '08)*, pp. 1276–1281, May 2008.

[13] H. Gattringer, *Starr-Elastische Robotersysteme: Theorie und Anwendungen*, Springer, Berlin, Germany, 2011.

[14] J. Swevers, C. Ganseman, J. De Schutter, and H. Van Brussel, "Experimental robot identification using optimised periodic trajectories," *Mechanical Systems and Signal Processing*, vol. 10, no. 5, pp. 561–577, 1996.

[15] D. Kostić, B. de Jager, M. Steinbuch, and R. Hensen, "Modeling and identification for high-performance robot control: an RRR-robotic arm case study," *IEEE Transactions on Control Systems Technology*, vol. 12, no. 6, pp. 904–919, 2004.

[16] R. Johanni, *Optimale Bahnplanung bei Industrierobotern*, Technische Universität München, München, Germany, 1988.

[17] R. Riepl and H. Gattringer, "An approach to optimal motion planning for robotic applications," in *Proceedings of the 9th International Conference on Motion and Vibration Control*, 2008.

[18] D. Leineweber, I. Bauer, A. Schäfer, H. Bock, and J. Schlöder, "An efficient multiple shooting based reduced SQP strategy

for large-scale dynamic process optimization (Parts I and II)," *Computers and Chemical Engineering*, vol. 27, pp. 157–174, 2003.

[19] J. Nocedal and S. J. Wright, *Numerical Optimization*, Springer Science+Business Media, LLC, New York, NY, USA, 2nd edition, 2006.

[20] J. T. Betts, *Practical Methods for Optimal Control Using Nonlinear Programming*, Society for Industrial and Applied Mathematics, Philadelphia, Pa, USA, 2001.

[21] M. Diehl, D. B. Leineweber, and A. Schäfer, *MUSCOD-II Users' Manual*, Universität Heidelberg, Heidelberg, Germany, 2001.

12

An Inventory Model with Finite Replenishment Rate, Trade Credit Policy and Price-Discount Offer

Biswajit Sarkar,[1] **Shib Sankar Sana,**[2] **and Kripasindhu Chaudhuri**[3]

[1] *Department of Applied Mathematics with Oceanology and Computer Programming, Vidyasagar University, Midnapore 721-102, India*
[2] *Department of Mathematics, Bhangar Mahavidyalaya, University of Calcutta, Kolkata 743-502, India*
[3] *Department of Mathematics, Jadavpur University, Kolkata 700-032, India*

Correspondence should be addressed to Biswajit Sarkar; bsbiswajitsarkar@gmail.com

Academic Editor: Paul C. Xirouchakis

When some suppliers offer trade credit periods and price discounts to retailers in order to increase the demand of their products, retailers have to face different types of discount offers and credits within which they have to take a decision which is the best offer for them to make more profit. The retailers try to buy perfect-quality items at a reasonable price, and also they try to invest returns obtained by selling those items in such a manner that their business is not hampered. In this point of view, we consider an *economic order quantity* (EOQ) model for various types of time-dependent demand when delay in payment and price discount are permitted by suppliers to retailers. The models of various demand patterns are discussed analytically. Some numerical examples and graphical representations are considered to illustrate the model.

1. Introduction

Many classical inventory models assume that demand is constant. In present marketing environment, few items follow constant demand. Many product's demands follow variable time-varying demand. The recent trend of the marketing system is to provide more buy opportunities to the retailer by the supplier by offering different discounts. To take the discount opportunity, retailers prefer to buy more beyond their capacity of buying. As a result, the supplier has the opportunity to sell more for better earning. This is the benefit of the supplier. The classical inventory model does not consider the delay time concept or variable demand. The proposed model considers time-varying demand and delay in payments along with finite replenishment rate.

The basic well-known *square root formula* for the EOQ of the item was formulated by Harris [1] based on constant demand. Donaldson [2] extended the constant demand to linear time-dependent demand model analytically with finite time horizon. Following Donaldson [2], significant contribution in this direction came out from researchers like Goyal [3], Goswami and Chaudhuri [4], Goyal et al. [5], and others.

Hariga and Benkherouf [6] discussed an optimal and heuristic replenishment model for deteriorating items with an exponentially time-varying demand. Wee [7] studied a deterministic lot size inventory model for deteriorating items with shortage and decline market. Khanra and Chaudhuri [8] extended an inventory model with quadratic increasing demand over a finite time horizon and shortages. Sana and Chaudhuri [9] studied an inventory model with linear trend demand incorporating shortages. Cárdenas-Barrón [10] discussed the derivation of inventory models by using analytic geometry and algebra. Sarkar et al. [11] explained an inventory model with quadratic time-varying demand by considering Euler-Lagrange method.

It is common to all that every customer prefers to buy more at reduced price. Some researchers like Abad [12], and Kim and Hwang [13] developed the traditional quantity discount model. In the traditional EOQ model, it was assumed that the retailer pays the purchasing cost when he received the items from a supplier. In trade-credit policy, the supplier allows a certain fixed period to pay the purchasing cost. This fixed period which is settled by the supplier is called the *credit period* to the retailer. During this credit period,

TABLE 1: Comparisons of this model with previous works.

Author/authors	Linear/quadratic demand	Exponential demand	Constant demand	Tread credit policy
Harris [1]			√	
Donaldson [2]	√			
Goyal [14]			√	√
Goyal [3]	√			
Goswami and Chaudhuri [4]	√			
Hariga and Benkherouf [6]		√		√
Teng [18]			√	√
Khanra and Chaudhuri [8]	√			
Arcelus et al. [19]			√	√
Sana and Chaudhuri [9]	√			
Huang [20]			√	√
Huang [21]			√	√
Cárdenas-Barrón [10]			√	√
Teng et al. [23]			√	√
Sarkar [24]	√			√
This paper	√	√	√	√

the supplier sells items to the retailer with different types of discounts to obtain more profit as early as possible during the credit period. Depending on this policy, Goyal [14] first developed an inventory model with permissible delay in payments. Aggarwal and Jaggi [15] developed an inventory model with an exponentially deteriorating rate by considering permissible delay in payments. Chu et al. [16] extended Goyal's [14] model by considering the case of deterioration. Jamal et al. [17] extended an inventory model with shortages.

Teng [18] developed an EOQ model for a retailer to order small lot size in order to take the benefit of permissible delay in payments. Arcelus et al. [19] developed an inventory model by considering the retailer's maximizing profit and inventory policies for vendor's trade promotion offer of price/credit on the purchase of perishable items. Huang [20] extended an inventory model of retailer's inventory system as a cost minimization model to determine the retailer's optimum inventory cycle time and optimal order quantity. Huang [21] developed an *economic production quantity* (EPQ) model of retailer's inventory system to investigate the optimal retailer's decisions under two levels of trade credit policy. Cárdenas-Barrón [22] extended optimal ordering policies in response to a discount offer. Teng et al. [23] explained optimal ordering decisions with returns and excess inventory. Sarkar [24] discussed an inventory model with delay in payments in the presence of imperfect production. Sarkar [25] developed an inventory model with delay in payments and time-varying deterioration rate. Forghani et al. [26] explained an inventory model in the single period inventory system with price adjustment.

This paper considers an inventory model for credit periods and price-discount offers with different types of time varying demand and constant supply rate K up to time $t = t_1$. During $[0, t_1]$, inventory piles up by adjusting the demand. The accumulated inventory level at time t_1 depletes gradually to meet the demand, and the level reaches zero level at time

T ($t_1 \leq T$). The agreement between the supplier and the retailer is such that total purchasing cost of whole amount (Kt_1) would be paid within the time R ($R > t_1$) with purchasing cost at discount rate ρ. The different delay periods with different discount rates on the purchasing cost are permitted by the supplier to the retailer. During the credit period, the retailer can earn interest by selling items whereas interest of purchasing cost is charged against the delay of excess time of credit of payment period by the retailer to the supplier.

The rest of the paper is designed as follows. The mathematical model is presented in Section 2. In Section 3, numerical examples are given. Finally, concluding remarks are explained in Section 4. See Table 1 for the comparison of this model with previous works.

2. Mathematical Model

We consider the following notation to develop the model.

T^*: the optimal length of inventory cycle (decision variable)

t_1^*: the optimal duration of replenishment (decision variable)

$Q_1(t)$: on-hand inventory at time t ($0 \leq t \leq t_1$)

$Q_2(t)$: on-hand inventory at time t ($t_1 \leq t \leq T$)

$D(t)$: time-varying demand rate

K: constant replenishment/supply rate

R: variable delay period

R_i: ith permissible delay period

ρ_i: discount rate on purchasing cost at ith permissible delay period

C_1: ordering cost per order

C_2: unit holding cost per unit time, excluding interest charge

C_3: purchasing cost per unit

C_4: maximum retail price per unit

P: selling price per unit

I_c: rate of interest gaining due to the credit balance

I_f: rate of interest due to financing inventory

T: length of the inventory cycle

t_1: duration of the replenishment

Z_{1i}: average profit of the system when $T \geq R_i$

Z_{2i}: average profit of the system when $T \leq R_i$.

The following assumptions are considered to develop this model.

(1) The inventory system involves only single type of product.

(2) The demand rate is constant or time dependent (quadratic, linear, and exponential).

(3) Different discount rates on the purchasing cost for different delay periods are considered.

(4) Replenishment rate is instantaneously infinite, but its size is finite.

(5) Time horizon is infinite.

(6) Lead time is negligible.

(7) Neither shortage nor backlogging is considered.

The cycle starts with zero inventory at supply rate K. The replenishment or supply continues up to time t_1. During the time span $[0, t_1]$, inventory piles up by adjusting the demand in the market. This accumulated inventory level at time t_1 depletes gradually to meet the demand and it reaches zero level at time $t = T$ $(T > t_1)$. Generally, the supplier offers delay period R $(R > t_1)$ to the retailer to pay the total purchasing cost $(C_3 K t_1)$ of items. For different delay periods R_i $(i = 1, 2, 3)$, different discounts ρ_i $(i = 1, 2, 3)$ of purchasing cost are offered to the retailer by the supplier. In this direction, we consider the purchasing cost of different delay periods as follows:

$$C_3 = \begin{cases} C_4 (1 - \rho_1) & \text{when } R = R_1, \\ C_4 (1 - \rho_2) & \text{when } R = R_2, \\ C_4 (1 - \rho_3) & \text{when } R = R_3, \\ \infty & \text{when } R > R_3, \end{cases} \quad (1)$$

where R_i's are the ith permissible delay to settle the purchasing cost at which the discount rate to the retailer is ρ_i. Also C_3 tends to be ∞ at $R > R_3$. That is, at infinite purchasing cost, the retailer never purchases any item from the supplier. Indirectly, the supplier would not supply the product to the retailer while delay period R exceeds R_3. In our model two cases may arise.

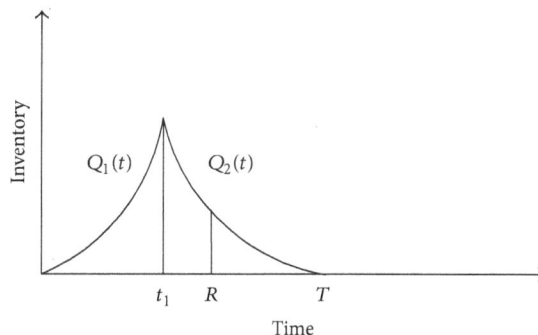

FIGURE 1: Inventory versus time (Case 1 $T \geq R$).

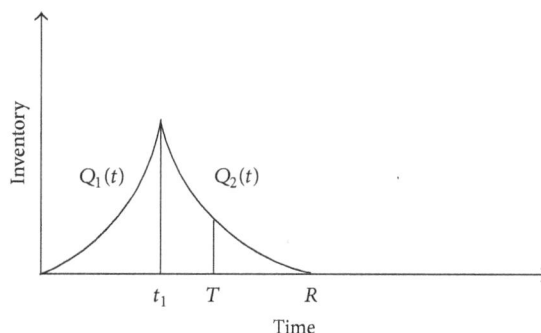

FIGURE 2: Inventory versus time (Case 2 $T \geq R$).

Case 1 $(T \geq R)$. That is, inventory cycle length T is larger or equal to the credit period R (see Figure 1). When $T \geq R$, there are some profits based on credit balance during the delay period and there is some interest charged due to financing inventory during $[R, T]$.

Case 2 $(T \leq R)$. That is, inventory cycle length T is smaller or equal to the credit period R (see Figure 2). When $T \leq R$, there are some profits based on credit balance during the delay period and there is no interest charged due to financing inventory.

The governing differential equations of this model are

$$\frac{dQ_1}{dt} = K - D(t) \quad \text{with } Q_1(0) = 0, \ 0 \leq t \leq t_1,$$
$$\frac{dQ_2}{dt} = -D(t) \quad \text{with } Q_2(T) = 0, \ t_1 \leq t \leq T. \quad (2)$$

From (2), we obtain

$$Q_1(t) = Kt - \int_0^t D(t) \, dt, \quad 0 \leq t \leq t_1, \quad (3)$$

$$Q_2(t) = \int_t^T D(t) \, dt, \quad t_1 \leq t \leq T. \quad (4)$$

By utilizing the continuity at t_1, $Q_1(t_1) = Q_2(t_1)$, we have

$$Kt_1 - \int_0^{t_1} D(t) \, dt = \int_{t_1}^T D(t) \, dt, \quad (5)$$

that is,

$$t_1 = \frac{1}{K} \int_0^T D(t)\,dt. \tag{6}$$

Now we formulate Cases 1 and 2 using (3) to (6).

Case 1 ($T \geq R$). When the inventory cycle (T) is larger or equal to the credit period R, the holding cost excluding the interest charges is $C_2 \{ \int_0^{t_1} Q_1(t)dt + \int_{t_1}^T Q_2(t)dt \}$.

The profit gains due to credit balance during the delay period $[0, R]$ are $I_c P \int_0^R (R-t)D(t)dt$.

The interest charged for financing inventory during $[R, T]$ is $I_f C_3 \int_R^T Q_2(t)dt$.

Therefore, the total profit is (see Figure 1)

$$P_{1i} = \Bigg[(P - C_3) K t_1$$

$$+ I_c P \left\{ \int_0^{t_1} (R_i - t) D(t)\,dt + \int_{t_1}^{R_i} (R_i - t) D(t)\,dt \right\}$$

$$- C_2 \left\{ \int_0^{t_1} Q_1(t)\,dt + \int_{t_1}^T Q_2(t)\,dt \right\}$$

$$- I_f C_3 \int_{R_i}^T Q_2(t)\,dt - C_1 \Bigg] \quad \text{for } i \in \{1, 2, 3\}. \tag{7}$$

Hence, the average profit is

$$Z_{1i} = \frac{P_{1i}}{T}$$

$$= \frac{1}{T} \Bigg[(P - C_3) K t_1$$

$$+ I_c P \left\{ \int_0^{t_1} (R_i - t) D(t)\,dt + \int_{t_1}^{R_i} (R_i - t) D(t)\,dt \right\}$$

$$- C_2 \left\{ \int_0^{t_1} Q_1(t)\,dt + \int_{t_1}^T Q_2(t)\,dt \right\}$$

$$- I_f C_3 \int_{R_i}^T Q_2(t)\,dt - C_1 \Bigg] \quad \text{for } i \in \{1, 2, 3\}. \tag{8}$$

Case 2 ($T \leq R$). When the inventory cycle (T) is smaller or equal to the credit period R, the holding cost excluding the interest charges is $C_2 \{ \int_0^{t_1} Q_1(t)dt + \int_{t_1}^T Q_2(t)dt \}$.

The profit gains due to credit balance during the time $[0, R]$ are $I_c P [\int_0^T (T-t)D(t)dt + Kt_1(R-T)]$.

Therefore, the total profit is (see Figure 2)

$$P_{2i} = \Bigg[(P - C_3) K t_1$$

$$+ I_c P \left\{ \int_0^{t_1} (T - t) D(t)\,dt \right.$$

$$\left. + \int_{t_1}^T (T - t) D(t)\,dt + K t_1 (R_i - T) \right\}$$

$$- C_2 \left\{ \int_0^{t_1} Q_1(t)\,dt + \int_{t_1}^T Q_2(t)\,dt \right\} - C_1 \Bigg]$$

$$\text{for } i \in \{1, 2, 3\}. \tag{9}$$

Hence, the average profit is

$$Z_{2i} = \frac{P_{2i}}{T}$$

$$= \frac{1}{T} \Bigg[(P - C_3) K t_1$$

$$+ I_c P \left\{ \int_0^{t_1} (T - t) D(t)\,dt \right.$$

$$\left. + \int_{t_1}^T (T - t) D(t)\,dt + K t_1 (R_i - T) \right\}$$

$$- C_2 \left\{ \int_0^{t_1} Q_1(t)\,dt + \int_{t_1}^T Q_2(t)\,dt \right\} - C_1 \Bigg]$$

$$\text{for } i \in \{1, 2, 3\}. \tag{10}$$

Now our objective is to maximize $Z = \mathrm{Sup}\{Z_{1i}, Z_{2i}\}$ for $i \in \{1, 2, 3\}$ and obtain the optimal replenishment period t_1^* and inventory cycle length T^*. We discuss various time-dependent or constant demands by utilizing this general formula.

2.1. Quadratic Demand Pattern. The quadratic time-dependent demand is of the form $D(t) = a + bt + ct^2$ where $a > 0$, $b > 0$, and $c > 0$. This trend of demand is applied to the products like essential commodities and seasonal goods.

From (2), we have

$$Q_1(t) = Kt - at - \frac{bt^2}{2} - \frac{ct^3}{3}, \tag{11}$$

$$Q_2(t) = a(T - t) + \frac{b}{2}(T^2 - t^2) + \frac{c}{3}(T^3 - t^3).$$

Using these $Q_1(t)$ and $Q_2(t)$ in (8) and (10), we obtain

$$Z_{1i} = \frac{1}{T}\left[(P - C_3)Kt_1\right.$$

$$+ I_cP\left\{\frac{aR_i^2}{2} + \frac{bR_i^3}{6} + \frac{cR_i^4}{12}\right\}$$

$$- C_2\left\{\frac{Kt_1^2}{2} + \frac{aT^2}{2} + \frac{bT^3}{3} + \frac{cT^4}{4}\right.$$

$$\left. - aTt_1 - \frac{bT^2t_1}{2} - \frac{cT^3t_1}{3}\right\}$$

$$- I_fC_3\left\{\frac{aT^2}{2} + \frac{bT^3}{3} + \frac{cT^4}{4}\right.$$

$$- a\left(TR_i - \frac{R_i^2}{2}\right) - \frac{b}{2}\left(T^2R_i - \frac{R_i^3}{3}\right)$$

$$\left.\left. - \frac{c}{3}\left(T^3R_i - \frac{R_i^4}{4}\right)\right\} - C_1\right],$$

$$Z_{2i} = \frac{1}{T}\left[(P - C_3)Kt_1\right.$$

$$+ I_cP\left\{\frac{aT^2}{2} + \frac{bT^3}{6} + \frac{cT^4}{12} + Kt_1(R_i - T)\right\}$$

$$- C_2\left\{\frac{Kt_1^2}{2} + \frac{aT^2}{2} + \frac{bT^3}{3}\right.$$

$$\left.\left. + \frac{cT^4}{4} - aTt_1 - \frac{bT^2t_1}{2} - \frac{cT^3t_1}{3}\right\} - C_1\right].$$

(12)

Using (6) in the above objective functions, we obtain

$$Z_{1i} = \left[(P - C_3)\left(a + \frac{bT}{2} + \frac{cT^2}{3}\right)\right.$$

$$+ \frac{I_cP}{T}\left\{\frac{aR_i^2}{2} + \frac{bR_i^3}{6} + \frac{cR_i^4}{12}\right\}$$

$$- C_2\left\{\frac{1}{2K}\left(a^2T + \frac{b^2T^3}{4} + \frac{c^2T^5}{9} + abT^2\right.\right.$$

$$\left. + \frac{bcT^4}{3} + \frac{2acT^3}{3}\right) + \frac{aT}{2} + \frac{bT^2}{3}$$

$$+ \frac{cT^3}{4} - \frac{a^2T}{K} - \frac{abT^2}{K}$$

$$\left. - \frac{2acT^3}{3K} - \frac{b^2T^3}{4K} - \frac{bcT^4}{3K} - \frac{c^2T^5}{9K}\right\}$$

$$- I_fC_3\left\{\frac{aT}{2} + \frac{bT^2}{3} + \frac{cT^3}{4} - aR_i\right.$$

$$+ \frac{aR_i^2}{2T} - \frac{bTR_i}{2} + \frac{bR_i^3}{6T}$$

$$\left.\left. - \frac{cT^2R_i}{3} + \frac{cR_i^4}{12T}\right\} - \frac{C_1}{T}\right],$$

$$Z_{2i} = \left[(P - C_3)\left(a + \frac{bT}{2} + \frac{cT^2}{3}\right)\right.$$

$$+ I_cP\left\{aR_i + \frac{bTR_i}{2} + \frac{cT^2R_i}{3} - \frac{aT}{2} - \frac{bT^2}{3} - \frac{cT^3}{4}\right\}$$

$$- C_2\left\{\frac{1}{2K}\left(a^2T + \frac{b^2T^3}{4} + \frac{c^2T^5}{9} + abT^2\right.\right.$$

$$\left. + \frac{bcT^4}{3} + \frac{2acT^3}{3}\right) + \frac{aT}{2} + \frac{bT^2}{3}$$

$$+ \frac{cT^3}{4} - \frac{a^2T}{K} - \frac{abT^2}{K} - \frac{2acT^3}{3K}$$

$$\left.\left. - \frac{b^2T^3}{4K} - \frac{bcT^4}{3K} - \frac{c^2T^5}{9K}\right\} - \frac{C_1}{T}\right].$$

(13)

To obtain the optimal cycle length, we construct two lemmas as follows.

Lemma 1. When $(dZ_{1i}/dT)|_{T=T^*} = 0$ exists for $T^* \in [R_i, \infty)$, then Z_{1i} has a maximum value at $T = T^*$ if $[4\xi_1(c)(T^*)^6 + \xi_2(b,c)(T^*)^5 + \xi_3(a,b,c)(T^*)^4 + \xi_4(a,b,c)(T^*)^3 - 2\xi_6(a,b,c)] < 0$. Otherwise, Z_{1i} has a maximum value at unique $T^* = \{T \mid (c/3)T^3 + (b/2)T^2 + aT - KR_i = 0, T > 0\}$ if $[\xi_1(c)R_i^6 + \xi_2(b,c)R_i^5 + \xi_3(a,b,c)R_i^4 + \xi_4(a,b,c)R_i^3 + \xi_5(a,b)R_i^2 + \xi_6(a,b,c)] > 0$ holds.

Proof. See Appendix A. □

Lemma 2. When $(dZ_{2i}/dT)|_{T=T^*} = 0$ exists for $T^* \in [0, R_i]$, then Z_{2i} has a maximum value at $T = T^*$ if $[5(T^*)^6\xi_1(c) + (T^*)^5\xi_2(b,c) + 2(T^*)^4\xi_7(a,b,c) + (T^*)^3\xi_8(a,b,c) - 2C_1] < 0$. Otherwise, Z_{2i} has a maximum value at $T = R_i$ if $[(R_i)^6\xi_1(c) + (R_i)^5\xi_2(b,c) + (R_i)^4\xi_7(a,b,c) + (R_i)^3\xi_8(a,b,c) + (R_i)^2\xi_9(a,b,c) + C_1] > 0$.

Proof. See Appendix A. □

2.2. *Linear Time-Dependent Demand Pattern.* We consider linear time-dependent demand which can be obtained if we substitute $c = 0$ in the form of quadratic demand. We get $D(t) = a + bt, a > 0, b > 0$ (see for instances Donaldson [2], Goyal [3], and Goswami and Chaudhuri [4]). Generally, in some computer games, computer android applications, we obtain this type of linear time-dependent demand.

Now from (2), we have

$$Q_1(t) = Kt - at - \frac{bt^2}{2}, \qquad (14)$$

and $Q_2(t) = a(T - t) + (b/2)(T^2 - t^2)$, respectively.

Using $Q_1(t)$ and $Q_2(t)$ in (8) and (10), we obtain

$$Z_{1i} = \frac{1}{T}\left[(P - C_3)Kt_1 \right.$$

$$+ I_c P \left\{ \frac{aR_i^2}{2} + \frac{bR_i^3}{6} \right\}$$

$$- C_2 \left\{ \frac{Kt_1^2}{2} + \frac{aT^2}{2} + \frac{bT^3}{3} - aTt_1 - \frac{bT^2 t_1}{2} \right\}$$

$$- I_f C_3 \left\{ \frac{aT^2}{2} + \frac{bT^3}{3} - a\left(TR_i - \frac{R_i^2}{2} \right) \right.$$

$$\left. \left. - \frac{b}{2}\left(T^2 R_i - \frac{R_i^3}{3} \right) \right\} - C_1 \right],$$

$$Z_{2i} = \frac{1}{T}\left[(P - C_3)Kt_1 \right.$$

$$+ I_c P \left\{ \frac{aT^2}{2} + \frac{bT^3}{6} + Kt_1(R_i - T) \right\}$$

$$\left. - C_2 \left\{ \frac{Kt_1^2}{2} + \frac{aT^2}{2} + \frac{bT^3}{3} - aTt_1 - \frac{bT^2 t_1}{2} \right\} - C_1 \right].$$

$$(15)$$

Using (6) in the above objective functions, we have

$$Z_{1i} = \left[(P - C_3)\left(a + \frac{bT}{2} \right) + \frac{I_c P}{T}\left\{ \frac{aR_i^2}{2} + \frac{bR_i^3}{6} \right\} \right.$$

$$- C_2 \left\{ \frac{1}{2K}\left(a^2 T + \frac{b^2 T^3}{4} + abT^2 \right) \right.$$

$$\left. + \frac{aT}{2} + \frac{bT^2}{3} - \frac{a^2 T}{K} - \frac{abT^2}{K} - \frac{b^2 T^3}{4K} \right\}$$

$$- I_f C_3 \left\{ \frac{aT}{2} + \frac{bT^2}{3} - aR_i + \frac{aR_i^2}{2T} \right.$$

$$\left. \left. - \frac{bTR_i}{2} + \frac{bR_i^3}{6T} \right\} - \frac{C_1}{T} \right],$$

$$Z_{2i} = \left[(P - C_3)\left(a + \frac{bT}{2} \right) \right.$$

$$+ I_c P \left\{ aR_i + \frac{bTR_i}{2} - \frac{aT}{2} - \frac{bT^2}{3} \right\}$$

$$- C_2 \left\{ \frac{1}{2K}\left(a^2 T + \frac{b^2 T^3}{4} + abT^2 \right) \right.$$

$$+ \frac{aT}{2} + \frac{bT^2}{3} - \frac{a^2 T}{K}$$

$$\left. \left. - \frac{abT^2}{K} - \frac{b^2 T^3}{4K} \right\} - \frac{C_1}{T} \right].$$

$$(16)$$

To obtain the optimal cycle length, we construct the following lemmas.

Lemma 3. *When $(dZ_{1i}/dT)|_{T=T^*} = 0$ exists for $T^* \in [R_i, \infty)$, then Z_{1i} has a maximum value at $T = T^*$ if $[2(T^*)^4 \xi_3(a, b, 0) + (T^*)^3 \xi_4(a, b, 0) - 2\xi_6(a, b, 0)] < 0$. Otherwise Z_{1i} has a global maximum value at $T = ((-a) + (\sqrt{a^2 + 2KbR_i}))/b$ if $R_i \leq 2(K - a)/b$ holds.*

Proof. See Appendix B.   □

Lemma 4. *When $(dZ_{2i}/dT)|_{T=T^*} = 0$ exists for $T^* \in [0, R_i]$, then Z_{2i} has a maximum value at $T = T^*$ if $[2(T^*)^4 \xi_3(a, b, 0) + (T^*)^3 \xi_8(a, b, 0) - 2C_1] < 0$. Otherwise Z_{2i} has a maximum value at $T = R_i$ if $[(R_i)^4 \xi_3(a, b, 0) + (R_i)^3 \xi_8(a, b, 0) + (R_i)^2 \xi_5(a, b) + C_1] > 0$.*

Proof. See Appendix B.   □

2.3. Constant Demand Pattern. We consider the demand as constant which can be found by substituting $b = 0$ in the linear demand. We obtain $D = a$, $a > 0$. See for instances Harris [1] and Cárdenas-Barrón [10, 22]. This type of demand is usually found in product's life cycle.

Now from (2), we have $Q_1(t) = (K - a)t$ and $Q_2(t) = a(T - t)$, respectively.

Using $Q_1(t)$ and $Q_2(t)$ in (8) and (10), we get

$$Z_{1i} = \frac{1}{T}\left[(P - C_3)Kt_1 + I_c P\left(\frac{aR_i^2}{2} \right) \right.$$

$$- C_2 \left\{ \frac{Kt_1^2}{2} + \frac{aT^2}{2} - aTt_1 \right\}$$

$$\left. - I_f C_3 \left\{ \frac{aT^2}{2} - a\left(TR_i - \frac{R_i^2}{2} \right) \right\} - C_1 \right], \quad (17)$$

$$Z_{2i} = \frac{1}{T}\left[(P - C_3)Kt_1 + I_c P\left\{ \frac{aT^2}{2} + Kt_1(R_i - T) \right\} \right.$$

$$\left. - C_2 \left\{ \frac{Kt_1^2}{2} + \frac{aT^2}{2} - aTt_1 \right\} - C_1 \right], \quad (18)$$

respectively. Using (6) in the above functions, we obtain

$$Z_{1i} = \left[(P - C_3)a + \frac{I_c P a R_i^2}{2T} - C_2 \left\{ \frac{aT}{2} - \frac{a^2 T}{2K} \right\} \right.$$

$$\left. - I_f C_3 \left\{ \frac{aT}{2} - aR_i + \frac{aR_i^2}{2T} \right\} - \frac{C_1}{T} \right],$$

$$Z_{2i} = \Bigg[(P - C_3)(a) + I_c P \left\{ aR_i - \frac{aT}{2} \right\}$$

$$- C_2 \left\{ \frac{aT}{2} - \frac{a^2 T}{2K} \right\} - \frac{C_1}{T} \Bigg].$$

$$(19)$$

To obtain optimal cycle length, we formulate the following lemmas.

Lemma 5. *If the conditions* $R_i^2 < K\{2C_1 + aR_i^2 (I_f C_3 - I_c P)\}/(C_2 a(K - a) + I_f C_3 aK) < K^2 R_i^2/a^2$ *and* $K\{2C_1 + aR_i^2(I_f C_3 - I_c P)\} > 0$ *hold, then* Z_{1i} *has a global maximum at* $T = T^* = [K\{2C_1 + aR_i^2(I_f C_3 - I_c P)\}/(C_2 a(K - a) + I_f C_3 aK)]^{1/2}$.

Proof. See Appendix C. □

Lemma 6. *If* $0 < 2C_1 K/(C_2 a(K-a) - I_c PaK) < R_i^2$ *hold, then* Z_{2i} *has a global maximum at* $T = T^* = [2C_1 K/(C_2 a(K - a) - I_c PaK)]^{1/2}$.

Proof. See Appendix C. □

2.4. Exponential Demand Pattern. We consider another important type of demand as exponential type, that is, $D(t)$ varies exponentially with time t. In this case, the demand rate increases very fast as in seasonal goods, new computer parts like RAM and data storage device. For this type of demand rate, we consider $D(t) = a \exp(bt)$ where $a > 0$, $b > 0$.

From (2), we have

$$Q_1(t) = Kt + \frac{a}{b}\{1 - \exp(bt)\},$$

$$Q_2(t) = \frac{a}{b}\{\exp(bT) - \exp(bt)\}. \quad (20)$$

Using these $Q_1(t)$ and $Q_2(t)$ in (8) and (10), we obtain

$$Z_{1i} = \frac{1}{T}\Bigg[(P - C_3)Kt_1 - I_c P\left\{\frac{aR_i}{b} + \frac{a}{b^2}(1 - \exp(bR_i))\right\}$$

$$- C_2\left\{\frac{Kt_1^2}{2} + \frac{a}{b}\left(t_1 + \frac{1}{b} + T\exp(bT)\right.\right.$$

$$\left.\left. - \frac{\exp(bT)}{b} - t_1\exp(bT)\right)\right\}$$

$$- I_f C_3 \frac{a}{b}\left\{(T - R_i)\exp(bT)\right.$$

$$\left. - \frac{\exp(bT) - \exp(bR_i)}{b}\right\} - C_1\Bigg],$$

$$Z_{2i} = \frac{1}{T}\Bigg[(P - C_3)Kt_1$$

$$- I_c P\left\{\frac{aT}{b} + \frac{a}{b^2}(1 - \exp(bT)) - Kt_1(R_i - T)\right\}$$

$$- C_2\left\{\frac{Kt_1^2}{2} + \frac{a}{b}\left(t_1 + \frac{1}{b} + T\exp(bT) - \frac{\exp(bT)}{b}\right.\right.$$

$$\left.\left. - t_1\exp(bT)\right)\right\} - C_1\Bigg].$$

$$(21)$$

From (6), we have

$$t_1 = \frac{1}{K}\int_0^T a\exp(bt)\,dt = \frac{a}{Kb}[\exp(bT) - 1]. \quad (22)$$

Substituting this in Z_{1i} and Z_{2i}, we have

$$Z_{1i} = \frac{a}{bT}\Bigg[(P - C_3)(\exp(bT) - 1)$$

$$- I_c P\left\{R_i + \frac{1}{b}(1 - \exp(bR_i))\right\}$$

$$- C_2\left\{\frac{1}{b} + T\exp(bT) - \frac{\exp(bT)}{b} + \frac{a\exp(bT)}{Kb}\right.$$

$$\left. - \frac{a\exp(2bT)}{2Kb} - \frac{a}{2Kb}\right\}$$

$$- I_f C_3\left\{(T - R_i)\exp(bT)\right.$$

$$\left. - \frac{\exp(bT) - \exp(bR_i)}{b}\right\} - \frac{bC_1}{a}\Bigg],$$

$$Z_{2i} = \frac{a}{bT}\Bigg[(P - C_3)(\exp(bT) - 1)$$

$$- I_c P\left\{\frac{1 - \exp(bT)}{b} + R_i + (T - R_i)\exp(bT)\right\}$$

$$- C_2\left\{\frac{1}{b} + T\exp(bT) - \frac{\exp(bT)}{b} + \frac{a\exp(bT)}{Kb}\right.$$

$$\left. - \frac{a\exp(2bT)}{2Kb} - \frac{a}{2Kb}\right)\right\} - \frac{bC_1}{a}\Bigg].$$

$$(23)$$

Lemma 7. *When* $(dZ_{1i}/dT)|_{T=T^*} = 0$ *exists for* $T^* \in [R_i, \infty)$, *then* Z_{1i} *has a maximum value at* $T = T^*$ *if* $(d^2 Z_{1i}/dT^2)|_{T=T^*} < 0$. *Otherwise,* Z_{1i} *has a unique maximum value* $T = T^* = (1/b)[\ln(1 + KbR_i/a)]$, *if* $(dZ_{1i}/dT)|_{T=R_i} > 0$.

Proof. See Appendix D. □

Lemma 8. *When* $(dZ_{2i}/dT)|_{T=T^*} = 0$ *exists for* $T^* \in [0, R_i]$, *then* Z_{2i} *has a maximum value at* $T = T^*$ *if* $(d^2 Z_{2i}/dT^2)|_{T=T^*} < 0$. *Otherwise,* Z_{2i} *has a maximum value* $T = R_i$, *if* $(dZ_{2i}/dT)|_{T=R_i} > 0$

Proof. See Appendix D. □

3. Numerical Examples

Example A. We consider the following parametric values in appropriate units: $C_1 = \$250$ per order, $C_2 = \$20$/unit/year,

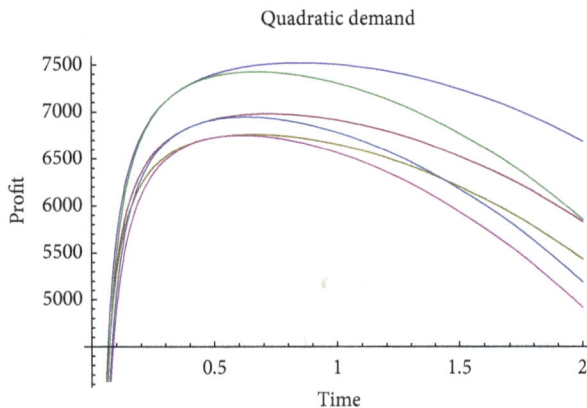

FIGURE 3: The figures of the average profit functions are in descending order as Z_{11}, Z_{21}, Z_{12}, Z_{22}, Z_{13}, and Z_{23}.

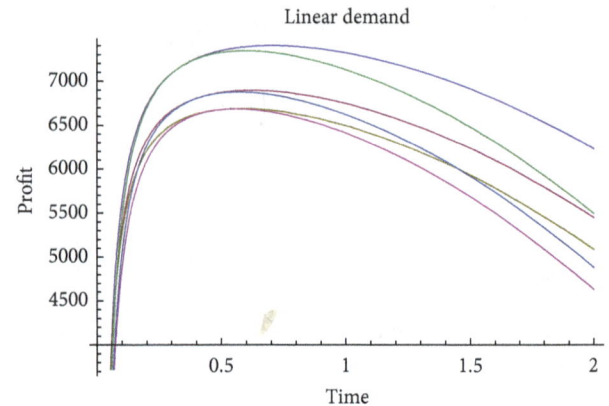

FIGURE 4: The figures of the average profit functions are in descending order as Z_{11}, Z_{21}, Z_{12}, Z_{22}, Z_{13}, and Z_{23}.

R_1 = 120/365 year, R_2 = 150/365 year, R_3 = 180/365 year, ρ_1 = 10%, ρ_2 = 5%, ρ_3 = 2%, a = 100 units, b = 50 units, c = 10 units, $D(t) = a + bt + ct^2$, C_4 = \$120/units, P = \$180/units, I_f = 0.16/\$/year, I_c = 0.13/\$/year, and K = 500 units.

Then the optimal solutions are {Z_{11} = \$7522.33, T^* = 0.85 year, t_1^* = 0.21 year}, {Z_{12} = \$6980.84, T^* = 0.72 year, t_1^* = 0.17 year}, {Z_{13} = \$6760.68, T^* = 0.66 year, t_1^* = 0.15 year}, {Z_{21} = \$7174.54, T^* = 0.33 year, t_1^* = 0.07 year}, {Z_{22} = \$6852.25, T^* = 0.41 year, t_1^* = 0.91 year}, {Z_{23} = \$6725.72, T^* = 0.49 year, t_1^* = 0.11 year}.

Among the above optimal solutions, the optimal solution is {Z_{11} = \$7522.33, T^* = 0.85 year, t_1^* = 0.21 year} global maximum. The average profit function is highly nonlinear. Figure 3 shows the concavity of the functions and comparisons of the cost functions.

Example B. We consider the following parametric values in appropriate units: C_1 = \$250 per order, C_2 = \$20/unit/year, R_1 = 120/365 year, R_2 = 150/365 year, R_3 = 180/365 year, ρ_1 = 10%, ρ_2 = 5%, ρ_3 = 2%, a = 100 units, b = 50 units, $D(t) = a + bt$, C_4 = \$120/units, P = \$180/units, I_f = 0.16/\$/year, I_c = 0.13/\$/year, and K = 500 units.

Then the optimal solutions are {Z_{11} = \$7404.27, T^* = 0.70 year, t_1^* = 0.17 year}, {Z_{12} = \$6895.63, T^* = 0.62 year, t_1^* = 0.14 year}, {Z_{13} = \$6689.02, T^* = 0.58 year, t_1^* = 0.13 year}, {Z_{21} = \$7149.18, T^* = 0.33 year, t_1^* = 0.07 year}, {Z_{22} = \$6816.19, T^* = 0.41 year, t_1^* = 0.09 year}, and {Z_{23} = \$6677, T^* = 0.49 year, t_1^* = 0.11 year}.

Among the above optimal solutions, the optimal solution is {Z_{11} = \$7404.27, T^* = 0.70 year, t_1^* = 0.17 year} global maximum. The average profit function is highly nonlinear. Figure 4 shows the concavity of the functions and comparisons of the cost functions.

Example C. We consider the following parametric values in appropriate units: C_1 = \$250 per order, C_2 = \$20/unit/year, R_1 = 120/365 year, R_2 = 150/365 year, R_3 = 180/365 year, ρ_1 = 10%, ρ_2 = 5%, ρ_3 = 2%, a = 100 units, $D(t) = a$, C_4 = \$120/unit, P = \$180/unit, I_f = 0.16/\$/year, I_c = 0.13/\$/year, and K = 500.

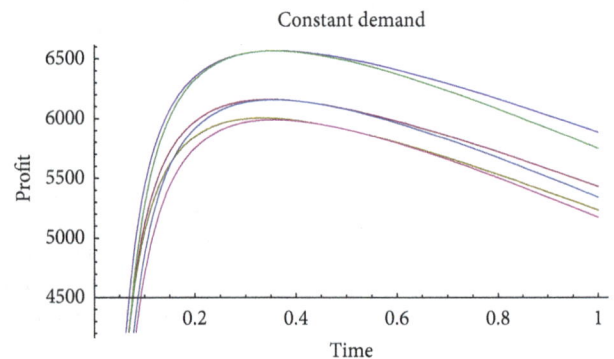

FIGURE 5: The figures of the average profit functions are in descending order as Z_{11}, Z_{21}, Z_{12}, Z_{22}, Z_{13}, and Z_{23}.

Then the optimal solutions are {Z_{11} = \$6566.5, T^* = 0.36 year, t_1^* = 0.07 year}, {Z_{12} = \$6160.63, T^* = 0.35 year, t_1^* = 0.07 year}, {Z_{13} = \$6004.87, T^* = 0.33 year, t_1^* = 0.07 year}, {Z_{21} = \$6561.20, T^* = 0.33 year, t_1^* = 0.07 year}, {Z_{22} = \$6143.73, T^* = 0.41 year, t_1^* = 0.08 year}, and {Z_{23} = \$5915.57, T^* = 0.49 year, t_1^* = 0.10 year}.

From the above optimal solutions, the optimal solution is {Z_{11} = \$6566.5, T^* = 0.36 year, t_1^* = 0.07 year} global maximum. We obtain a closed type formula for T^*, and Figure 5 shows the concavity of the functions and comparisons of the cost functions.

Example D. We consider the following parametric values in appropriate units: C_1 = \$250 per order, C_2 = \$10/unit/month, R_1 = 1 month, R_2 = 2 months, R_3 = 3 months, ρ_1 = 20%, ρ_2 = 15%, ρ_3 = 10%, a = 50 units, b = 0.16 units, $D(t) = a \exp(bt)$, C_4 = \$200/unit, P = \$265/unit, I_f = 0.16\$/month, I_c = 0.13\$/month, and K = 500 units.

Then the optimal solutions are {Z_{11} = \$5840.84, T^* = 5.86 months, t_1^* = 0.97 month}, {Z_{12} = \$5227.55, T^* = 4.51 months, t_1^* = 0.66 month}, {Z_{13} = \$4712.47, T^* = 3.32 months, t_1^* = 0.44 month}, {Z_{21} = \$5270.15, T^* = 1 month, t_1^* = 0.11 month}, {Z_{22} = \$5081.96, T^* = 2 months,

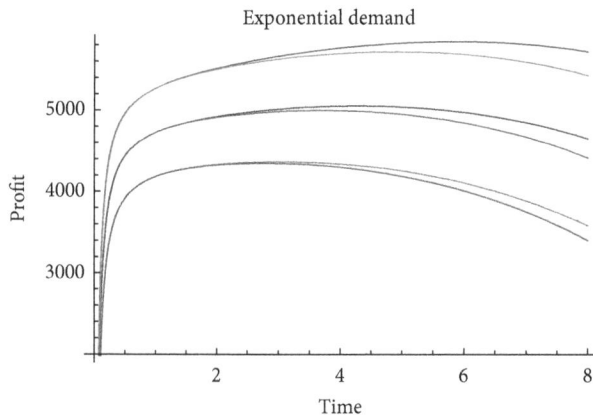

FIGURE 6: The figures of the average profit functions are in descending order as Z_{11}, Z_{21}, Z_{12}, Z_{22}, Z_{13}, and Z_{23}.

$t_1^* = 0.24$ month}, and {$Z_{23} = \$4710.04$, $T^* = 3$ months, $t_1^* = 0.39$ month}.

From the above optimal solutions, the optimal solution is {$Z_{11} = \$5840.84$, $T^* = 5.86$ months, $t_1^* = 0.97$ month} global maximum. The average profit function is highly non-linear. Figure 6 shows the concavity of the functions and comparisons of the cost functions.

Sensitivity Analysis. The sensitivity analysis of the key parameters is given in Table 2 for quadratic demand, Table 3 for linear demand, Table 4 for constant demand, and Table 5 for exponential demand.

(i) If ordering cost increases, then material handlling cost, shipping cost, and placing order's cost increase; as a result the total relevant profit decreases.

(ii) If the unit holding cost per unit item increases, the total profit of the system decreases.

(iii) Increasing value of maximum retailer price increases the total purchasing cost of the whole system which decreases the total profit.

(iv) Increasing value of selling price per item increases the total selling price per lot which reduces the total cost of the whole system. Therefore, the total profit of the system increases.

(v) If we increase the replenishment rate then the holding cost increases which results the decreasing value of the total profit.

4. Concluding Remarks

In most EOQ models, the retailer pays the purchasing cost of items to the supplier as soon as the items are received. In the competitive marketing environment, the supplier offers the retailer a delay period called the *trade credit period* to stimulate the retailer to buy more items. Before the end of the trade credit period, the retailer can sell his products, accumulate revenue, and earn interest. Besides it, a higher interest is

TABLE 2: Sensitivity analysis for the quadratic demand pattern.

Parameters	%	Z_1	Z_2	Z_3
	Quadratic demand Case 1			
C_0	−50%	2.14	2.80	3.24
	−25%	1.02	1.31	1.49
	+25%	−0.95	−1.20	−1.35
	+50%	−1.86	−2.32	−2.59
C_H	−50%	8.17	6.89	6.27
	−25%	3.34	2.89	2.66
	+25%	−2.50	−2.25	−2.11
	+50%	−4.50	−4.11	−3.87
C_R	−50%	18.02	25.83	29.23
	−25%	55.72	58.33	59.46
	+25%	−42.46	−47.02	−49.56
	+50%	−82.40	−92.16	−97.71
P	−50%	−137.21	−148.03	−154.16
	−25%	−70.15	−75.12	−77.87
	+25%	78.03	166.01	221.24
	+50%	54.07	223.77	221.87
K	−50%	4.76	3.76	3.34
	−25%	1.24	1.05	0.95
	+25%	−0.66	−0.57	0.52
	+50%	−1.06	−0.92	0.85
	Quadratic demand Case 2			
C_0	−50%	5.30	4.44	3.77
	−25%	2.65	2.22	1.88
	+25%	−2.65	−2.22	−1.88
	+50%	−5.30	−4.44	−3.77
C_H	−50%	2.01	2.70	3.37
	−25%	1.01	1.35	1.69
	+25%	−1.01	−1.35	−1.69
	+50%	−2.01	−2.70	−3.37
C_R	−50%	81.72	92.20	98.91
	−25%	40.86	46.00	49.45
	+25%	−40.86	−46.00	−49.45
	+50%	−81.72	−92.20	−98.91
P	−50%	−139.04	−149.34	−156.05
	−25%	−69.52	−74.67	−78.03
	+25%	69.52	74.67	78.03
	+50%	139.04	149.34	156.05
K	−50%	1.08	1.47	1.88
	−25%	0.36	0.49	0.62
	+25%	−0.22	−0.29	−0.37
	+50%	−0.36	−0.49	−0.62

charged if the payment is not settled by the end of the trade credit period. Several papers discussing the topic have been mentioned in the literature that investigates inventory problems under various conditions. Many articles in this direction consider the replenishment rate to be instantaneously infinite, but its size is finite and the lot is placed in the beginning of the cycle. Some models consider finite replenishment/supply rate for constant demand rate. Our models

TABLE 3: Sensitivity analysis for the linear demand pattern.

Parameters	%	Z_1	Z_2	Z_3
		Linear demand Case 1		
C_0	−50%	2.67	3.32	3.75
	−25%	1.26	1.54	1.72
	+25%	−1.16	−1.40	−1.54
	+50%	−2.24	−2.69	−2.96
C_H	−50%	5.84	5.29	4.97
	−25%	2.54	2.34	2.21
	+25%	−2.07	−1.95	−1.86
	+50%	−3.82	−3.63	−3.47
C_R	−50%	112.63	121.07	124.66
	−25%	47.64	51.69	53.57
	+25%	−41.94	−46.70	−49.32
	+50%	−82.21	−92.11	−97.72
P	−50%	−137.74	−148.48	−154.55
	−25%	−69.87	−74.97	−77.78
	+25%	77.20	80.96	82.68
	+50%	169.90	176.81	179.62
K	−50%	3.14	2.80	2.60
	−25%	0.92	0.84	0.78
	+25%	−0.51	−0.47	−0.44
	+50%	−0.83	−0.77	−0.73
		Linear demand Case 2		
C_0	−50%	5.32	4.46	3.80
	−25%	2.66	2.23	1.90
	+25%	−2.66	−2.23	−1.90
	+50%	−5.32	−4.46	−3.80
C_H	−50%	2.38	3.24	4.13
	−25%	1.01	1.35	1.68
	+25%	−1.01	−1.35	−1.68
	+50%	−2.01	−2.69	−3.37
C_R	−50%	81.74	92.22	98.92
	−25%	40.87	46.11	49.46
	+25%	−40.87	−46.11	−49.46
	+50%	−81.74	−92.22	−98.92
P	−50%	−139.07	−149.37	−78.08
	−25%	−69.54	−74.69	−78.04
	+25%	69.54	74.69	78.04
	+50%	139.07	149.37	156.08
K	−50%	1.08	1.47	1.86
	−25%	0.36	0.49	0.62
	+25%	−0.22	−0.29	−0.37
	+50%	1.08	1.47	1.86

TABLE 4: Sensitivity analysis for the constant demand pattern.

Parameters	%	Z_1	Z_2	Z_3
		Constant demand Case 1		
C_0	−50%	6.39	7.18	7.80
	−25%	2.86	3.18	3.42
	+25%	−2.74	−2.73	−2.90
	+50%	−4.68	−5.15	−5.46
C_H	−50%	2.35	2.40	2.37
	−25%	1.13	1.16	3.79
	+25%	−1.07	−1.10	−3.60
	+50%	−2.08	−2.14	−4.60
C_R	−50%	82.26	92.65	93.61
	−25%	41.13	46.31	45.47
	+25%	−41.13	−46.30	−50.49
	+50%	−82.25	−92.59	−98.38
P	−50%	−139.56	−150.26	−156.00
	−25%	−69.82	−75.23	−78.76
	+25%	69.92	75.51	74.24
	+50%	139.96	151.45	159.23
K	−50%	1.13	1.16	1.15
	−25%	0.37	0.38	0.37
	+25%	−0.22	−0.22	−0.22
	+50%	−0.36	−0.37	−0.37
		Constant demand Case 2		
C_0	−50%	5.80	4.95	4.28
	−25%	2.90	2.48	2.14
	+25%	−2.90	−2.48	−2.14
	+50%	−5.80	−4.95	−4.28
C_H	−50%	2.00	2.68	3.33
	−25%	1.00	1.34	1.67
	+25%	−1.00	−1.34	−1.67
	+50%	−2.00	−2.68	−3.33
C_R	−50%	82.30	92.78	99.40
	−25%	41.15	46.39	49.70
	+25%	−41.15	−46.39	−49.70
	+50%	82.30	−92.78	−99.40
P	−50%	−140.10	−150.40	−157.02
	−25%	−70.05	−75.20	−78.51
	+25%	70.05	75.20	78.51
	+50%	140.10	150.40	157.02
K	−50%	1.00	1.34	1.67
	−25%	0.33	0.44	0.56
	+25%	−0.20	−0.27	−0.33
	+50%	−0.33	−0.44	−0.56

consider the demand of the products as an increasing function of time. From the view point of the supplier, price discount on the purchasing cost of the items by retailer is given at a different delay period to motivate the retailer to buy more. Generally, suppliers consider maximum delay period, after which they would not take any risk of getting back money from retailers. As a result, the purchasing cost C_3 is infinite when R is greater than R_3; that is, retailer never purchases

items at infinite cost. In addition, we establish eight effective and easy-to-use lemmas and figures to help the retailer to take optimal strategy for his marketing policy. We solved the model analytically. We found global optimal solutions for different types of demand patterns. We illustrate the numerical results graphically. As far as the knowledge of authors goes, such type of model has not yet been discussed in the EOQ literature. The application of our model in a different

TABLE 5: Sensitivity analysis for the exponential demand pattern.

Parameters	%	Z_1	Z_2	Z_3
		Exponential demand Case 1		
C_0	−50%	0.37	0.54	0.83
	−25%	0.18	−0.27	0.41
	+25%	−0.18	−0.26	−0.39
	+50%	−0.36	−0.52	−0.77
C_H	−50%	—	—	—
	−25%	18.16	13.01	9.39
	+25%	−6.93	−5.23	−3.78
	+50%	−10.30	−7.89	−5.80
C_R	−50%	—	—	—
	−25%	—	—	—
	+25%	−46.01	−46.01	−50.15
	+50%	−84.03	−84.03	−95.06
P	−50%	—	—	—
	−25%	−71.44	−76.39	−83.42
	+25%	—	—	—
	+50%	—	—	—
K	−50%	—	—	5.03
	−25%	2.82	1.80	1.16
	+25%	−1.25	−0.86	−0.57
	+50%	−1.97	−1.37	−0.92
		Exponential demand Case 2		
C_0	−50%	2.37	1.23	0.88
	−25%	1.86	0.61	0.44
	+25%	−1.18	−0.61	−0.44
	+50%	−2.37	−1.23	−0.88
C_H	−50%	2.36	5.42	9.72
	−25%	1.18	2.71	4.86
	+25%	−1.18	−2.71	−4.86
	+50%	−2.36	−2.36	−9.72
C_R	−50%	82.31	98.56	122.63
	−25%	41.15	49.28	61.31
	+25%	−41.15	−49.28	−61.31
	+50%	−82.31	−98.56	—
P	−50%	—	—	—
	−25%	−68.52	−77.61	−91.62
	+25%	68.52	77.61	91.62
	+50%	137.04	155.21	183.23
K	−50%	0.56	1.37	2.62
	−25%	0.18	0.46	0.87
	+25%	−0.11	−0.27	−0.52
	+50%	−0.18	−0.46	−0.87

— indicates the value of the parameter is either imaginary or the value does not exist at that point.

field of inventory models where the assumptions are valid. Further extension of the model may be generalized by considering shortages, deterioration, and stochastic demand. One immediate extension of this model is to consider partial backlogging.

Appendices

A. Derivation of Lemmas 1 and 2

Differentiating (13), the equations become

$$
\frac{dZ_{1i}}{dT} = \left[(P - C_3)\left(\frac{b}{2} + \frac{2cT}{3}\right) - \frac{I_c P}{T^2}\left\{\frac{aR_i^2}{2} + \frac{bR_i^3}{6} + \frac{cR_i^4}{12}\right\} \right.
$$

$$
- C_2\left\{\frac{1}{2K}\left(a^2 + \frac{3b^2T^2}{4} + \frac{5c^2T^4}{9} + 2abT\right.\right.
$$

$$
\left.\left. + \frac{4bcT^3}{3} + 2acT^2\right) + \frac{a}{2} + \frac{2bT}{3}\right.
$$

$$
+ \frac{3cT^2}{4} - \frac{a^2}{K} - \frac{2abT}{K} - \frac{2acT^2}{K}
$$

$$
\left. - \frac{3b^2T^2}{4K} - \frac{4bcT^3}{3K} - \frac{5c^2T^4}{9K}\right\}
$$

$$
- I_f C_3 \left\{\frac{a}{2} + \frac{2bT}{3} + \frac{3cT^2}{4} - \frac{aR_i^2}{2T^2} - \frac{bR_i}{2} - \frac{bR_i^3}{6T^2}\right.
$$

$$
\left.\left. - \frac{2cTR_i}{3} - \frac{cR_i^4}{12T^2}\right\} + \frac{C_1}{T^2}\right]
$$

$$
= \frac{1}{T^2}\left[\xi_1(c)T^6 + \xi_2(b,c)T^5 + \xi_3(a,b,c)T^4\right.
$$

$$
\left. + \xi_4(a,b,c)T^3 + \xi_5(a,b)T^2 + \xi_6(a,b,c)\right],
$$

(A.1)

where

$$
\xi_1(c) = \frac{5c^2C_2}{18K}, \qquad \xi_2(b,c) = \frac{2bcC_2}{3K},
$$

$$
\xi_3(a,b,c) = \left\{\left(2ac + \frac{3b^2}{4}\right)\frac{C_2}{2K} - \left(C_2 + I_f C_3\right)\frac{3c}{4}\right\},
$$

$$
\xi_4(a,b,c) = \left\{(P - C_3)\frac{2c}{3} + \frac{C_2 ab}{K}\right.
$$

$$
\left. - \left(C_2 + I_f C_3\right)\frac{2b}{3} + \frac{2I_f C_3 cR_i}{3}\right\},
$$

$$
\xi_5(a,b) = \left\{(P - C_3)\frac{b}{2} + \frac{C_2 a^2}{K}\right.
$$

$$
\left. - \left(C_2 + I_f C_3\right)\frac{a}{2} + \frac{I_f C_3 bR_i}{2}\right\},
$$

$$
\xi_6(a,b,c) = C_1 + \left\{\frac{aR_i^2}{2} + \frac{bR_i^3}{6} + \frac{cR_i^4}{12}\right\}\left(I_f C_3 - I_c P\right),
$$

$$\frac{d^2 Z_{1i}}{dT^2} = \left[(P - C_3) \left(\frac{2c}{3} \right) + \frac{2I_c P}{T^3} \left\{ \frac{aR_i^2}{2} + \frac{bR_i^3}{6} + \frac{cR_i^4}{12} \right\} \right.$$

$$- C_2 \left\{ \frac{1}{2K} \left(\frac{3b^2 T}{2} + \frac{20c^2 T^3}{9} + 2ab + 4bcT^2 \right. \right.$$

$$\left. + 4acT \right) + \frac{2b}{3} + \frac{3cT}{2} - \frac{2ab}{K}$$

$$- \frac{4acT}{K} - \frac{3b^2 T}{2K} - \frac{4bcT^2}{K} - \frac{20c^2 T^3}{9K} \right\}$$

$$- I_f C_3 \left\{ \frac{2b}{3} + \frac{3cT}{2} + \frac{aR_i^2}{T^3} + \frac{bR_i^3}{3T^3} \right.$$

$$\left. - \frac{2cR_i}{3} + \frac{cR_i^4}{6T^3} \right\} - \frac{2C_1}{T^3} \right]$$

$$= \frac{1}{T^3} \left[4\xi_1(c) T^6 + \xi_2(b,c) T^5 + \xi_3(a,b,c) T^4 \right.$$

$$\left. + \xi_4(a,b,c) T^3 - 2\xi_6(a,b,c) \right].$$

(A.2)

Now,

$$\frac{dZ_{2i}}{dT} = \left[(P - C_3) \left(\frac{b}{2} + \frac{2cT}{3} \right) \right.$$

$$- I_c P \left\{ \frac{2cTR_i}{3} + \frac{bR_i}{2} - \frac{a}{2} - \frac{2bT}{3} - \frac{3cT^2}{4} \right\}$$

$$- C_2 \left\{ \frac{1}{2K} \left(a^2 + \frac{3b^2 T^2}{4} + \frac{5c^2 T^4}{9} \right. \right.$$

$$\left. + 2abT + \frac{4bcT^3}{3} + 2acT^2 \right) + \frac{a}{2} + \frac{2bT}{3}$$

$$+ \frac{3cT^2}{4} - \frac{a^2}{K} - \frac{2abT}{K} - \frac{2acT^2}{K}$$

$$\left. - \frac{3b^2 T^2}{4K} - \frac{4bcT^3}{3K} - \frac{5c^2 T^4}{9K} \right\} + \frac{C_1}{T^2} \right]$$

$$= \frac{1}{T^2} \left[\xi_1(c) T^6 + \xi_2(b,c) T^5 + \xi_7(a,b,c) T^4 \right.$$

$$\left. + \xi_8(a,b,c) T^3 + \xi_9(a,b) T^2 + C_1 \right],$$

(A.3)

where

$$\xi_1(c) = \frac{5c^2 C_2}{18K}, \qquad \xi_2(b,c) = \frac{2bcC_2}{3K},$$

$$\xi_7(a,b,c) = \left\{ \left(2ac + \frac{3b^2}{4} \right) \frac{C_2}{2K} - (C_2 + I_c P) \frac{3c}{4} \right\},$$

$$\xi_8(a,b,c) = \left\{ (P - C_3) \frac{2c}{3} + \frac{C_2 ab}{K} \right.$$

$$\left. - (C_2 + I_c P) \frac{2b}{3} + \frac{2I_c PcR_i}{3} \right\},$$

$$\xi_9(a,b) = \left\{ (P - C_3) \frac{b}{2} + \frac{C_2 a^2}{K} - (C_2 + I_c P) \frac{a}{2} + \frac{I_c PbR_i}{2} \right\},$$

$$\frac{d^2 Z_{2i}}{dT^2} = \left[(P - C_3) \left(\frac{2c}{3} \right) - I_c P \left\{ \frac{2cR_i}{3} - \frac{2b}{3} - \frac{3cT}{2} \right\} \right.$$

$$- C_2 \left\{ \frac{1}{2K} \left(\frac{3b^2 T}{2} + \frac{20c^2 T^3}{9} \right. \right.$$

$$\left. + 2ab + 4bcT^2 + 4acT \right)$$

$$+ \frac{2b}{3} + \frac{3cT}{2} - \frac{2ab}{K} - \frac{4acT}{K}$$

$$\left. - \frac{3b^2 T}{2K} - \frac{4bcT^2}{K} - \frac{20c^2 T^3}{9K} \right\} - \frac{2C_1}{T^3} \right]$$

$$= \frac{1}{T^3} \left[5\xi_1(c) T^6 + \xi_2(b,c) T^5 \right.$$

$$\left. + 2\xi_7(a,b,c) T^4 + \xi_8(a,b,c) T^3 - 2C_1 \right].$$

(A.4)

To obtain the optimal cycle length, proof of two lemmas are as follows.

Proof of Lemma 1. The first part of the lemma is obvious from the above calculations of dZ_{1i}/dT and $d^2 Z_{1i}/dT^2$. For the second part, if Z_{1i} does not have any stationary points in $[R_i, \infty)$, then Z_{1i} is either monotonic increasing or monotonic decreasing function of $T \in [R_i, \infty)$. Here, $dZ_{1i}/dT \to \infty$ as $T \to \infty$ because $\xi_1(c) > 0$. Hence, Z_{1i} will be monotonic increasing if $(dZ_{1i}/dT)|_{T=R_i} > 0$ as Z_{1i} does not have stationary points in $[R_i, \infty)$. Here, $t_1 = (1/K)(aT + (b/2)T^2 + (c/3)T^3)$, t_1 increase with increasing value of T. As in our models $t_1 \leq R_i$; hence, for the feasibility of the model, $(t_1)^* = R_i$ and T^* is obtained from $(c/3)(T^*)^3 + (b/2)(T^*)^2 + a(T^*) = KR_i$. By Descarte's rule, $(c/3)(T^*)^3 + (b/2)(T^*)^2 + a(T^*) - KR_i = 0$ may have atmost one positive root and two negative roots. Therefore, Z_{1i} has a maximum value at $T^* = \{T \mid (c/3)T^3 + (b/2)T^2 + aT - KR_i = 0, T > 0\}$ if $(1/R_i^2)[\xi_1(c)R_i^6 + \xi_2(b,c)R_i^5 + \xi_3(a,b,c)R_i^4 + \xi_4(a,b,c)R_i^3 + \xi_5(a,b)R_i^2 + \xi_6(a,b,c)] > 0$. Here T^* is unique if it exists, by Descarte's rule, hence the proof. □

Proof of Lemma 2. The first part of the lemma is obvious (see the expressions of dZ_{1i}/dT and $d^2 Z_{1i}/dT^2$). For the second part, if Z_{2i} does not have any stationary points in $[0, R_i]$, then Z_{2i} is either monotonic increasing or monotonic decreasing function of $T \in [0, R_i]$. Here $dZ_{2i}/dT \to \infty$ as $T \to 0$ because $C_1 > 0$. Hence, Z_{2i} is monotonic increasing if $(dZ_{2i}/dT)|_{T=R_i} > 0$ as Z_{2i} does not have stationary points in $[0, R_i]$. Hence Z_{2i} has a maximum value at $T = R_i$ if

$[(R_i)^6\xi_1(c) + (R_i)^5\xi_2(b,c) + (R_i)^4\xi_7(a,b,c) + (R_i)^3\xi_8(a,b,c) + (R_i)^2\xi_9(a,b,c) + C_1] > 0$, hence the proof. $\qquad\square$

B. Derivation of Lemmas 3 and 4

Now differentiating (16), we obtain

$$\frac{dZ_{1i}}{dT} = \left[(P - C_3)\left(\frac{b}{2}\right) - \frac{I_c P}{T^2}\left\{\frac{aR_i^2}{2} + \frac{bR_i^3}{6}\right\}\right.$$

$$- C_2\left\{\frac{1}{2K}\left(a^2 + \frac{3b^2T^2}{4} + 2abT\right) + \frac{a}{2}\right.$$

$$\left. + \frac{2bT}{3} - \frac{a^2}{K} - \frac{2abT}{K} - \frac{3b^2T^2}{4K}\right\}$$

$$\left. - I_f C_3\left\{\frac{a}{2} + \frac{2bT}{3} - \frac{aR_i^2}{2T^2} - \frac{bR_i}{2} - \frac{bR_i^3}{6T^2}\right\} + \frac{C_1}{T^2}\right]$$

$$= \frac{1}{T^2}\left[\xi_3(a,b,0)\,T^4 + \xi_4(a,b,0)\,T^3\right.$$

$$\left. + \xi_5(a,b)\,T^2 + \xi_6(a,b,0)\right],$$

$$\text{(B.1)}$$

where

$$\xi_3(a,b,0) = \left(\frac{3b^2}{4}\right)\frac{C_2}{2K},$$

$$\xi_4(a,b,0) = \left\{\frac{C_2 ab}{K} - (C_2 + I_f C_3)\frac{2b}{3}\right\},$$

$$\xi_5(a,b)$$

$$= \left\{(P - C_3)\frac{b}{2} + \frac{C_2 a^2}{K} - (C_2 + I_f C_3)\frac{a}{2} + \frac{I_f C_3 b R_i}{2}\right\},$$

$$\xi_6(a,b,0) = C_1 + \left\{\frac{aR^2}{2} + \frac{bR^3}{6}\right\}(I_f C_3 - I_c P),$$

$$\frac{d^2 Z_{1i}}{dT^2} = \left[\frac{2I_c P}{T^3}\left\{\frac{aR_i^2}{2} + \frac{bR_i^3}{6}\right\}\right.$$

$$- C_2\left\{\frac{1}{2K}\left(\frac{3b^2T}{2} + 2ab\right)\right.$$

$$\left. + \frac{2b}{3} - \frac{2ab}{K} - \frac{3b^2T}{2K}\right\}$$

$$\left. - I_f C_3\left\{\frac{2b}{3} + \frac{aR_i^2}{T^3} + \frac{bR_i^3}{3T^3}\right\} - \frac{2C_1}{T^3}\right]$$

$$= \frac{1}{T^3}\left[2\xi_3(a,b,0)\,T^4 + \xi_4(a,b,0)\,T^3 - 2\xi_6(a,b,0)\right].$$

$$\text{(B.2)}$$

Now,

$$\frac{dZ_{2i}}{dT} = \left[(P - C_3)\left(\frac{b}{2}\right) - I_c P\left\{\frac{bR_i}{2} - \frac{a}{2} - \frac{2bT}{3}\right\}\right.$$

$$- C_2\left\{\frac{1}{2K}\left(a^2 + \frac{3b^2T^2}{4} + 2abT\right) + \frac{a}{2} + \frac{2bT}{3}\right.$$

$$\left. - \frac{a^2}{K} - \frac{2abT}{K} - \frac{3b^2T^2}{4K}\right\} + \frac{C_1}{T^2}\right]$$

$$= \frac{1}{T^2}\left(\xi_3(a,b,0)\,T^4 + \xi_8(a,b,0)\,T^3 + \xi_5(a,b)\,T^2 + C_1\right),$$

$$\text{(B.3)}$$

where

$$\xi_8(a,b,0) = \left\{\frac{C_2 ab}{K} - (C_2 + I_c P)\frac{2b}{3}\right\},$$

$$\frac{d^2 Z_{2i}}{dT^2} = \left[I_c P\left\{\frac{2b}{3}\right\}\right.$$

$$- C_2\left\{\frac{1}{2K}\left(\frac{3b^2T}{2} + 2ab\right)\right.$$

$$\left. + \frac{2b}{3} - \frac{2ab}{K} - \frac{3b^2T}{2K}\right\} - \frac{2C_1}{T^3}\right]$$

$$= \frac{1}{T^3}\left[2\xi_3(a,b,0)\,T^4 + \xi_8(a,b,0)\,T^3 - 2C_1\right].$$

$$\text{(B.4)}$$

Proof of Lemma 3. The first part of the lemma is obvious from the formulas of dZ_{1i}/dT and d^2Z_{1i}/dT^2. For the second part, if Z_{1i} does not have any stationary points in $[R_i, \infty)$, then Z_{1i} is either a monotonic increasing or monotonic decreasing function of $T \in [R_i, \infty)$. Here, $dZ_{1i}/dT \to \infty$ as $T \to \infty$ because $\xi_3(a,b,0) > 0$. Hence, Z_{1i} is monotonic increasing if $[(R_i)^4\xi_3(a,b,0) + (R_i)^3\xi_4(a,b,0) + (R_i)^2\xi_5(a,b) + \xi_6(a,b,0)] > 0$ as Z_{1i} does not have stationary points in $[R_i, \infty)$. Here $t_1 = (1/K)(aT + (b/2)T^2)$ and t_1 increases with increasing value of T. According to our model, $t_1 < R_i$; hence for the feasibility of the model, $t_1^* = R_i$ and T^* is obtained from $(b/2)(T^*)^2 + a(T^*) = KR_i$. By Descarte's rule $(b/2)(T^*)^2 + a(T^*) - KR_i = 0$ may have atmost one positive root and one negative root. Now the positive root of the above equation is $T^* = (-a + \sqrt{a^2 + 2KbR_i})/b$. Here T^* is unique by Descarte's rule. For feasibility of our model, T^* must be greater than or equal to R_i. Therefore, $T^* \geq R_i \Rightarrow -a + \sqrt{a^2 + 2KbR_i} \geq bR_i \Rightarrow \sqrt{a^2 + 2KbR_i} \geq a + bR_i \Rightarrow a^2 + 2KbR_i \geq (a + bR_i)^2 \Rightarrow a^2 + 2KbR_i - a^2 - 2abR_i - b^2(R_i)^2 \geq 0 \Rightarrow 2(K-a)/b \geq R_i$. Thus Z_{1i} has a unique maximum value at $T = (-a + \sqrt{a^2 + 2KbR_i})/b$ if $R_i \leq 2(K - a)/b$ holds, hence the proof. $\qquad\square$

Proof of Lemma 4. The first part of the lemma is obvious (see the expressions of dZ_{1i}/dT and d^2Z_{1i}/dT^2). For the second part, if Z_{2i} does not have any stationary points in $[0, R_i]$, then Z_{2i} is either monotonic increasing or monotonic decreasing function of $T \in [0, R_i]$. Here, $dZ_{2i}/dT \to \infty$ as $T \to 0$ because $C_1 > 0$. So, Z_{2i} is monotonic increasing if $(1/(R_i)^2)[(R_i)^4\xi_3(a, b, 0) + (R_i)^3\xi_8(a, b, 0) + (R_i)^2\xi_5(a, b) + C_1] > 0$ as Z_{2i} does not have stationary points in $[0, R_i]$. Hence, Z_{2i} has a maximum value when $T = R_i$ if $[(R_i)^4\xi_3(a, b, 0) + (R_i)^3\xi_8(a, b, 0) + (R_i)^2\xi_5(a, b) + C_1] > 0$. Hence the proof. $\qquad\square$

C. Derivation of Lemmas 5 and 6

Differentiating (19), we have

$$\frac{dZ_{1i}}{dT} = \left[-\frac{I_cP}{T^2}\left\{\frac{aR_i^2}{2}\right\} - C_2\left\{\frac{a}{2} - \frac{a^2}{2K}\right\} \right.$$
$$\left. -I_fC_3\left\{\frac{a}{2} - \frac{aR_i^2}{2T^2}\right\} + \frac{C_1}{T^2} \right],$$

$$\frac{d^2Z_{1i}}{dT^2} = \left[\frac{I_cP}{T^3}\{aR_i^2\} - I_fC_3\left\{\frac{aR_i^2}{T^3}\right\} - \frac{2C_1}{T^3} \right], \qquad \text{(C.1)}$$

$$\frac{dZ_{2i}}{dT} = \left[I_cP\left\{\frac{a}{2}\right\} - C_2\left\{\frac{a}{2} - \frac{a^2}{2K}\right\} + \frac{C_1}{T^2} \right],$$

$$\frac{d^2Z_{2i}}{dT^2} = \left[-\frac{2C_1}{T^3} \right].$$

Proof of Lemma 5. We have $dZ_{1i}/dT = [-(I_cP/T^2)\{(aR_i^2)/2\} - C_2\{a/2 - a^2/2K\} - (I_fC_3)\{a/2 - aR_i^2/2T^2\} + C_1/T^2]$ and $d^2Z_{1i}/dT^2 = [(I_cX/T^3)\{aR_i^2\} - I_fC_3\{aR^2/T^3\} - 2C_1/T^3]$. For maximization, $dZ_{1i}/dT = 0$. This implies that $T = [K\{2C_1 + aR_i^2(I_fC_3 - I_cP)\}/(C_2a(K - a) + I_fC_3aK)]^{1/2} = T^*$(say). At $T = T^*$, $d^2Z_{1i}/dT^2 = -(1/K)[(C_2a(K - a) + I_fC_3aK)/K\{2C_1 + aR_i^2(I_fC_3 - I_cP)\}]^{1/2}$. If $K\{2C_1 + aR_i^2(I_fC_3 - I_cP)\} > 0$, then $d^2Z_{1i}/dT^2 < 0$. Again for the feasibility of the model, $t_1 = aT/K < R_i \Rightarrow T < (KR_i)/a$ and $T > R_i$. Combining these, we have $R_i < T < KR_i/a$. That is, $R_i^2 < T^2 < (KR_i)^2/a^2$. Hence at $T = T^* = [K\{2C_1 + aR_i^2(I_fC_3 - I_cP)\}/(C_2a(K - a) + I_fC_3aK)]^{1/2}$, Z_{1i} has a maximum value if $R_i^2 < K\{2C_1 + aR_i^2(I_fC_3 - I_cP)\}/(C_2a(K - a) + I_fC_3aK) < K^2R_i^2/a^2$ hold, hence the proof. $\qquad\square$

Proof of Lemma 6. We have $dZ_{2i}/dT = [I_cP(a/2) - C_2\{a/2 - a^2/2K\} + C_1/T^2]$ and $d^2Z_{2i}/dT^2 = [-2C_1/T^3]$. For optimization $dZ_{2i}/dT = 0$. This gives $T = [2C_1K/(C_2a(K - a) - I_cPaK)]^{1/2}$. For the real values of T, $C_2a(K - a) - I_cPaK > 0$ and also for the feasibility of the model, $T < R_i$ which imply that $T^2 < (R_i)^2 \Rightarrow 0 < 2C_1K/(C_2a(K - a) - I_cPaK) < R_i^2$. The second derivative of Z_{2i} at $T = T^*$ is always negative. Therefore Z_{2i} has a global maximum at $T = T^* = [2C_1K/(C_2a(K - a) - I_cPaK)]^{1/2}$ if $0 < 2C_1K/(C_2a(K - a) - I_cPaK) < R_i^2$ hold, hence the lemma. $\qquad\square$

D. Derivation of Lemmas 7 and 8

Now differentiating Z_{1i} and Z_{2i} with respect to T, we obtain

$$\frac{dZ_{1i}}{dT} = \frac{a}{bT}\left[(P - C_3)\left(b\exp(bT) - \frac{(\exp(bT) - 1)}{T}\right) \right.$$

$$+ \frac{I_cP}{T}\left\{R_i + \frac{1}{b}(1 - \exp(bR_i))\right\}$$

$$-C_2\left\{\frac{a\exp(bT)}{K} - \frac{a\exp(2bT)}{K}\right.$$

$$+ Tb\exp(bT) - \frac{a\exp(bT)}{KbT}$$

$$+ \frac{a\exp(2bT)}{2TKb} - \frac{1}{bT}$$

$$+ \frac{a}{2KbT} - \exp(bT) + \frac{\exp(bT)}{bT}\right\}$$

$$-I_fC_3\left\{(T - R_i)b\exp(bT)\right.$$

$$- \frac{(T - R_i)\exp(bT)}{T}$$

$$+ \frac{\exp(bT) - \exp(bR_i)}{bT}\right\} + \frac{bC_1}{aT}\right].$$
$$\text{(D.1)}$$

Using $dZ_{1i}/dT = 0$,

$$\frac{d^2Z_{1i}}{dT^2}$$

$$= \frac{a}{bT}\left[(P - C_3) \right.$$

$$\times \left(b^2\exp(bT) - \frac{Tb\exp(bT) - \exp(bT) + 1}{T^2}\right)$$

$$- \frac{I_cP}{T^2}\left\{R_i + \frac{1}{b}(1 - \exp(bR_i))\right\}$$

$$-C_2\left\{\frac{ab\exp(bT)}{K} - \frac{2ab\exp(2bT)}{K}\right.$$

$$+ Tb^2\exp(bT) - \frac{a\exp(bT)}{KT} + \frac{a\exp(bT)}{KbT^2}$$

$$+ \frac{a\exp(2bT)}{KT} - \frac{a\exp(2bT)}{2T^2Kb} + \frac{1}{bT^2}$$

$$- \frac{a}{2KbT^2} + \frac{\exp(bT)}{T} - \frac{\exp(bT)}{bT^2}\right\}$$

$$-I_f C_3 \left\{ (T - R_i)\, b^2 \exp(bT) \right.$$

$$- \frac{R_i \left(Tb \exp(bT) - \exp(bT) \right)}{T^2}$$

$$\left. + \frac{Tb \exp(bT) - \exp(bT) + \exp(bR_i)}{bT^2} \right\}$$

$$- \frac{bC_1}{aT^2} \Bigg].$$

$$(D.2)$$

From Case 2, we have

$$\frac{dZ_{2i}}{dT} = \frac{a}{bT} \left[(P - C_3) \left(b \exp(bT) - \frac{(\exp(bT) - 1)}{T} \right) \right.$$

$$- \frac{I_c P}{T} \left\{ Tb \exp(bT) - R_i b \exp(bT) \right.$$

$$- \frac{1}{bT} \left(1 - \exp(bT) \right)$$

$$\left. + \frac{R_i \exp(bT)}{T} - \frac{R_i}{T} - \exp(bT) \right\}$$

$$- C_2 \left\{ \frac{a \exp(bT)}{K} - \frac{a \exp(2bT)}{K} \right.$$

$$+ Tb \exp(bT) - \frac{a \exp(bT)}{KbT}$$

$$+ \frac{a \exp(2bT)}{2TKb} - \frac{1}{bT} + \frac{a}{2KbT}$$

$$\left. \left. - \exp(bT) + \frac{\exp(bT)}{bT} \right\} + \frac{bC_1}{aT} \right].$$

$$(D.3)$$

Using $dZ_{2i}/dT = 0$,

$$\frac{d^2 Z_{2i}}{dT^2}$$

$$= \frac{a}{bT} \left[(P - C_3) \right.$$

$$\times \left(b^2 \exp(bT) - \frac{Tb \exp(bT) - \exp(bT) + 1}{T^2} \right)$$

$$- I_c P \left\{ Tb^2 \exp(bT) - R_i b^2 \exp(bT) - \frac{\exp(bT)}{T} \right.$$

$$+ \frac{1}{bT^2} - \frac{\exp(bT)}{bT^2} + \frac{R_i b \exp(bT)}{T}$$

$$\left. - \frac{R_i \exp(bT)}{T^2} + \frac{R_i}{T^2} \right\}$$

$$- C_2 \left\{ \frac{ab \exp(bT)}{K} - \frac{2ab \exp(2bT)}{K} + Tb^2 \exp(bT) \right.$$

$$- \frac{a \exp(bT)}{KT} + \frac{a \exp(bT)}{KbT^2} + \frac{a \exp(2bT)}{KT}$$

$$- \frac{a \exp(2bT)}{2T^2 Kb} + \frac{1}{bT^2} - \frac{a}{2KbT^2}$$

$$\left. + \frac{\exp(bT)}{T} - \frac{\exp(bT)}{bT^2} \right\} - \frac{bC_1}{aT^2} \Bigg].$$

$$(D.4)$$

Proof of Lemma 7. The first part of the lemma is obvious. For the second part, if Z_{1i} does not have any stationary points in $[R_i, \infty)$, then Z_{1i} is either monotonic increasing or monotonic decreasing function of $T \in [R_i, \infty)$. Here, $dZ_{1i}/dT \to \infty$ as $T \to \infty$ because

$$\frac{dZ_{1i}}{dT}$$

$$= \frac{a}{bT} \left[(P - C_3) \left(b \exp(bT) - \frac{(\exp(bT) - 1)}{T} \right) \right.$$

$$+ \frac{I_c P}{T} \left\{ R_i + \frac{1}{b} \left(1 - \exp(bR_i) \right) \right\}$$

$$- C_2 \left\{ \frac{a \exp(bT)}{K} - \frac{a \exp(2bT)}{K} + Tb \exp(bT) \right.$$

$$- \frac{a \exp(bT)}{KbT} + \frac{a \exp(2bT)}{2TKb} - \frac{1}{bT}$$

$$\left. + \frac{a}{2KbT} - \exp(bT) + \frac{\exp(bT)}{bT} \right\}$$

$$- I_f C_3 \left\{ (T - R_i)\, b \exp(bT) - \frac{(T - R_i) \exp(bT)}{T} \right.$$

$$\left. \left. + \frac{\exp(bT) - \exp(bR_i)}{bT} \right\} + \frac{bC_1}{aT} \right]$$

$$= \frac{a \exp(bT)}{bT} \left[(P - C_3) \left(b - \frac{1}{T} \right) \right.$$

$$- C_2 \left\{ \frac{a}{K} + Tb - \frac{a}{KbT} - 1 + \frac{1}{bT} \right\}$$

$$\left. - I_f C_3 \left\{ (T - R_i)\, b + \frac{R_i}{T} + \frac{1}{bT} \right\} \right]$$

$$+ \frac{a}{bT} \left[\frac{(P - C_3)}{T} + \frac{I_c P}{T} \left\{ R_i + \frac{1}{b} \right\} \right.$$

$$\left. - C_2 \left\{ \frac{a}{2KbT} - \frac{1}{bT} \right\} + I_f C_3 \frac{\exp(bR_i)}{bT} \right]$$

$$+ \frac{a}{bT} \left[-C_2 \left\{ -\frac{a}{K} + \frac{a}{2TKb} \right\} \right] \exp(2bT) + \frac{C_1}{T^2}.$$

$$(D.5)$$

Using L'Hospital's rule

$$\lim_{T \to \infty} \frac{dZ_{1i}}{dT}$$

$$= \lim_{T \to \infty} \frac{a \exp(bT)}{bT^2}$$

$$\times \left[(P - C_3)(bT - 1) \right.$$

$$- C_2 \left\{ \frac{aT}{K} + T^2 b - \frac{a}{Kb} - T + \frac{1}{b} \right\}$$

$$\left. - I_f C_3 \left\{ (T - R_i) Tb + R_i + \frac{1}{b} \right\} \right]$$

$$+ \lim_{T \to \infty} \frac{a}{bT^2} \left[(P - C_3) + I_c P \left\{ R_i + \frac{1}{b} \right\} \right.$$

$$\left. - C_2 \left\{ \frac{a}{2Kb} - \frac{1}{b} \right\} + I_f C_3 \frac{\exp(bR_i)}{b} \right]$$

$$+ \lim_{T \to \infty} \frac{a}{bT^2} \exp(2bT) \left[-C_2 \left\{ -\frac{aT}{K} + \frac{a}{2Kb} \right\} \right]$$

$$+ \lim_{T \to \infty} \frac{C_1}{T^2} = \frac{a}{b} \lim_{T \to \infty} \frac{F(T)}{T^2} + \frac{a}{b} \lim_{T \to \infty} \frac{G(T)}{T^2}$$

$$= \frac{a}{b} \lim_{T \to \infty} \frac{F(T) + G(T)}{T^2}$$

$$\simeq \frac{a}{b} \lim_{T \to \infty} \frac{G(T)}{T^2} \quad \text{since } |G(T)| > |F(T)|$$

$$= \frac{a}{b} \lim_{T \to \infty} \frac{G'(T)}{2T} = \frac{a}{2b} \lim_{T \to \infty} G''(T) \longrightarrow +\infty,$$

$$\tag{D.6}$$

where

$$F(T) = \left[\exp(bT) \left[(P - C_3)(bT - 1) \right. \right.$$

$$- C_2 \left\{ \frac{aT}{K} + T^2 b - \frac{a}{Kb} - T + \frac{1}{b} \right\}$$

$$\left. - I_f C_3 \left\{ (T - R_i) Tb + R_i + \frac{1}{b} \right\} \right]$$

$$+ \left[\frac{(P - C_3)}{T} + I_c P \left\{ R_i + \frac{1}{b} \right\} \right.$$

$$\left. - C_2 \left\{ \frac{a}{2Kb} - \frac{1}{b} \right\} + I_f C_3 \frac{\exp(bR_i)}{b} \right]$$

$$+ \frac{C_1}{T^2} \right],$$

$$G(T) = -\frac{aC_2}{2Kb} \exp(2bT)(1 - 2bT).$$

$$\tag{D.7}$$

Now,

$$G(T) = -\frac{aC_2}{2Kb} \exp(2bT)(1 - 2bT) \Longrightarrow G'(T)$$

$$= \frac{2abTC_2}{K} \exp(2bT) \Longrightarrow G''(T)$$

$$= \frac{(2b + 1) 2abC_2}{K} \exp(2bT) \Longrightarrow G''(T) \longrightarrow +\infty.$$

$$\tag{D.8}$$

Hence, Z_{1i} is monotonic increasing if $(dZ_{1i}/dT)|_{T=R_i} > 0$ as Z_{1i} does not have stationary points in $[R_i, \infty)$. Here, $t_1 = (a/Kb)(\exp(bT) - 1)$ and t_1 increases with increasing value of T. As in our model, $t_1 < R_i$; hence, for the feasibility of the model, $t_1^* = R_i$ and T^* is obtained from $(a/b)(\exp(bT^*) - 1) = KR_i$. Therefore, Z_{1i} has a maximum value at $T^* = (1/b) \ln(1 + KbR_i/a)$ if $(dZ_{1i}/dT)|_{T=R_i} > 0$. Here T^* is unique, hence the proof. $\square$

Proof of Lemma 8. The first part of the lemma is obvious. For the second part, if Z_{2i} does not have any stationary points in $[0, R_i]$, then Z_{2i} is either monotonic increasing or monotonic decreasing function of $T \in [0, R_i]$. Here $dZ_{2i}/dT \to \infty$ as $T \to 0$ because

$$\frac{dZ_{2i}}{dT}$$

$$= \frac{a}{bT} \left[(P - C_3) \left(b \exp(bT) - \frac{(\exp(bT) - 1)}{T} \right) \right.$$

$$- \frac{I_c P}{T} \left\{ Tb \exp(bT) - R_i b \exp(bT) \right.$$

$$- \frac{1}{bT} (1 - \exp(bT)) + \frac{R_i \exp(bT)}{T}$$

$$\left. - \frac{R_i}{T} - \exp(bT) \right\}$$

$$- C_2 \left\{ \frac{a \exp(bT)}{K} - \frac{a \exp(2bT)}{K} \right.$$

$$+ Tb \exp(bT) - \frac{a \exp(bT)}{KbT}$$

$$+ \frac{a \exp(2bT)}{2TKb} - \frac{1}{bT} + \frac{a}{2KbT}$$

$$\left. \left. - \exp(bT) + \frac{\exp(bT)}{bT} \right\} + \frac{bC_1}{aT} \right].$$

$$\tag{D.9}$$

Using L'Hospital's Rule, we obtain

$$\lim_{T \to 0} \frac{dZ_{2i}}{dT}$$

$$= \lim_{T \to 0} \frac{a}{bT^3}$$

$$\times \left[(P - C_3) \{bT \exp(bT) - (\exp(bT) - 1)\} T \right.$$

$$- I_c P \left\{ T \left(Tb \exp(bT) - R_i b \exp(bT) \right) \right.$$

$$- \frac{1}{b} (1 - \exp(bT)) + R_i \exp(bT)$$

$$\left. - R_i - T \exp(bT) \right\}$$

$$- C_2 \left\{ T \left(\frac{a \exp(bT) - a \exp(2bT)}{K} + Tb \exp(bT) \right) \right.$$

$$- \frac{2a \exp(bT) - a \exp(2bT)}{2Kb} - \frac{1}{b} + \frac{a}{2Kb}$$

$$\left. \left. - T \exp(bT) + \frac{\exp(bT)}{b} \right\} + \frac{bC_1 T}{a} \right]$$

$$= \lim_{T \to 0} \frac{H(T)}{T^3} = \lim_{T \to 0} \frac{H'(T)}{3T^2} \longrightarrow +\infty,$$

$$(D.10)$$

where

$$H(T)$$

$$= \frac{a}{b} \left[(P - C_3) \right.$$

$$\times \{bT \exp(bT) - (\exp(bT) - 1)\} T$$

$$- I_c P \left\{ T \left(Tb \exp(bT) - R_i b \exp(bT) \right) \right.$$

$$- \frac{1}{b} (1 - \exp(bT)) + R_i \exp(bT)$$

$$\left. - R_i - T \exp(bT) \right\}$$

$$- C_2 \left\{ T \left(\frac{a \exp(bT) - a \exp(2bT)}{K} + Tb \exp(bT) \right) \right.$$

$$- \frac{2a \exp(bT) - a \exp(2bT)}{2Kb} - \frac{1}{b} + \frac{a}{2Kb}$$

$$\left. \left. - T \exp(bT) + \frac{\exp(bT)}{b} \right\} + \frac{bC_1 T}{a} \right].$$

$$(D.11)$$

Now,

$$H'(T)$$

$$= \frac{a}{b} \left[(P - C_3) \right.$$

$$\times \{bT \exp(bT) - \exp(bT) + 1 + T^2 b^2 \exp(bT)\}$$

$$- I_c P \left\{ Tb \exp(bT) + T^2 b^2 \exp(bT) \right.$$

$$\left. - R_i Tb^2 \exp(bT) \right\}$$

$$- C_2 \left\{ T \left(\frac{a \exp(bT) - a \exp(2bT)}{K} - \exp(bT) \right) \right.$$

$$- T^2 \left(\frac{ab \exp(bT) - 2ab \exp(2bT)}{K} \right.$$

$$\left. + 2b \exp(bT) \right) + T^3 b^2 \exp(bT)$$

$$- \frac{2a \exp(bT) - a \exp(2bT)}{2Kb}$$

$$\left. \left. - \frac{1}{b} + \frac{a}{2Kb} + \frac{\exp(bT)}{b} \right\} + \frac{bC_1}{a} \right].$$

$$(D.12)$$

Here, $\lim_{T \to 0} H'(T) = C_1$ so that Z_{1i} is monotonic increasing if $(dZ_{1i}/dT)|_{T=R_i} > 0$ as Z_{1i} does not have stationary points in $[0, R_i]$. Here, $t_1 = (a/Kb)(\exp(bT) - 1)$, and t_1 increases with increasing value of T. According to our model, $t_1 < R_i$; hence, for the feasibility of the model, $t_1^* = R_i$ and T^* is obtained from $(a/b)(\exp(bT^*) - 1) = KR_i$. Therefore, Z_{1i} has a maximum value at $T^* = (1/b) \ln(1 + KbR_i/a)$ if $(dZ_{1i}/dT)|_{T=R_i} > 0$. Here, T^* is unique, hence the proof. $\square$

Acknowledgments

The authors want to acknowledge the reviewers for their constructive comments to revise the paper. The first author would like to express his heartiest gratitude to his parents, wife, and son for their valuable support during the research. This research is supported by the University Grant Commission, Delhi, India, from the research grant of the Minor Project (no. 41-1433/2012(SR)). The corresponding author wants to acknowledge the infrastructural assistance available at Vidyasagar University, West Bengal, India.

References

[1] F. W. Harris, "How many parts to make at once factory," *The Magazine of Management*, vol. 10, pp. 135–136, 1913.

[2] W. A. Donaldson, "Inventory replenishment policy for a linear trend in demand: an analytical solution," *Operational Research Quarterly*, vol. 28, pp. 663–670, 1977.

[3] S. K. Goyal, "On improving replenishment policies for linear trend in demand," *Engineering Costs and Production Economics*, vol. 10, no. 1, pp. 73–76, 1986.

[4] A. Goswami and K. S. Chaudhuri, "EOQ model for deteriorating items with shortages and a linear trend in demand," *Journal of the Operational Research Society*, vol. 42, no. 12, pp. 1105–1110, 1991.

[5] S. K. Goyal, D. Morin, and F. Nebebe, "Finite horizon trended inventory replenishment problem with shortages," *Journal of the Operational Research Society*, vol. 43, no. 12, pp. 1173–1178, 1992.

[6] M. A. Hariga and L. Benkherouf, "Optimal and heuristic inventory replenishment models for deteriorating items with exponential time-varying demand," *European Journal of Operational Research*, vol. 79, no. 1, pp. 123–137, 1994.

[7] H. M. Wee, "A deterministic lot-size inventory model for deteriorating items with shortages and a declining market,"

Computers and Operations Research, vol. 22, no. 3, pp. 345–356, 1995.

[8] S. Khanra and K. S. Chaudhuri, "A note on an order-level inventory model for a deteriorating item with time-dependent quadratic demand," *Computers and Operations Research*, vol. 30, no. 12, pp. 1901–1916, 2003.

[9] S. Sana and K. S. Chaudhuri, "On a volume flexible production policy for a deteriorating item with time-dependent demand and shortages," *Advanced Modeling and Optimization*, vol. 6, no. 1, pp. 57–74, 2004.

[10] L. E. Cárdenas-Barrón, "Optimal ordering policies in response to a discount offer: corrections," *International Journal of Production Economics*, vol. 122, pp. 783–789, 2009.

[11] B. Sarkar, S. S. Sana, and K. Chaudhuri, "An imperfect production process for time varying demand with inflation and time value of money—an EMQ model," *Expert Systems with Applications*, vol. 38, no. 11, pp. 13543–13548, 2011.

[12] P. L. Abad, "Determining optimal selling price and lot size when suppliers offers all unit quantity discounts," *Decision Science*, vol. 19, pp. 622–634, 1988.

[13] K. H. Kim and H. Hwang, "An incremental discount pricing schedule with multiple customers and single price break," *European Journal of Operational Research*, vol. 35, no. 1, pp. 71–79, 1988.

[14] S. K. Goyal, "Economic order quantity under conditions of permissible delay in payments," *Journal of the Operational Research Society*, vol. 36, no. 4, pp. 335–338, 1985.

[15] S. P. Aggarwal and C. K. Jaggi, "Ordering policies of deterioration items under permissible delay in payments," *Journal of the Operational Research Society*, vol. 46, pp. 658–662, 1995.

[16] P. Chu, K.-J. Chung, and S.-P. Lan, "Economic order quantity of deteriorating items under permissible delay in payments," *Computers and Operations Research*, vol. 25, no. 10, pp. 817–824, 1998.

[17] A. M. M. Jamal, B. R. Sarker, and S. Wang, "Optimal payment time for a retailer under permitted delay of payment by the wholesaler," *International Journal of Production Economics*, vol. 66, no. 1, pp. 59–66, 2000.

[18] J.-T. Teng, "On the economic order quantity under conditions of permissible delay in payments," *Journal of the Operational Research Society*, vol. 53, no. 8, pp. 915–918, 2002.

[19] F. J. Arcelus, N. H. Shah, and G. Srinivasan, "Retailer's pricing, credit and inventory policies for deteriorating items in response to temporary price/credit incentives," *International Journal of Production Economics*, vol. 81-82, pp. 153–162, 2003.

[20] Y.-F. Huang, "Economic order quantity under conditionally permissible delay in payments," *European Journal of Operational Research*, vol. 176, no. 2, pp. 911–924, 2007.

[21] Y.-F. Huang, "Optimal retailer's replenishment decisions in the EPQ model under two levels of trade credit policy," *European Journal of Operational Research*, vol. 176, no. 3, pp. 1577–1591, 2007.

[22] L. E. Cárdenas-Barrón, "The derivation of EOQ/EPQ inventory models with two backorders costs using analytic geometry and algebra," *Applied Mathematical Modelling*, vol. 35, no. 5, pp. 2394–2407, 2011.

[23] H.-M. Teng, P.-H. Hsu, Y. Chiu, and H. M. Wee, "Optimal ordering decisions with returns and excess inventory," *Applied Mathematics and Computation*, vol. 217, no. 22, pp. 9009–9018, 2011.

[24] B. Sarkar, "An EOQ model with delay in payments and stock dependent demand in the presence of imperfect production," *Applied Mathematics and Computation*, vol. 218, no. 17, pp. 8295–8308, 2012.

[25] B. Sarkar, "An EOQ model with delay in payments and time varying deterioration rate," *Mathematical and Computer Modelling*, vol. 55, no. 3-4, pp. 367–377, 2012.

[26] K. Forghani, A. Mirzazadeh, and M. Rafiee, "A price-dependent demand model in the single period inventory system with price adjustment," *Journal of Industrial Engineering*, vol. 2013, Article ID 593108, 9 pages, 2013.

13

Fuzzy Quality Function Deployment: An Analytical Literature Review

Mohammad Abdolshah[1] and Mohsen Moradi[2]

[1] Department of Engineering Faculty, Islamic Azad University, Semnan Branch, P.O. Box 35136-93688, Semnan, Iran
[2] Department of Industrial Engineering, Semnan University, Semnan, Iran

Correspondence should be addressed to Mohammad Abdolshah; abdolshah@gmail.com

Academic Editor: Hsin-Hung Wu

This paper presents an analytical literature review on fuzzy quality function deployment (FQFD) of papers published between 2000 and 2011. In this review, publications were divided into two main groups. First group included publications which proposed some models to develop FQFD. The second one was related to new applications of FQFD models. Next, publications were analyzed and research gaps and future directions were presented. We reached some conclusions including the following. (i) Most of studies were focused on quantitative methods to accomplish phase 1 of QFD or House of Quality (HoQ). The most employed techniques were multicriteria decision making (MCDM) methods. (ii) Although main purpose of using QFD was product development, other factors such as risk and competiveness analysis should be considered in product development process. (iii) A promising approach is using of metaheuristic methods for solving complicated problems of FQFD. (iv) There are a few studies on completing all phases of FQFD.

1. Introduction

Quality function deployment (QFD) is a customer-driven product development tool to achieve higher customer satisfaction through translating customer needs (CNs) into design requirements (DRs), part characteristics (PCs), and production plans and control [1]. Chan and Wu [2] defined QFD as "a system to assure that customer needs drive product design and production process." QFD is used essentially in order to design product according to customer favorites. A general QFD process consists of 4 phases. First phase, which is called House of Quality (HoQ), is an important stage in deploying QFD process. In this stage, after determining CNs and technical characteristics (TCs), relationships between CNs (Whats) and TCs (Hows) as well as their interdependencies are established and their importance weight is calculated [1]. In second phase TCs are translated to important PCs. Critical parameters of process are established in third stage and finally production requirements are specified (fourth phase) [3].

Most of required data in QFD processes and activities are expressed in natural language. Customers, for example, say their expectations from product by using expressions such as "easy to use," "safe," and "comfortable" which all of them have ambiguity. Computing these ambiguities in a requirement is an important issue [4]. Using tools from fuzzy sets and their concepts, we can approximate linguistic data to a numeric precision [5]. This review, consisting of a bank with more than 70 papers, divided publications in two main groups. First was publications combining FQFD with other methods to develop its efficiency and effectiveness, and latter was publications in new and major applications of FQFD. Each of these two groups was divided into subgroups. In the 2nd section we discuss these two groups, respectively. Discussed papers were analyzed in the fourth section and the literature vacancies were described.

2. Proposed Models for Developing FQFD

There are lots of models proposed to develop FQFD. In fact, according to wide aspects of deploying fuzzy QFD, it may be used in combination with many other methods and models. All of these models are common in using fuzzy logic.

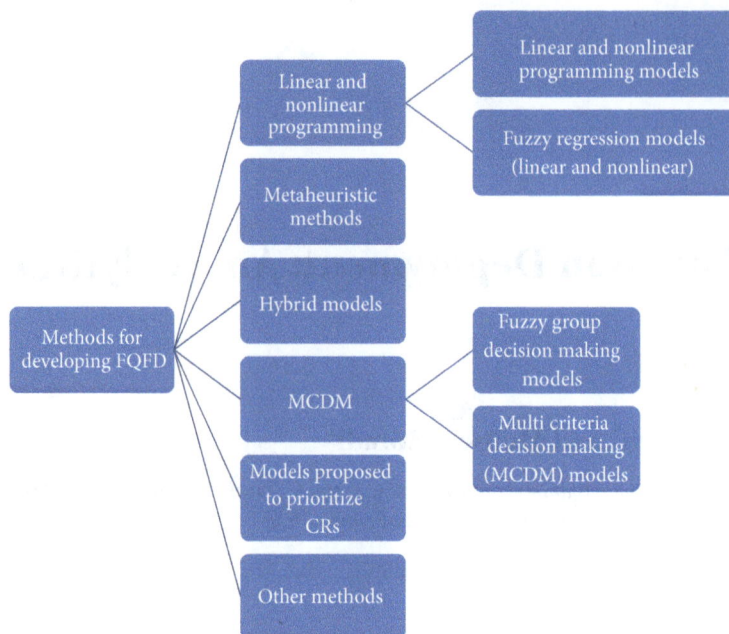

FIGURE 1: Categorizing the models for developing FQFD.

Here, according to the literature, this group is divided into 5 following subgroups to be discussed. Then as shown in Figure 1 these 5 groups can be discussed in 8 different groups.

2.1. Fuzzy Linear and Nonlinear Programming Models. Linear programming methods have been widely used in FQFD. These models are mostly employed to compute fulfillment levels of process parameters (e.g., TCs, PCs) respecting some constraints like budget, technical difficulty, and technology. Followings are some presented models in the literature.

Ko and Chen [6] presented a fuzzy linear programming (FLP) using QFD and fuzzy normal relations evaluation method for new product development (NPD). They used FLP to determine fulfillment levels of engineering characteristics and design requirements to maximize customer satisfaction with respect to company's sources, technical difficulties, and market competition constraints. In this model, in order to identify relationships between CRs and DRs as well as interdependencies within DRs, Wassermann's equation has been used. Also relationships between characteristics were represented by fuzzy number to cope with fuzziness. In that model, percentage increase of costs corresponding to increase of fulfillment levels of each TC was specified and total costs could not be more than a given value. Minimum fulfillment level of each DR was known and fulfillment level of DR couldn't be less.

Luo et al. [7] proposed an optimization method based on FQFD for part selecting. Their model became an integer linear programming model at the end. This model considered CNs as fuzzy numbers and translated them to TCs and finally to PCs. Their final goal was minimizing the differences between CNs and end product. Chen and Ko [1] proposed fuzzy linear programming models to determine the fulfillment levels of PCs and failure modes and effect analyses

(FMEA) for risk analysis. This paper proposed FLP models for determining fulfillment levels of PCs to achieve given levels of DRs to satisfy customers and considered both phases 1 and 2 of NPD. Furthermore, to reduce risk FMEA as a constraint for models in QFD process was introduced. To assure customer satisfaction, this paper considered fulfillment level of Jth DR in phase 2 more than or equal to that in phase 1.

Chen and Ko [3], in other study, considered close relationships of 4 phases of QFD by means of end chain concept (MEC) and proposed a set of FLP models to determine contribution level of HOWs in customer satisfaction. They also used FMEA in phases 2 and 3 and risk evaluation in phase 4, to solve potential risk problem in NPD project. In addition to DRs, in these 4 phases approach PCs, process planning (PP), and process requirements (PRs) have been considered. Solving risk problem in NPD, the authors used fuzzy FMEA as a model constraint for risk evaluation of DRs, PCs, PPs, and PRs in 4 phases. Reliability of models' results depended on reliability of relations between variables and parameters in HoQ, as they said. It was proposed to apply other concepts such as Kano's model instead of MEC in QFD modeling.

In addition to FLP models, one can use fuzzy nonlinear programming (FNLP) models in QFD. Although using these models makes it more difficult, they have a better efficiency in sophisticated problem, because in such problems, relations are often non-linear.

Chen and Ko [6] presented an FNLP model in combination with Kano's model to specify performance levels of DR. They classified DRs in three groups of exciting, functional, and basic using Kano's model. This FNLP was based on FLP model of Chen and Weng (2003). They illustrated with an example that their model was better than corresponding NLP model. Sener and Karsak [8] proposed a hybrid model discussed in Sections 2.2–2.8.

2.2. Multicriteria Decision Making (MCDM) Models. It can be claimed that among quantitative models used in FQFD, MCDM models are the most used ones. These models in QFD matrixes (often HoQ) are applied in determining importance of parameters.

Kim et al. [9] used MCDM models for calculating target values of TCs as an optimization problem. Target function was finding target value levels of TCs in such a way that customer satisfaction became maximized for target function values.

Kwong and Bay [10] employed fuzzy analytical hierarchy process (FAHP) to weight customer needs. They proposed a fuzzy model for weighting CNs based on fuzzy measurements and traditional AHP. They used triangular fuzzy numbers (TFNs). Two advantages of their model was (1) using fuzzy numbers to deal with ambiguous nature of human judgments and (2) adoption of fuzzy numbers allowed designing team to have freedom of estimation regarding overall goal of customer satisfaction. These estimations could be optimistic or pessimistic.

Erol and Ferrell [11] proposed a method to select among finite alternatives when there are more than one object and data are both quantitative and qualitative. FQFD was used to convert qualitative data to quantitative ones; then this data was combined with other quantitative one to construct multiobjective mathematical model. A modified version of preemptive goal programming was used to solve the model and some solutions close to ideal solution were provided for decision makers to make final decision.

Büyüközkan et al. [12] employed fuzzy analytical network process (FANP) to determine DR weights. They used TFNs in their model. Although FANP proposed in this paper has more calculations than other methods such as FAHP and is more time consuming, results are more precise and, considering long-term competiveness, it is more useful for companies.

Kahraman et al. [5] purposed a fuzzy optimization model for specifying product's technical requirement which should be considered during design process in QFD. Considering interdependencies, target function's coefficient were calculated using FANP. Doing the same with FAHP, they compared results of both methods. Although results were close because of relations within TCs, apparently, FANP had better credibility.

Chen and Weng [13] presented an approach for evaluating of TCs in FQFD using goal programming models. They employed fuzzy goal programming to identify fulfillment levels of DRs according to constraints such as cost, customer satisfaction, and technical difficulties. Coefficients of goal programming were fuzzy to reflect ambiguity in linguistic data. This model also considered business competiveness in determining minimum fulfillment level of DRs. This model used α-Cuts to determine membership function of fuzzy goals and fulfillment levels of DRs.

Gunasekaran et al. [14] purposed an MCDM method using Monte Carlo simulation and FQFD for optimization of supply chain management (SCM). Here, customer needs were simulated by FQFD, Monte Carlo simulation, and a multiobjective model for optimizing customer preferences.

By presenting some novel criteria such as design complexity, speed of construction, environment, and aesthetic and construction complexity, Mousavi et al. [15] developed a hybrid FQFD model using fuzzy technique for order performance by similarity to Ideal Solution (TOPSIS) for bridge scheme selecting, which was a complex engineering project.

Lin and Lee [16] presented a model of QFD and FANP to solve NPD problem in TFT-LCD industry. They used fuzzy Delphi method to simplify number of factors in the model and made paired comparisons easier.

Celik et al. [17] proposed an integrated FQFD model for investment decisions of shipping routing in crude oil tanker market. This model used FQFD principles for shipping investment process through substituting HoQ with ship of quality (SoQ). In addition, FAHP and fuzzy axiomatic design (FAD) algorithms were integrated in SoQ in order to reach quantitative results. SoQ framework was tested by a set of periodic data and current trends in tankers market for big companies like Suezmaxes and Aftermaxes.

Khademi-Zare et al. [18] provided two prioritizing models based on FQFD for ranking strategic actions (SAs) of mobile cellular telecommunication in Iran. Considering the gap between current situation and positive ideal situation for customer attribute (CAs), they used TOPSIS for ranking CAs in first model. Using AHP in the second model, more factors have been considered to rank CAs. A fuzzy utility factor, B_j, was introduced for ranking CAs in both models. Both models were able to consider ambiguity in human judgments and allowed customer-oriented companies like Mobile Communication of Iran (MCI) to use voice of customers for extracting benefits in a more expressive way.

Lin et al. [19] presented a model to analyze environmental requirements of products with linguistic preferences using FQFD and ANP with interdependence relations between environmental production requirements (EPRs) and sustainable product indicators (SPIs). At first, to facilitate main issue of QFD problem, Whats questions of EPRs and Hows questions of SPIs have been made, which were two main components of HoQ. In conjunction with fuzzy set theory and ANP, the systemic analytical approaches were proposed.

Mousavi [15] proposed a systematic decision making process for evaluating conceptual bridge design and selecting the best idea through the new methodology based on integrated optimization. In the first phase, QFD has been employed to translate project requirements to design requirements. Then, the best structure as an alternative was selected by TOPSIS based on weighted criteria achieved from first phase. In this study the rating values regarding each alternative and criteria throughout the phases were described in a fuzzy environment by means of linguistic variables. Finally, a case study was provided to illustrate the implementation process of the integrated methodology for bridge superstructure design.

In his work, Wang [20] utilized FQFD and MCDM methods for optimized modular design. This article integrated FAHP and fuzzy decision making trial and evaluation laboratory (DEMATEL) to achieve weights of DRs and created constructing blocks of optimum modular design.

Nepal et al. [21] proposed an FAHP framework according to fuzzy set theory, to prioritize customer satisfaction (CS) attributes in target planning for vehicle design. In addition, unlike previous studies, they considered a broad range of technical and strategic factors to evaluate weights. Then, these weights were introduced in target planning by identifying the gap in current CS level. This framework was deployed in Microsoft Excel, so it could be utilized with a limited training in organization. Unlike traditional AHP, FAHP gave QFD team members freedom of estimation, since judgments could vary from very optimistic to very pessimistic.

Zheng et al. [22] presented an evaluation method for integrating fourth-party logistics (4PL) supply chain based on FQFD and AHP. At first, they studied 4PL requirements and introduced them into supply chain design process by QFD. Then, nonvalue processes were eliminated using benchmarking method and supply chain was rebuilt by supply chain fuzzy evaluation theory. Utilizing multiobjective optimizing theory, optimum model of 4PL was achieved by evaluating value. Finally, they used FAHP to present evaluation method of 4PL integration.

Gungor et al. [23] proposed a fuzzy decision making system (FDMS) based on fuzzy control rules for new product design using FQFD. Customer needs, as inputs of FDMS, were fuzzified and prioritized using membership function concept. DRs were rated by if-then rules. This paper determined fulfillment levels of DRs by FANP and its results were compared with those of proposed FDMS.

Lee and Lin [24] proposed a fuzzy model consisting of FANP and QFD in new product development. Parameters of each phase of QFD were determined through literature review and interviewing with experts and fuzzy Delphi method was utilized for screening important factors. Also for constructing relations within HoQ, fuzzy interpretive structural modeling with FANP was used.

Liu et al. [25] prepared a model for product design and selection using FQFD and MCDM methods. This paper integrated FQFD and product prototype selection and proposed a method for product design selection. α-Cuts operation was utilized in QFD for fuzzy set calculations of each component. Competitive analyses as well as relations between TCs were considered too. Considering TCs and product development factors in product prototype selection, an MCDM method was proposed for prototype selection. This method provided product developers with useful information and precise analysis. It is a useful decision aid tool.

Liu and Wang [26] proposed an advanced QFD model with FANP to consider interdependencies within and relations between QFD components. To extend applicability of QFD, authors were seeking to extend QFD from product planning phase for part deploying phase.

Zandi and Tavana [27] proposed a methodology for evaluation and selection of the best e-CRM framework in agile production using FQFD, fuzzy group real option analysis (ROA), and fuzzy TOPSIS method. First, e-CRM frameworks were prioritized according to financially oriented characteristics using an ROA model. Next, the e-CRM frameworks were ranked according to their customer-oriented characteristics using a hybrid fuzzy group permutation and a four-phase fuzzy quality function deployment (QFD) model with respect to three main perspectives of agile manufacturing (i.e., strategic, operational, and functional agilities). Finally, the best agile e-CRM framework was selected using a TOPSIS model.

Zarei et al. [28] proposed a hybrid methodology of AHP-QFD to increase leanness of food supply chain. Linking lean attributes (LAs) and lean enablers (LEs), this study used quality function deployment (QFD) to identify viable LEs to be practically implemented in order to increase the leanness of the food chain. Furthermore, triangular fuzzy numbers were used to deal with linguistic judgments expressing relationships and correlations required in QFD. FAHP was employed to prioritize LAs.

Yousefie et al. [29] proposed an original approach for the management tools selection based on the quality function deployment (QFD) approach. Specifically, the research addressed the issue of how to deploy the house of quality (HOQ) to effectively and efficiently improve management tools selection processes and thus company satisfaction about its excellence achievement. Entropy method was used to perform competitive analysis and calculation weights of competitive priorities. Then, FAHP was employed to prioritize management tools. Presented hybrid models and methodologies by Karsak and Özogul [30], Huang and Li [31], Wang and Chin [32], Tolga and Alptekin [33], Karsak et al. [34], and Ozdemir and Ayag [35] were discussed in Section 2.8.

2.3. Fuzzy Group Decision Making Models. Zhang and Chu [36] introduced a model for selection of multiple design schemes of complex products based on QFD and fuzzy group decision making. They divided the scheme to parts and evaluated weights of each part using group decision making in QFD and rank the designs. In addition to selection of optimized design, QFD was employed to integrate customer requirements into design. This paper also used group decision making for determining relations between characteristics.

Büyüközkana et al. [37] proposed a new fuzzy group decision making method for product design by QFD. In their model, inputs of QFD process were both quantitative and qualitative. Quantitative data was translated to qualitative one; then fuzzified and finally weights of TCs were computed.

Liu [38] used a fuzzy group decision making with risk-taking attribute to deploy QFD, because they thought decision makers' opinions about risk taking were different (optimistic, normal, and conservative). Finally, TCs were prioritized and HoQ was completed.

Zhang and Chu [36] proposed a fuzzy group decision making to aggregate multiformat and multigranularity judgments of decision makers based on two optimizing models (logarithmic least square and weighted least square) for constructing HoQ. Logarithmic least square model and two normalized formulae were utilized to solve fuzzy paired comparisons matrices and normalizing evaluations in linguistics format. Then, weighted least square model was used to aggregate ultimate normalized multigranularity evaluations. Sanayei et al. [39] utilized MCDM model of Vlse Kriterijumska Optimizacija I Kompromisno Resenje (VIKOR), which means multicriteria optimization and compromise solution, in association with fuzzy group decision making for supplier selection. VIKOR method is a multicriteria decision making method for solving problems which have inconsistent criteria. It selects an alternative which has the least distance from ideal solution.

Lin et al. [40] employed QFD and fuzzy group decision making for service innovation. This study developed a scientific framework for tourism service management from epistemology perspective. This article used FQFD and fuzzy group decision making to analyze various service evaluation criteria. In particular, the methodology allowed the identification of service attributes perceived to affect service design performances from the tourist's point of view, enabling the assessment of possible gaps between tourists' and hotel's perception of service delivery. To assess viable strategic designs, in the proposed approach they introduced a utility factor, considering the costs of implementation for each "How." Wanga and Xionga [41] provided an integrated approach of group decision making based on linguistic variables for QFD applications. Proposed methodology does its computations with words—without converting to fuzzy numbers—so information loss risk is low. Hybrid model of Liu et al. [26] is discussed in Section 2.8.

2.4. Metaheuristic Methods.
According to meta-heuristic algorithms' ability to solve complex problems, they can be used in different stages of QFD process. Discussed articles often use meta-heuristic methods to identify relations between CNs and TCs.

Hsiao and Liu [42] proposed a neurofuzzy evolutionary approach for product design. Their model was based on artificial intelligence including fuzzy theory, backpropagation neural network (BPN), and genetic algorithm (GA), along with morphological analysis to synthesize, evaluate, and optimizate of product design. Fuzzy logic was utilized for modeling imprecise market information, BPN for determining relations between CNs and design parameters, morphological analysis for constructing design alternatives, and finally GA was used to select optimum design.

Lin et al. [40] prepared an intelligence model to estimate product design time (PDT) using intelligence method. Due to lack of information in early stage of product development, this model utilized fuzzy logic to complete HoQ. This model employed QFD to extract DRs from CNs. Then a fuzzy neural network was built to combine data and estimated PDT, which made use of fuzzy comprehensive evaluation to simplify structure. In a word, the whole estimation method consisted of four steps: time factors identification, product characteristics extraction by QFD and function mapping pattern, FNN learning, and PDT estimation. Finally, to illustrate the procedure of the estimation method, the case of injection mold design was studied. This model had some limitations. It was inapplicable for developing brand new products, because the influencing weights of linguistic variables obtained by experience or experiment were important for the parsimonious FNN model.

As the functional relationships between customer requirements and engineering characteristics in QFD are uncertain, unclear, and fuzzy, Huang and Li [31] proposed radial basis function (RBF) to determine the functional relationships for QFD and QFD functional relationships model based on RBF. According to RBF neural network, nonlinear mapping space from the input space to the output can be realized, and optimal relationships pattern of the input and output would be obtained. The customer requirements and engineering characteristics in QFD constituted the input and output of the RBF.

The optimal relationships were constructed through the neural network training. Wanga and Xionga [41] analyzed the limitations of traditional methods by using the product planning HoQ; the available linguistic terms based on experts' knowledge, and with the artificial neural network, were introduced to realize neural network-based fuzzy reasoning. The final importance of the technical requirements was evaluated reasonably and effectively. This model considered market competitiveness and technical competitiveness. The importance of technical requirement was determined through four steps: (1) acquisition of customer requirements and ranking of their importance measures, (2) establishment of technical requirements and their relations with customer requirements, (3) establishment of market competitiveness and technical competitiveness, and (4) evaluation of technical requirement importance with competitiveness.

Lee and Lin [24] proposed an ANN-based dynamic FQFD method in order to solve the dynamic and fuzzy nature of QFD; they solved the problem of the dynamic nature by using neural network method, while trapezoidal fuzzy number for its ambiguity was introduced. Firstly, a combined method with neural networks and FQFD was established; after learning and training, the method could quickly and effectively deliver customer requirements to the products designers. This model could solve a non-linear problem and met the non-linear changes in customer needs. Hybrid model of Liu [25] and Karsak and Özogul [30] paper were described in Section 2.8.

2.5. Fuzzy Regression Models (Linear and Nonlinear).
These models were used to complete HoQ and their main focus was on finding relations between parameters of HoQ. Chen and Chen [13] utilized a fuzzy nonlinear possibilistic regression approach to model product planning (first phase of QFD). The model was able to incorporate both qualitative and quantitative data in determining relations between CRs and DRs as well as within DRs. Using linear regression, some coefficients became crisp because of linear programming nature. Therefore, they used nonlinear regression in their model. TFNs were used in this model. Sener and Karsak [8] and Karsak et al. [34] studies were discussed in Section 2.8.

2.6. Models Proposed to Prioritize CRs.
Identifying CNs and their importance weight is first step of QFD process. Due to its importance, some papers have been focused on determining CNs and their importance weight. Lai et al. [43] proposed a new methodology in their articles which considered competitors information. This paper considered competitive environment and product's current performance.

The proposed algorithm was complicated, but the authors claimed that its solutions were more accurate than similar models. Mehdizadeh [44] used fuzzy centroid-based method and considered competitive environment to rank CRs. He employed both normal and nonnormal numbers. All of previous studies employed normal fuzzy numbers which resulted in misleading solutions in case of non-normal fuzzy numbers. Using fuzzy centroid-based method, this paper solved the problem.

2.7. Hybrid Models. Some of discussed models encompass more than one model of above ones, so we discussed them under subgroup of hybrid models. Using fuzzy set theory and Euclidean distance, Guo et al. [45] put forward a methodology called Euclidean space distances weighting ranking method. They applied this method in fuzzy AHP. This method satisfied additive consistent fuzzy matrix. In addition to designing an algorithm for calculating weights, they developed a module to design new product based on customer needs weight calculation model. Sener and Karsak [8] used an optimization model and fuzzy regression on the basis of non-linear programming to determine target values of TR. Using linear models, regression coefficients approached to zero and, due to this, they used nonlinear models to determine functional relationships between CRs and TRs and within them. Finally, using a fuzzy mathematical programming model, target levels of TRs have been achieved. This fuzzy mathematical programming model incorporated both center values and spread values of parametric estimations of functional relationships in optimization process and, thus, avoided loss of information in design phase.

Liu [25] used a fuzzy group decision making model in association with genetic algorithm in QFD process to complete HoQ and calculation of weights of TRs. The ranking was done in two phases according to budget and time constraints.

Huang and Li [31] employed BP neural network and AHP in QFD fuzzy to determine key technology of product planning and designing. First, they used AHP to determine weights of CRs and then, as the relationships between TRs and CRs, TRs and part characteristics in QFD were uncertain, nonlinear and fuzzy BP neural network was presented to determine these relationships. After determining weights, part characteristics were identified using QFD and key technology of production was selected.

Karsak and Özogul [30], in a paper titled an integrated decision making approach for ERP system selection, they presented a new decision framework for selecting ERP system based on QFD, fuzzy linear regression, and zero-one goal programming (ZOGP). Proposed framework considered both company requirements and features of ERP system and provided a tool not only to link company requirements and features of ERP system but also to interact between them through applying QFD principles. Using fuzzy linear regression, target level of ERP characteristics and reachable maximum values of CNs were achieved. Finally, ZOGP was employed to select ERP system which minimized weights aggregate of deviations from reachable maximum values of organizational needs.

Wang and Chin [32], used fuzzy AHP to prioritize characteristics in QFD. This model utilized linear goal programming (LGP) to calculate normalized fuzzy weights and pairwise comparisons matrices of AHP. Proposed LGP method was tested by three numerical examples including new product development (NPD). The results showed that LGP method can derive precise fuzzy weights for perfectly consistent fuzzy comparison matrices and normalized optimal fuzzy weights for inconsistent fuzzy comparison matrices on the basis of minimum deviation. It was also shown that fuzzy AHP can be used as a very useful decision support tool for NPD project screening.

Mu et al. [46] presented an integrated model of fuzzy multiobjective model and Kano's model to determine nonlinear relationships in HoQ matrix. They considered firm's budget as a constraint for the model and utilized fuzzy multiobjective model to achieve maximum customer satisfaction considering the constraint.

Karsak et al. [34] utilized ANP and GP in QFD to select TRs, emphasizing design process according to given targets. This algorithm included two phases. At the first phase, HoQ was constructed using ANP, and at the second phase, TRs that should be focused were determined using ZOGP.

Tolga and Alptekin [33] integrated QFD with fuzzy compromise-based goal programming to identify how much product features should be improved. ANP is utilized to evaluate inner dependencies within customer needs, product attributes, and also the relationships between them. The constraints of compromise-based goal programming included manufacturer budget and product competitive performance in market.

Ozdemir and Ayag [35] proposed a model for NPD multiobjective problem and its solution with intelligent approach. They combined multiattribute method of TOPSIS and multiobjective method of goal programming (GP) as well as economic analysis. Economic analysis was used for demand changes during product life cycle. Fuzzy equivalent worth of each new product alternative was determined using fuzzy life cycle monetary input, and results were input to the multicriteria analysis. The selected multicriteria analysis tool was TOPSIS in the paper and some additional judgmental criteria were also considered to rank new product alternatives. The preference weights used to rank alternatives were fed to a goal programming model which made an ultimate selection of new product(s) to be produced in the manufacturing system under capacity, sales potential, and workforce balance constraints. The goal programming model had an objective of minimizing the weighted sum of positive deviation from target of total cost and negative deviation of target total preference weights. The proposed approach was implemented using real data of a continuous production system for the need of introducing new products into market.

Wang [20] integrated QFD with multiattribute decision making (MCDM) models for optimal modular design. This paper combined fuzzy AHP and fuzzy DEMATEL to develop marketing-driven product. Fuzzy AHP was used to determine important weight of CNs; then fuzzy DEMATEL was applied to achieve design requirements and to construct common/specific building blocks for achieving an optimal modular design.

Gungor et al. [23] developed a fuzzy decision making system (FDMS) based on fuzzy control rules to design new product using QFD which considered CNs as factors. Customer needs were determined as input variables and fuzzified using membership function concept. Weights of these factors were fuzzified to ensure the consistency of the decision maker while assigning the importance of each factor over another. By applying IF-THEN decision rules, DRs of the firm were scored. This paper also used fuzzy analytic-network process (FANP) to determine the fulfillment levels of DRs of the firm and its results were compared with FDMS's ones.

Sanayei et al. [39] used MCDM model of VIKOR (Vlse Kriterijumska Optimizacija I Kompromisno Resenje means multicriteria optimization and compromise solution) along with group decision making methods for selecting supplier. VIKOR was an MCDM method for solving problems with conflicting and noncommensurable (different units) criteria and choosing the alternative that was closest to ideal solution.

2.8. Other Methods. Kahraman et al. [5] used fuzzy ranking methods for evaluating DRs. Fuzzy ranking methods were rarely used in previous studies. This paper used three fuzzy ranking methods in PVC windows industry and finally compared them using sensitivity analysis.

3. Applications of FQFD

Some of papers discussed FQFD applications. Although QFD had numerous applications and traditional QFD has been deployed in lots of firms, here we discussed most important applications of FQFD. The most important application of QFD was supply chain management (SCM). Moreover, there were some papers in product design and other applications.

3.1. Supply Chain Management (SCM). In a paper about selecting supplier, Bevilacqua et al. [47] proposed a new method which translated HoQ as problem of a huge company of clutch coupling production.

Zhang and Chu [48] put forward a new method for contingency management of 3rd party logistics (3PL). They suggested a multiobject framework using FQFD and group decision making. In addition to contingency management, risk management was also considered.

Wang et al. [49] proposed a practicable method for determining customization place of service products logistics to improve capability of 3PL customized servicing. First, they used FQFD to import CNs in locating process. Then, a multiobject integer programming model was developed to determine optimum locating program under constraints of 3PL interior sources and payment values of buyer firm for outsourcing.

Rau and Fang [50] proposed a combined model of FQFD and TRIZ to solve conflicts within design characteristics for packaging design in notebooks logistic. In this model, first, design requirements and design characteristics were specified. Then their relationships were identified based on

language variables using experts' opinions using HoQ. Weight of design characteristics was calculated using fuzzy integral method. Finally, modifications and innovation principles are determined for solving conflicts of product pack design problem corresponding to priority of design characteristics.

Bottani and Rizzi [51] suggested a method for customer service management in supply chain using FQFD. The main intention was using HoQ to improve logistics processes effectively and efficiently. They estimated distance between firm's performance in logistic services and customer needs. According to customer importance, CNs were weighted to identify key factors for services improvement. The study began with identifying characteristics which purchased product for satisfying customers (what variables). Then it created criteria for supplier evaluation using a final ranking based on fuzzy proportion index for conclusion. All of process was deployed using fuzzy numbers. Using a fuzzy algorithm allowed company to define importance weights of What, relationship rates between Whats and Hows, and effect of each potential supplier by language variables. This paper attended specifically to different personal evaluations in HoQ accomplishment process and fuzzy triangular numbers were suggested to consider ambiguity in language evaluations.

Amin and Razmi [52] suggested a fuzzy integrated model based on firm's strategy for managing, evaluation, and supplier development. In the first phase, QFD was used to rank the best ISPs based on qualitative criteria. In the first phase, QFD was employed for ranking best ISPs based on qualitative criteria. Then, a quantitative model was used to consider quantitative criteria. Finally, two models were developed and the best ISP was selected. In the next phase, a new algorithm was developed to evaluate selected ISPs in three features: customer, performance, and competition. Fuzzy logic and triangular fuzzy numbers were used for dealing with ambiguity in human thinking. A case study brought forward to show evaluation and selection phases of ISPs.

Sohn and Choi [53] used QFD for SCM considering reliability. They developed an FQFD model to determine relationships between customer needs and design characteristics according to reliability test. They employ QFD to determine relationships between CNs and design characteristics in each supply chain. Then, they assumed these relationships were reversible and considered end user needs as a function of reliability test performance variables in the last chain of product design process. Finally, fuzzy MCDM was utilized to find an optimum solution set regarding demanded reliability performance. Other works including Gunasekaran et al. [14], Lin et al. [40], Zheng et al. [22], and Zarei et al. [28] were discussed in Section 2.

3.2. Product Design. Due to law restrictions and public pressures, many companies were making products consistent with environmental concerns. Many of these products were produced and sent to market so far, but most of them were rejected by customers and couldn't gain a market share. It seemed that they faced this problem because they just considered environmental conditions and neglected customer needs [54]. Recently, some studies have been focused on this problem to solve it by considering both environmental and customer requirements simultaneously using FQFD.

In a paper titled as "Integration of environmental considerations in quality function deployment by using fuzzy logic," Kuo et al. [54] proposed a developed Eco-QFD model to form a design team considering both environmental concerns and customer satisfaction by using QFD. A fuzzy group method was applied to Eco-QFD for product development planning to reduce the vagueness and uncertainty in a group decision making process. This fuzzy multiobjective model not only considered the overall customer satisfaction but also encouraged enterprises to produce an environmentally friendly product. With an interactive approach, the optimal balance between environmental acceptability and overall customer satisfaction could be obtained. Finally, a case study illustrating the application of the proposed model was also provided. Kuo and Hung [55] proposed a fuzzy Eco-QFD model to design products based on environmental considerations. They used a fuzzy multiobjective model to aid the design team in choosing target levels for engineering characteristics. Paper of Lin et al. [40] about environmental production was discussed in Section 2.2.

Lin et al. [40] employed FQFD to analyze Island hotels management by using lingual preferences. They proposed an approach for managing internal and external services as well as service innovations based on an FQFD framework. That was a methodology adapted to service development properly. The paper addressed how to apply HoQ to improve hotel services innovation process and tourists satisfaction efficiently and effectively. Fuzzy logic provided a methodology to deal with lingual judgments nature in HoQ. At the end a case study was discussed to examine model accuracy. Work of Lin et al. [40] on service innovation was discussed in Section 2.3.

3.3. Other Applications. Jia and Bai [56] suggested a method to develop manufacturing strategy using QFD. This method consisted of 11 steps and used QFD as a transferring tool to relate competitive factors with manufacturing decision groups (such as structural and infrastructural decision groups) and a main tool in different stages of developing manufacturing strategy. This paper also integrated fuzzy set theory with HoQ to deal with vagueness of decision process inputs.

Şen and Baraçlı [57] presented a methodology to find software selection requirements for organizations based on FQFD. They proposed a QFD approach to determine which nonfunctional requirements reported by recent studies were important in software selection decision making and combined it with functional requirements. The proposed solution not only helped decision makers to determine software selection's characteristics and criteria, but also provided their importance weights.

Zhang and Chu [48], using rough set theory, proposed a new methodology for FQFD to facilitate decision making in early stages of product development. At the end, they compared their proposed method with traditional FQFD and found result of proposed method more accurate.

TABLE 1: The frequency of proposed methodologies for FQFD.

ID	Method	Number of research
1	Fuzzy linear and nonlinear programming models	53
2	Fuzzy regression models (linear and nonlinear)	3
3	Metaheuristic methods	6
4	Hybrid models	13
5	Fuzzy group decision making models	8
6	Multicriteria decision making (MCDM) models	29
7	Models proposed to prioritize CRs	2
8	Other methods	1

Yan et al. [58] employed FQFD to design decision support system for hazardous material road transportation accidents. First, they identified problem of decision support system plan selection including requirements of governmental organization and transportation companies. Then, they used QFD to translate these requirements to technical characteristics of decision support system.

4. Conclusion

As mentioned, nowadays FQFD is used as a powerful tool in designing and developing products and decision making from supplier selection to Eco-design product development. In comparison with traditional QFD, using fuzzy logic is unavoidable. Actually, regarding application of lingual variables in paired comparisons, ratings, and weightings, using crisp numbers leads to lose information. Combining fuzzy logic and QFD has made a new methodology named fuzzy QFD. Using many case studies showed that this methodology is effective and reliable. In this review, after analyzing papers published between 2000 and 2011, we categorized papers in two major groups of (i) proposed models of FQFD and (ii) QFD employed applications. The first consisted of 8 subgroups and the second topic included 3 subgroups. Literature review showed that FQFD has been used mostly in supply chain management. After discussing and analyzing papers, we identified models' weakness, strength points, and literature vacancies and proposed some directions for future research. The frequency of proposed methodologies for FQFD (Table 1) showed that fuzzy linear and nonlinear programming models have been used 53 times with the first rank. Therefore it seems that since linear and nonlinear programming models can find the optimum solution, mostly researches tend to use this methodology.

5. Future Research

Most of discussed papers only studied first phase of QFD (i.e., quality matrix) and just completed HoQ. This was maybe due to importance of this stage, but to deliver a product according to customer requirements in target market, we should complete all phases of QFD. Results of first phase were weighted (or prioritized) TCs which should be translated to part characteristics and process requirements, respectively, so we could produce a product appropriate for target market. Hence it seems essential to extend proposed models for all phases of QFD. Regarding HoQ, other problem is its big size and complexity in many NPD problems. This issue, especially in fuzzy environment, caused greater complexity. Metaheuristic algorithms such as artificial neural networks (ANN), which were used to determine complex and non-liner relations, were one of feasible solutions. For this problem, most of researchers assumed that relations between TCs and CNs as well as relations within TCs were linear in HoQ. On the other hand, model's applicability and being user-friendly were more important than model. Introducing fuzzy logic with its comprehensive, long, and time-consuming calculation of the models has stopped many models in the theory phase and does not allow these models to become practicable. A good solution to this problem is to design some modules and software which make calculations easier and allow users to get their outputs by a little change in inputs. Another way to increase users' interest in proposed models is using expert and decision support systems which decrease calculations. It's noteworthy that designing databases is an effective way for gathering, maintenance, and easier access to CNs for next developments of products design and process risk analysis is another concept which is less studied in papers. Even the most optimistic people accept a percentage of risk for process. We should consider process risk in addition to QFD process and method execution because running the project in organization needs capital. This area needs many studies to be carried out, too. Other issue, which is less considered in papers, is use of appropriate fuzzy number. We should investigate and find suitable kind of fuzzy numbers for each especial application. Most of researches have used triangular fuzzy number (TFN) because their calculations were easier than other fuzzy numbers. Nevertheless, using TFNs might cause losing information in some cases. Also most of studies use normal fuzzy numbers. In this case, appropriate fuzzy number—normal or abnormal—should be determined. In some cases, algebraic operations are used for working with fuzzy sets which can lead to wrong results.we should use fuzzy operation such as x-cuts in these cases. Also, defuzzifying of fuzzy numbers many lead to lose information, so we should try not to defuzzify in early stages of calculations and use fuzzy operations as far as we can. Other proposed method to deal with this issue is using lingual variables and avoiding translating them into fuzzy numbers [59].

Although QFD has been invented to design product according to customer needs, we cannot just consider CNs and forget about market competiveness. It is possible to increase your market share even with a product lower than customer expectation because of market low competitiveness and vice versa. Hence, in addition to attention to customer needs, one should use competition analysis to determine strengths, weaknesses, target market situation, and sale points. It should be noticed that customer preferences may change after introducing product to market due to different factors [8]. So, models should be developed in a way that can anticipate changes in customer needs and put this anticipating in product design after initial introducing of product. Although fuzzy linear and nonlinear programming models have been used frequently in FQFD process, most of their application is in determining of satisfaction level of process characteristics according to existing constraints. Most of papers used linear methods and there are a few papers used nonlinear models. In case of multiattribute models, one must notice differences between AHP and ANP methods. Although fuzzy AHP is easier and less time consuming than fuzzy ANP, but when there are interdependencies between parameters, one must use fuzzy ANP. Increasing application of FAHP in papers, in spite of interdependencies between TCs and CNs, is concerning and should be reviewed. Goal programming is one of the MCDM methods and used widely in FQFD process. Goal programming includes several methods such as zero-one goal programming and weighting goal programming. Most of studies use zero-one goal programming; it should be investigated when we can choose one of GP methods. Using GP in combination with other methods is another important point. We cannot fulfill all needs of design team with just these methods. Also, number of employed constraints in this decision making methods should be selected according to company situation, kind of problem and its complexity, and market competitiveness situation, so its results could be applied in real world. While many GP models consider coefficients as crisp numbers, they should be considered as fuzzy numbers, according to fuzzy environment and operations. It decreases risk of information lost.

Many of discussed papers used hybrid models. In spite of big size of calculations, it seems that combination of different methods is a promising trend in FQFD. They can be combined to cover weaknesses of each other and use their strengths. It's better to compare results of all used methods to reach the best method. Using fuzzy group decision making models is another promising trend which seems necessary because of several decision makers (QFD team) who were involved in QFD process.

Metaheuristic algorithms are another new method employed in FQFD. The most frequent meta-heuristic algorithms used in FQFD are genetic algorithm (GA) and artificial neural networks (ANN). According to increasing trend of using these algorithms, it is expected that other meta-heuristic algorithms will be used in QFD, too. These methods were mostly used to determine relations between characteristics especially in quality matrix (or HoQ). They can be used to calculate target values of TCs and determine fulfillment level of these characteristics.

In the case of QFD applications, QFD can be used in all industries (both manufacturing and service), but, among papers published in different databases and discussed here in this review, QFD was used SCM and logistics more than other discussed applications. Using QFD to develop products according to environmental considerations is a new, promising trend. QFD applications in designing expert systems and decision support systems have a long way to go. In these two last cases, there is a mutual relation between QFD and them. We can use QFD to select and develop DSS and expert systems; on the other side, these systems can be used to deploy QFD's proposed models easier to make them user-friendly by decreasing calculations size.

References

[1] L. H. Chen and W. C. Ko, "Fuzzy approaches to quality function deployment for new product design," *Fuzzy Sets and Systems*, vol. 160, no. 18, pp. 2620–2639, 2009.

[2] L. K. Chan and M. L. Wu, "A systematic approach to quality function deployment with a full illustrative example," *Omega*, vol. 33, no. 2, pp. 119–139, 2005.

[3] L. H. Chen and W. C. Ko, "Fuzzy linear programming models for NPD using a four-phase QFD activity process based on the means-end chain concept," *European Journal of Operational Research*, vol. 201, no. 2, pp. 619–632, 2010.

[4] X. F. Liu and J. Yen, "Analytic framework for specifying and analyzing imprecise requirements," in *Proceedings of the 18th International Conference on Software Engineering*, pp. 60–69, IEEE Computer Society Press, Los Alamitos, Calif, USA, March 1996.

[5] C. Kahraman, T. Ertay, and G. Büyüközkan, "A fuzzy optimization model for QFD planning process using analytic network approach," *European Journal of Operational Research*, vol. 171, no. 2, pp. 390–411, 2006.

[6] L. H. Chen and W. C. Ko, "A fuzzy nonlinear model for quality function deployment considering Kano's concept," *Mathematical and Computer Modelling*, vol. 48, no. 3-4, pp. 581–593, 2008.

[7] X. G. Luo, J. F. Tang, and D. W. Wang, "An optimization method for components selection using quality function deployment," *International Journal of Advanced Manufacturing Technology*, vol. 39, no. 1-2, pp. 158–167, 2008.

[8] Z. Sener and E. E. Karsak, "A decision model for setting target levels in quality function deployment using nonlinear programming-based fuzzy regression and optimization," *International Journal of Advanced Manufacturing Technology*, vol. 48, no. 9–12, pp. 1173–1184, 2010.

[9] K. J. Kim, H. Moskowitz, A. Dhingra, and G. Evans, "Fuzzy multicriteria models for quality function deployment," *European Journal of Operational Research*, vol. 121, no. 3, pp. 504–518, 2000.

[10] C. K. Kwong and H. Bai, "A fuzzy AHP approach to the determination of importance weights of customer requirements in quality function deployment," *Journal of Intelligent Manufacturing*, vol. 13, no. 5, pp. 367–377, 2002.

[11] I. Erol and W. G. Ferrell, "A methodology for selection problems with multiple, conflicting objectives and both qualitative and quantitative criteria," *International Journal of Production Economics*, vol. 86, no. 3, pp. 187–199, 2003.

[12] G. Büyüközkan, T. Ertay, C. Kahraman, and D. Ruan, "Determining the importance weights for the design requirements in the house of quality using the fuzzy analytic network approach," *International Journal of Intelligent Systems*, vol. 19, no. 5, pp. 443–461, 2004.

[13] L. H. Chen and M. C. Weng, "An evaluation approach to engineering design in QFD processes using fuzzy goal programming models," *European Journal of Operational Research*, vol. 172, no. 1, pp. 230–248, 2006.

[14] N. Gunasekaran, S. Rathesh, S. Arunachalam, and S. C. L. Koh, "Optimizing supply chain management using fuzzy approach,"

Journal of Manufacturing Technology Management, vol. 17, no. 6, pp. 737–749, 2006.

[15] S. M. Mousavi, H. Malekly, H. Hashemi, and S. M. H. Mojtahedi, "A two-phase fuzzy decision making methodology for Bridge Scheme Selection," in *Proceedings of theIEEE International Conference on Industrial Engineering and Engineering Management (IEEM '08)*, pp. 415–419, December 2008.

[16] C. Y. Lin and A. H. I. Lee, "Preliminary study of a fuzzy integrated model for new product development of TFT-LCD," in *Proceedings of the IEEE International Conference on Service Operations and Logistics, and Informatics (IEEE/SOLI '08)*, pp. 2689–2695, October 2008.

[17] M. Celik, S. Cebi, C. Kahraman, and I. D. Er, "An integrated fuzzy QFD model proposal on routing of shipping investment decisions in crude oil tanker market," *Expert Systems with Applications*, vol. 36, no. 3, pp. 6227–6235, 2009.

[18] H. Khademi-Zare, M. Zarei, A. Sadeghieh, and M. Saleh Owlia, "Ranking the strategic actions of Iran mobile cellular telecommunication using two models of fuzzy QFD," *Telecommunications Policy*, vol. 34, no. 11, pp. 747–759, 2010.

[19] L. Z. Lin, W. C. Chen, and T. J. Chang, "Using FQFD to analyze island accommodation management in fuzzy linguistic preferences," *Expert Systems with Applications*, vol. 38, no. 6, pp. 7738–7745, 2011.

[20] C. H. Wang, "An integrated fuzzy multi-criteria decision making approach for realizing the practice of quality function deployment," in *Proceedings of the IEEE International Conference on Industrial Engineering and Engineering Management (IEEM '10)*, pp. 13–17, December 2010.

[21] B. Nepal, O. Yadav, and A. Murat, "A fuzzy-AHP approach to prioritization of CS attributes in target planning for automotive product development," *Expert Systems with Applications*, vol. 37, pp. 6775–6786, 2010.

[22] L. Zheng, T. Pan, and G. Yan, "The process integration evaluation method of the fourth party logistics using fuzzy theory," in *Proceedings of the 4th International Conference on Management of e-Commerce and e-Government (ICMeCG '10)*, pp. 313–316, October 2010.

[23] Z. Gungor, E. K. Delice, and S. E. Kesen, "New product design using FDMS and FANP under fuzzy environment," *Applied Soft Computing*, vol. 11, no. 4, pp. 3347–3356, 2011.

[24] A. H. I. Lee and C. Y. Lin, "An integrated fuzzy QFD framework for new product development," *Flexible Services and Manufacturing Journal*, vol. 23, no. 1, pp. 26–47, 2011.

[25] H. T. Liu, "Product design and selection using fuzzy QFD and fuzzy MCDM approaches," *Applied Mathematical Modelling*, vol. 35, no. 1, pp. 482–496, 2010.

[26] H. T. Liu and C. H. Wang, "An advanced quality function deployment model using fuzzy analytic network process," *Applied Mathematical Modelling*, vol. 34, no. 11, pp. 3333–3351, 2010.

[27] F. Zandi and M. Tavana, "A fuzzy group quality function deployment model for e-CRM framework assessment in agile manufacturing," *Computers and Industrial Engineering*, vol. 61, no. 1, pp. 1–19, 2011.

[28] M. Zarei, M. B. Fakhrzad, and M. Jamali Paghaleh, "Food supply chain leanness using a developed QFD model," *Journal of Food Engineering*, vol. 102, no. 1, pp. 25–33, 2011.

[29] S. Yousefie, M. Mohammadi, and J. H. Monfared, "Selection effective management tools on setting European Foundation for Quality Management (EFQM) model by a quality function

deployment (QFD) approach," *Expert Systems with Applications*, vol. 38, no. 8, pp. 9633–9647, 2011.

[30] E. E. Karsak and C. O. Özogul, "An integrated decision making approach for ERP system selection," *Expert Systems with Applications*, vol. 36, no. 1, pp. 660–667, 2009.

[31] L. Huang and X. Li, "Research on determining the key technology of new product plan and design," in *Proceedings of the IEEE International Conference on Robotics and Biomimetics (ROBIO '08)*, pp. 1532–1537, Bangkok, Thailand, February 2009.

[32] Y. M. Wang and K. S. Chin, "A linear goal programming priority method for fuzzy analytic hierarchy process and its applications in new product screening," *International Journal of Approximate Reasoning*, vol. 49, no. 2, pp. 451–465, 2008.

[33] E. Tolga and S. E. Alptekin, "Product evaluation and development process using a fuzzy compromise-based goal programming approach," *Journal of Intelligent and Fuzzy Systems*, vol. 19, no. 4-5, pp. 285–301, 2008.

[34] E. E. Karsak, S. Sozer, and S. E. Alptekin, "Product planning in quality function deployment using a combined analytic network process and goal programming approach," *Computers and Industrial Engineering*, vol. 44, no. 1, pp. 171–190, 2003.

[35] R. G. Ozdemir and Z. Ayag, "Fuzzy ANP-based modified TOPSIS for machine tool selection problem," in *Proceedings of the 16th International Working Seminar on Production Economics (IWSPE '10)*, Innsbruck, Austria, March 2010.

[36] Z. Zhang and X. Chu, "Fuzzy group decision-making for multi-format and multi-granularity linguistic judgments in quality function deployment," *Expert Systems with Applications*, vol. 36, no. 5, pp. 9150–9158, 2009.

[37] G. Büyüközkana, O. Feyzioğlu, and D. Ruanb, "Fuzzy group decision-making to multiple preference formats in quality function deployment," *Computers in Industry*, vol. 58, no. 5, pp. 392–402, 2007.

[38] H. T. Liu, "The extension of fuzzy QFD: from product planning to part deployment," *Expert Systems with Applications*, vol. 36, no. 8, pp. 11131–11144, 2009.

[39] A. Sanayei, S. Farid Mousavi, and A. Yazdankhah, "Group decision making process for supplier selection with VIKOR under fuzzy environment," *Expert Systems with Applications*, vol. 37, no. 1, pp. 24–30, 2010.

[40] L. Z. Lin, L. C. Huang, and H. R. Yeh, "Fuzzy group decision-making for service innovations in quality function deployment," *Group Decision and Negotiation*, vol. 21, no. 4, pp. 495–517, 2011.

[41] X.-T. Wanga and W. Xionga, "An integrated linguistic-based group decision-making approach for quality function deployment," *Expert Systems with Applications*, vol. 38, no. 12, pp. 14428–14438, 2011.

[42] S. W. Hsiao and E. Liu, "A neurofuzzy-evolutionary approach for product design," *Integrated Computer-Aided Engineering*, vol. 11, no. 4, pp. 323–338, 2004.

[43] F. Lai, D. Li, Q. Wang, and X. Zhao, "The information technology capability of third-party logistics providers: a resource-based view and empirical evidence from China," *Journal of Supply Chain Management*, vol. 44, no. 3, pp. 22–38, 2008.

[44] E. Mehdizadeh, "Ranking of customer requirements using the fuzzy centroid-based method," *International Journal of Quality and Reliability Management*, vol. 27, no. 2, pp. 201–216, 2010.

[45] M. Guo, J. B. Yang, K. S. Chin, H. W. Wang, and X. B. Liu, "Evidential reasoning approach for multiattribute decision analysis under both fuzzy and Interval uncertainty," *IEEE Transactions on Fuzzy Systems*, vol. 17, no. 3, pp. 683–697, 2009.

[46] T. Mu, A. K. Nandi, and R. M. Rangayyan, "Classification of breast masses via nonlinear transformation of features based on a kernel matrix," *Medical and Biological Engineering and Computing*, vol. 45, no. 8, pp. 769–780, 2007.

[47] M. Bevilacqua, F. E. Ciarapica, and G. Giacchetta, "A fuzzy-QFD approach to supplier selection," *Journal of Purchasing and Supply Management*, vol. 12, no. 1, pp. 14–27, 2006.

[48] Z. Zhang and X. Chu, "A selection model for multiple design schemes of complex product," in *Proceedings of the 4th International Conference on Fuzzy Systems and Knowledge Discovery (FSKD '07)*, pp. 483–487, August 2007.

[49] F. Wang, X. H. Li, W. N. Rui, and Y. Zhang, "A fuzzy QFD-based method for customizing positioning of logistics Service products of 3PLS," in *Proceedings of the International Conference on Wireless Communications, Networking and Mobile Computing (WiCOM '07)*, pp. 3331–3334, September 2007.

[50] H. Rau and Y. T. Fang, "Conflict resolution of product package design for logistics using the triz method," in *Proceedings of the International Conference on Machine Learning and Cybernetics*, pp. 2891–2896, July 2009.

[51] E. Bottani and A. Rizzi, "Strategic management of logistics service: a fuzzy QFD approach," *International Journal of Production Economics*, vol. 103, no. 2, pp. 585–599, 2006.

[52] S. H. Amin and J. Razmi, "An integrated fuzzy model for supplier management: a case study of ISP selection and evaluation," *Expert Systems with Applications*, vol. 36, no. 4, pp. 8639–8648, 2009.

[53] S. Y. Sohn and I. S. Choi, "Fuzzy QFD for supply chain management with reliability consideration," *Reliability Engineering and System Safety*, vol. 72, no. 3, pp. 327–334, 2001.

[54] I. H. Kuo, S. J. Horng, T. W. Kao, T. L. Lin, C. L. Lee, and Y. Pan, "An improved method for forecasting enrollments based on fuzzy time series and particle swarm optimization," *Expert Systems with Applications*, vol. 36, no. 3, pp. 6108–6117, 2009.

[55] T. C. Kuo, H. H. Wu, and J. I. Shieh, "Integration of environmental considerations in quality function deployment by using fuzzy logic," *Expert Systems with Applications*, vol. 36, no. 3, pp. 7148–7156, 2009.

[56] G. Z. Jia and M. Bai, "An approach for manufacturing strategy development based on fuzzy-QFD," *Computers and Industrial Engineering*, vol. 60, no. 3, pp. 445–454, 2011.

[57] C. G. Şen and H. Baraçlı, "Fuzzy quality function deployment based methodology for acquiring enterprise software selection requirements," *Expert Systems with Applications*, vol. 37, no. 4, pp. 3415–3426, 2010.

[58] Y. Yan, H. Liu, Y. Zhang, and Z. Zhou, "A Fuzzy-QFD approach to design decision support system for emergency response of hazardous materials road transportation accidents," in *Proceedings of the IEEE International Conference on Automation and Logistics (ICAL '10)*, pp. 506–510, August 2010.

[59] G. Wang, P. Shi, and C. Wen, "Fuzzy approximation relations on fuzzy *n*-cell number space and their applications in classification," *Information Sciences*, vol. 181, no. 18, pp. 3846–3860, 2011.

14

Joint Optimal Pricing and Inventory Control for Deteriorating Items under Inflation and Customer Returns

Maryam Ghoreishi, Alireza Arshsadi khamseh, and Abolfazl Mirzazadeh

Industrial Engineering Department, Kharazmi University, University Square, Dr. Beheshti Street, Karaj 31979-37551, Tehran, Iran

Correspondence should be addressed to Alireza Arshsadi khamseh; alireza.arshadikhamseh@gmail.com

Academic Editor: Ilkyeong Moon

This paper studies the effect of inflation and customer returns on joint pricing and inventory control for deteriorating items. We adopt a price and time dependent demand function, also the customer returns are considered as a function of both price and demand. Shortage is allowed and partially backlogged. The main objective is determining the optimal selling price, the optimal replenishment cycles, and the order quantity simultaneously such that the present value of total profit in a finite time horizon is maximized. An algorithm has been presented to find the optimal solution. Finally, we solve a numerical example to illustrate the solution procedure and the algorithm.

1. Introduction

Recently, many researchers have studied the problem of joint pricing and inventory control for deteriorating items. Generally, deterioration is defined as decay, damage, spoilage, evaporation, and loss of utility of the product. Most physical goods undergo decay or deterioration over time such as medicines, volatile liquids, blood banks, and others [1]. The first attempt to describe optimal ordering policies for deteriorating items was made by Ghare and Schrader [2]. Later, Covert and Philip [3] derived the model with variable deteriorating rate of two-parameter Weibull distribution. Goyal and Giri [4] presented a detailed review of deteriorating inventory literatures. Abad [5, 6] considered a pricing and lot-sizing problem for a perishable good under exponential decay and partial backlogging. Dye [7] proposed the joint pricing and ordering policies for a deteriorating inventory with price-dependent demand and partial backlogging. Dye et al. [8] developed an inventory and pricing strategy for deteriorating items with shortages when demand and deterioration rate are continuous and differentiable function of price and time, respectively. Chang et al. [9] introduced a deteriorating inventory model with price-time-dependent demand and partial backlogging. Nakhai and Maihami [10] developed the joint pricing and ordering policies for deteriorating items

with partial backlogging where the demand is considered as a function of both price and time. Tsao and Sheen [11] proposed the problem of dynamic pricing and replenishment for a deteriorating item under the supplier's trade credit and the retailer's promotional effort. Sarkar [12] extended the model with finite replenishment rate, stock-dependent demand, imperfect production, and delay in payments with two progressive periods. Sarkar [13] proposed an EOQ (economic order quantity) model for finite replenishment rate with delay in payments. In this model, deterioration and demand of the item have been considered as a time-dependent function. Sett et al. [14] considered a two-warehouse inventory model with quadratic increasing demand and time-varying deterioration. This model is derived with a finite replenishment rate and unequal length of the cycle time. Sarkar et al. [15] developed an economic production quantity model with stochastic demand in an imperfect production system. This model is derived for both continuous and discrete random demands.

In all the above models, the inflation and the time value of money were disregarded, but most of the countries have suffered from large-scale inflation and sharp decline in the purchasing power of money during years. As a result, while determining the optimal inventory policies, the effects of inflation and time value of money cannot be ignored. First, Buzacott [16] presented the EOQ model with inflation.

Following Buzacott [16], several researchers (Misra [17], Jolai et al. [18], etc.) have extended their approaches to distinguish the inventory models by considering the time value of money, the different inflation rates for the internal and external costs, finite replenishment, shortages, and so forth. Park [19] derived the economic order quantity in terms of purchasing credit. Datta and Pal [20] discussed a model with shortages and time-dependent demand rates to study the effects of inflation and time value of money on a finite time horizon. Goel et al. [21] developed the model economic discount value for multiple items with restricted warehouse space and the number of orders under inflationary conditions. Hall [22] presented a new model with the increasing purchasing price over time. Sarker and Pan [23] surveyed the effects of inflation and the time value of money on the optimal ordering quantities and the maximum allowable shortage in a finite replenishment inventory system. Hariga and Ben-Daya [24] have presented time-varying lot-sizing models with a time-varying demand pattern, taking into account the effects of inflation and time value of money. Horowitz [25] discussed an EOQ model with a normal distribution for the inflation. Moon and Lee [26] developed an EOQ model under inflation and discounting with a random product life cycle. Mirzazadeh and Sarfaraz [27] presented a multiple-item inventory system with a budget constraint and the uniform distribution function for the external inflation rate. Dey et al. [28] developed the model for a deteriorating item with time-dependent demand rate and interval-valued lead time under inflationary conditions. Mirzazadeh et al. [29] considered stochastic inflationary conditions with variable probability density functions (pdfs) over the time horizon and the demand rate is dependent on the inflation rates. Wee and Law [30] developed a deteriorating inventory model taking into account the time value of money for a deterministic inventory system with price-dependent demand. Hsieh and Dye [31] presented pricing and inventory control model for deterioration items taking into account the time value of money. In their model, shortage was allowed and partially backlogged and the demand was assumed as a function of price and time. Sarkar and Moon [32] developed a production inventory model for stochastic demand with inflation in an imperfect production system. Sarkar et al. [33] presented an EMQ (economic manufacturing quantity) model for time varying demand with inflation in an imperfect production process. Sarkar et al. [34] considered an economic order quantity model for various types of deterministic demand patterns in which the delay periods and different discount rates on purchasing cost are offered by the supplier to the retailers in the presence of inflation.

In the classical EOQ models, customer returns have not been considered, while in supply chain retailers can return some or all unsold items at the end of the selling season to the manufacturer and receive a full or partial refund. Hess and Mayhew [35] studied the problem of customer return by using regression methods to model the returns for a large direct market. Anderson et al. [36] found that the quantity sold has a strong positive linear relationship with number of returns. Same as Hess and Mayhew, they used regression models to show that as the price increases,

both the number of returns and the return rate increase. These empirical investigations provide evidence to support the view that customer returns increase with both the quantity sold and the price set for the product. Chen and Bell [37] considered the customer returns as a function of price and demand simultaneously. Pasternack [38] studied the newsvendor problem framework for a seasonal product where a percentage of the order quantity could be returned from the retailers to the manufacturer. Zhu [39] presented a single-item periodic-review model for the joint pricing and inventory replenishment problem with returns and expediting. Yet, only a few authors have investigated the effect of customer returns on joint pricing and inventory control.

In the previous research that considered the impact of customer returns on pricing and inventory control for deteriorating items, the effect of time value of money has not been considered. However, in order to consider the realistic circumstances, the effect of time value of money should be considered. On the other hand, in nearly all papers that consider the impact of customer returns on pricing and inventory control, the return functions are dependent on price or demand, separately. But the empirical findings of Anderson et al. [36] provide evidence to support the view that customer returns increase with both the quantity sold and the price set for the product. The present paper studies the effect of inflation and customer returns on the joint pricing and inventory control for deteriorating items. We assume that the customer returns increase with both the quantity sold and the product price. The demand is deterministic and depends on time and price simultaneously. Shortages are allowed and partially backlogged. An optimization procedure is presented to derive the optimal time with positive inventory, selling price, and the number of replenishments and then obtains the optimal order quantity when the total present value of profits is maximized. Thus, the replenishment and price policies are appropriately developed. Numerical examples are provided to illustrate the proposed model.

The rest of the paper is organized as follows. In Section 2, assumptions and notations throughout this paper are presented. In Section 3, we establish the mathematical model. Next, in Section 4, an algorithm is presented to find the optimal selling price and inventory control variables. In Section 5, we use a numerical example and, finally, summary and some suggestions for the future are presented in Section 6.

2. Notations and Assumptions

The following notations and assumptions are used throughout the paper.

Notations

A: Constant purchasing cost per order

c: Purchasing cost per unit

c_1: Holding cost per unit per unit time

c_2: Backorder cost per unit per unit time

c_3: Cost of lost sale per unit

p: Selling price per unit, where $p > c$

θ: Constant deterioration rate

r: Constant representing the difference between the discount (cost of capital) and the inflation rate

Q: Order quantity

T: Length of replenishment cycle time

t_1: Length of time in which there is no inventory shortage

SV: Salvage value per unit

H: Length of planning horizon

N: Number of replenishments during the time horizon H

T^*: Optimal length of the replenishment cycle time

Q^*: Optimal order quantity

t_1^*: Optimal length of time in which there is no inventory shortage

p^*: Optimal selling price per unit

$I_1(t)$: Inventory level at time $t \in [0, t_1]$

$I_2(t)$: Inventory level at time $t \in [t_1, T]$

I_0: Maximum inventory level

S: Maximum amount of demand backlogged

PWTP(p, t_1, T): The present-value of total profit over the time horizon.

Assumptions. In this paper, the following assumptions are considered.

(1) There is a constant fraction of the on-hand inventory deteriorates per unit of time and there is no repair or replacement of the deteriorated inventory.

(2) The replenishment rate is infinite and the lead time is zero.

(3) The demand rate, $D(p, t) = (a - bp)e^{\lambda t}$ (where $a > 0, b > 0$) is a linearly decreasing function of the price and decreases (increases) exponentially with time when $\lambda < 0$ $(\lambda > 0)$ [11].

(4) Shortages are allowed. The unsatisfied demand is backlogged, and the fraction of shortage backordered is $\beta(x) = k_0 e^{-\delta x}$, $(\delta > 0, 0 < k_0 \leq 1)$, where x is the waiting time up to the next replenishment, δ is a positive constant, and $0 \leq \beta(x) \leq 1, \beta(0) = 1$ [5].

(5) The time horizon is finite.

(6) Following the empirical findings of Anderson et al. [36], we assume that customer returns increase with both the quantity sold and the price using the following general form: $R(p, t) = \alpha D(p, t) + \beta p$ ($\beta \geq 0$, $0 \leq \alpha < 1$).

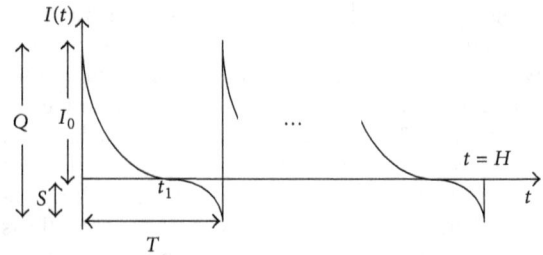

FIGURE 1: Graphical representation of inventory system.

3. Model Formulation

We use Nakhai and Maihami inventory shortage model [10]. According to this model; the inventory system is as follows: I_0 units of an item arrive at the inventory system at the beginning of each cycle and during the time interval $[0, t_1]$, drop to zero due to demand and deterioration. Finally, a shortage occurs due to demand and partial backlogging during the time interval $[t_1, T]$ (see Figure 1).

The equation representing the inventory status in system for the first interval [10] is as follows:

$$\frac{dI_1(t)}{dt} + \theta dI_1(t) = -D(p, t) \quad 0 \leq t \leq t_1, \quad (1)$$

when $I_1(t_1) = 0$, inventory level in the $(0, t_1)$ interval yields as the following:

$$I_1(t) = \frac{(a - bp)e^{-\theta t}}{\lambda + \theta} \left[e^{(\lambda + \theta)t_1} - e^{(\lambda + \theta)t} \right], \quad 0 \leq t \leq t_1. \quad (2)$$

Also in this interval with the condition $I_1(0) = I_0$, the maximum inventory level (I_0) yields as follows:

$$I_0 = \frac{(a - bp)}{\lambda + \theta} \left[e^{(\lambda + \theta)t_1} - 1 \right]. \quad (3)$$

In the second interval (t_1, T), shortage is partially backlogged according to fraction $\beta(T - t)$. Therefore, the inventory level at time t is obtained by the following:

$$\frac{dI_2}{dt} = -D(p, t)\beta(T - t) = \frac{-D(p, t)}{e^{\delta(T-t)}}, \quad t_1 \leq t \leq T. \quad (4)$$

The solution of the above differential equation, after applying the boundary conditions $I_2(t_1) = 0$, is

$$I_2(t) = \frac{(a - bp)e^{-\delta T} \left(e^{(\delta + \lambda)t_1} - e^{(\lambda + \delta)t} \right)}{\lambda + \delta}, \quad t_1 \leq t \leq T. \quad (5)$$

If we put $t = T$ into $I_2(t)$, the maximum amount of demand backlogging (S) will be obtained:

$$S = -I_2(T) = -\frac{(a - bp)e^{-\delta T} \left(e^{(\delta + \lambda)t_1} - e^{(\lambda + \delta)T} \right)}{\lambda + \delta}. \quad (6)$$

Order quantity per cycle (Q) is the sum of S and I_0, that is:

$$Q = S + I_0 = \frac{(a - bp)\, e^{-\delta T} \left(e^{(\delta + \lambda)t_1} - e^{(\lambda + \delta)T} \right)}{\lambda + \delta}$$
$$+ \frac{(a - bp)}{\lambda + \theta} \left[e^{(\lambda + \theta)t_1} - 1 \right]. \tag{7}$$

Now, we can obtain the present value of inventory costs and sales revenue for the first cycle, which consists of the following elements.

(1) Since replenishment in each cycle has been done at the start of each cycle, the present value of replenishment cost for the first cycle will be A, which is a constant value.

(2) Inventory occurs during period t_1; therefore, the present value of holding cost (HC) for the first cycle is

$$HC = c_1 \left(\int_0^{t_1} I_1(t) \cdot e^{-r \cdot t} dt \right). \tag{8}$$

(3) The present value of shortage cost (SC) due to backlog for the first cycle is

$$SC = c_2 \left(e^{-r \cdot t_1} \int_{t_1}^{T} -I_2(t) \cdot e^{-r \cdot t} dt \right). \tag{9}$$

(4) The present value of opportunity cost due to lost sales (OC) for the first cycle is

$$OC = c_3 \left(e^{-r \cdot t_1} \int_{t_1}^{T} D(p,t)(1 - \beta(T - t)) \cdot e^{-r \cdot t} dt \right). \tag{10}$$

(5) The present value of purchase cost (PC) for the first cycle is

$$PC = c \left(I_0 + S e^{-r \cdot T} \right). \tag{11}$$

(6) The present value of return cost (RC) for the first cycle is

$$RC = (p - SV) \int_0^{t_1} (\alpha D(p,t) + \beta p) e^{-r \cdot t} dt. \tag{12}$$

(7) The present value of sales revenue (SR) for the first cycle is

$$SR = p \left(\int_0^{t_1} D(p,t) \cdot e^{-r \cdot t} dt + S \cdot e^{-r \cdot T} \right). \tag{13}$$

There are N cycles during the planning horizon. Since inventory is assumed to start and end at zero, an extra replenishment at $t = H$ is required to satisfy the backorders of the last cycle in the planning horizon. Therefore, the total number of replenishment will be $N + 1$ times; the first replenishment lot size is I_0, and the 2nd, 3rd, …, Nth replenishment lot size is as follows:

$$Q = S + I_0. \tag{14}$$

Finally, the last or $(N + 1)$th replenishment lot size is S.

Therefore, the present value of total profit during planning horizon (denoted by $PWTP(p, t_1, T)$) is derived as follows:

$$PWTP(p, t_1, T)$$
$$= \sum_{i=0}^{N-1} (SR - A - HC - SC - OC - PC - RC)\, e^{-r \cdot i \cdot T}$$
$$- A \cdot e^{-r \cdot H}. \tag{15}$$

The value of the variable T can be replaced by the equation $T = H/N$ which uses Maclaurin's approximation for $\sum_{i=0}^{N-1} e^{-r \cdot i \cdot T} \cong 1 - e^{-r \cdot N \cdot T}/1 - e^{-r \cdot T}$. Thus, the objective of this paper is determining the values of t_1, p, and N that maximize $PWTP(p, t_1, T)$ subject to $p > 0$ and $0 < t_1 < T$, where N is a discrete variable and p and t_1 are continuous variables. For a given value of N, the necessary conditions for finding the optimal p^* and t_1^* are given as follows:

$$\frac{\partial PWTP}{\partial p}(p, t_1, N)$$

$$= -\frac{1}{(-\lambda + r)(\delta + \lambda) r \left(-1 + e^{-(rH/N)} \right)}$$
$$\times \Bigg(\left(-r e^{-(\delta H/N)} e^{-(rH/N)} (-\lambda + r)(a - bp) e^{((\delta + \lambda)H)/N} \right.$$
$$+ r e^{-(\delta H/N)} e^{(\delta + \lambda)t_1} (-\lambda + r)(a - bp) e^{-(rH/N)}$$
$$+ (\delta + \lambda) \left(r(a - bp) e^{-t_1(-\lambda + r)} \right.$$
$$+ \Bigg(\Big(((-SV + 2p)\beta + \alpha(a - bp)) r$$
$$-2 \left(-\frac{1}{2} SV + p \right) \lambda \beta \Big) e^{rt_1}$$
$$- r\alpha(a - bp) e^{\lambda t_1}$$
$$-2 \left(-\frac{1}{2} SV + p \right)(-\lambda + r)\beta \Big) e^{-rt_1}$$
$$\left. \left. -r(a - bp) \right) \right) \left(-1 + e^{-rH} \right) \Bigg) = 0, \tag{16}$$

$$\frac{\partial PWTP}{\partial t_1}(p, t_1, N)$$

$$= \frac{1}{(r - \delta - \lambda)(-\lambda + r)(\theta + r)(\delta + \lambda) r \left(-1 + e^{-(rH/N)} \right)}$$
$$\times \Bigg(\left(-2(\delta + \lambda) r(\theta + r) e^{-rt_1}(a - bp) c_3 \left(r - \frac{1}{2}\delta - \frac{1}{2}\lambda \right) \right.$$
$$\times (-\lambda + r) e^{-(rH/N)} e^{(-t_1(r - \delta - \lambda)N + H(r - \delta))/N}$$

TABLE 1: Optimal solution of the example.

N	p	Time interval		Q	PWTP
		t_1	T		
10	257.5664	2.5798	4	391.0126	32372.5220
11^*	254.8477^*	2.7126^*	3.63^*	392.2260^*	35919.3928^*
12	694.1629	0.2142	3.33	107.4832	-161403.4113

$$+ \left(2 (\delta + \lambda) \left(r - \frac{1}{2}\delta - \frac{1}{2}\lambda \right) e^{(t_1(\delta + \lambda)N + rH)/N} \right.$$

$$- r^2 e^{((\delta + \lambda)H + rt_1 N)/N} + e^{t_1(r + \delta + \lambda)} (r - \delta - \lambda)^2 \right)$$

$$\times (\theta + r) e^{-rt_1} (a - bp) (-\lambda + r) c_2$$

$$\times e^{(H(-r-\delta) - rt_1 N)/N} + (\delta + \lambda)$$

$$\times \left(2 (r - \delta - \lambda)(\theta + r) \right.$$

$$\times \left(\frac{1}{2}\lambda + r \right) e^{-rt_1} (a - bp)$$

$$\times c_3 e^{-(rH/N)} e^{(-t_1(-\lambda + r)N + rH)/N}$$

$$+ \left(e^{t_1(\delta + \lambda)} (-\lambda + r) \right.$$

$$\times (r - \delta - \lambda)(-p + c)$$

$$\times e^{-(\delta H/N)} + r\delta c_3 e^{-rt_1} e^{H\lambda/N} \right)$$

$$\times (\theta + r)(a - bp) e^{-(rH/N)} - (r - \delta - \lambda)$$

$$\times \left(-(rp + p\theta + c_1)(a - bp) e^{-t_1(-\lambda + r)} \right.$$

$$+ (a - bp)(c\theta + c_1 + rc) e^{t_1(\theta + \lambda)}$$

$$+ (p - SV) \left(\alpha (a - bp) e^{\lambda t_1} + \beta p \right)$$

$$\left. \times (\theta + r) e^{-rt_1} \right) (-\lambda + r) \right) r \right)$$

$$\times \left(-1 + e^{-rH} \right) \right) = 0.$$

$$(17)$$

4. Optimal Solution Procedure

The objective function has three variables. The number of replenishments (N) is a discrete variable, the length of time in which there is no inventory shortage (t_1), and the selling price per unit (p). The following algorithm is used to obtain the optimal amount of t_1, p, N [30].

Step 1. let $N = 1$.

Step 2. For different integer N values, derive t_1^* and p^* from (16) and (17). Substitute (p^*, t_1^*, N^*) to (15) to derive PWTP(p^*, t_1^*, N^*).

Step 3. Add one unit to N and repeat Step 2 for new N. If there is no increase in the last PWPT, then show the last one.

The (p^*, t_1^*, N^*) and PWTP(p^*, t_1^*, N^*) values constitute the optimal solution and satisfy the following conditions:

$$\Delta\text{PWTP}\left(p^*, t_1^*, N^*\right) < 0 < \Delta\text{PWTP}\left(p^*, t_1^*, N^* - 1\right),$$

$$(18)$$

where

$$\Delta\text{PWTP}\left(p^*, t_1^*, N^*\right) = \text{PWTP}\left(p^*, t_1^*, N^* + 1\right)$$

$$- \text{PWTP}\left(p^*, t_1^*, N^*\right).$$

$$(19)$$

Substitute (p^*, t_1^*, N^*) to (7) to derive the Nth replenishment lot size.

If the objective function is concave, the following sufficient conditions must be satisfied:

$$\left(\frac{\partial^2 \text{PWTP}}{\partial p \partial t_1} \right)^2 - \left(\frac{\partial^2 \text{PWTP}}{\partial t_1^2} \right) \left(\frac{\partial^2 \text{PWTP}}{\partial p^2} \right) < 0, \quad (20)$$

and any one of the following:

$$\frac{\partial^2 \text{PWTP}}{\partial t_1^2} < 0, \qquad \frac{\partial^2 \text{PWTP}}{\partial p^2} < 0. \quad (21)$$

Since PWTP is a very complicated function due to high-power expression of the exponential function, it is not possible to show analytically the validity of the above sufficient conditions. Thus, the sign of the above quantity in (21) is assessed numerically. The computational results are shown in the following illustrative example.

5. Numerical Example

To illustrate the solution procedure and the results, let us apply the proposed algorithm to solve the following numerical examples. The results can be found by applying Maple 13. This example is based on the following parameters and functions:

$D(p,t) = (500 - 0.5p)e^{-0.98t}$, $c_1 = \$50$/per unit/per unit time, $c_2 = \$30$/per unit/per unit time, $c_3 = \$30$/per unit, $c = \$100$/per unit, $A = \$30$/per order, $\theta = 0.08$, $\beta(x) = e^{0.2x}$, $r = 0.12$, $R(p,t) = 0.2 * D(p,t) + 0.3p$, $SV = \$200$/per unit, and $H = 40$ unit time.

From Table 1, if all the conditions and constraints in (15)–(21) are satisfied, optimal solution can be derived. In this example, the maximum present value of total profit is found

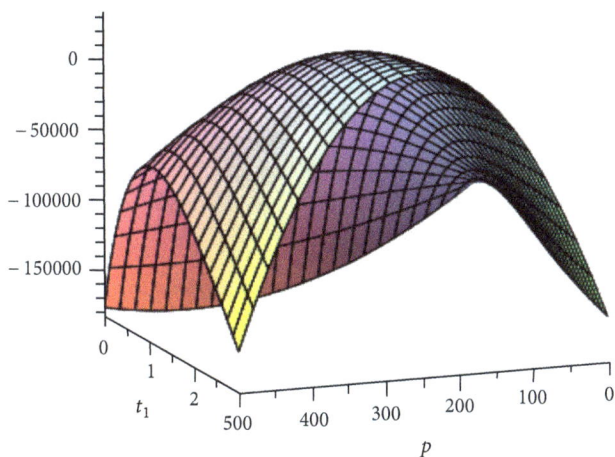

FIGURE 2: The graphical representation of the concavity of the present worth of total profit function $\text{PWTP}(p, t_1, 11)$.

in the 11th cycle. The total number of order is therefore $(N+1)$ or 12. With twelve orders, the optimal solution is as follows:

$$p^* = 254.8477, \qquad t_1^* = 2.7126,$$

$$T^* = 3.63, \qquad \text{PWTP}^* = 35919.3928, \qquad (22)$$

$$Q^* = 392.2260.$$

By substituting the optimal values of N^*, p^*, and t_1^* to (21), it will be shown that PWTP is strictly concave (Figure 2):

$$\frac{\partial^2 \text{PWTP}}{\partial t_1^2} = -11764.95287, \qquad \frac{\partial^2 \text{PWTP}}{\partial p^2} = -5.7433. \qquad (23)$$

6. Conclusion

In this work, we addressed the problem of joint pricing and inventory control model for deteriorating items taking into account the time value of money and customer returns. The demand is deterministic and depends on time and price simultaneously. Also, the customer returns assumed as a function of both the quantity sold and the price. Shortage is allowed and partially backlogged. An algorithm is presented for deriving the optimal replenishment and pricing policy that wants to maximize the present value of total profit. Finally, a numerical example is provided to illustrate the algorithm and the solution procedure.

This paper can be extended in several ways. For instance, the constant deterioration rate could be extended to a time-dependent function. Also, the deterministic demand function could be extended to the stochastic demand function. Finally, we could extend the model to incorporate some more realistic features such as quantity discounts and permissible delay in payments.

Acknowledgment

The authors thank the editors and anonymous referees for their constructive suggestions in the improvement of the paper.

References

[1] H. M. Wee, "Economic production lot size model for deteriorating items with partial back-ordering," *Computers and Industrial Engineering*, vol. 24, no. 3, pp. 449–458, 1993.

[2] P. M. Ghare and G. H. Schrader, "A model for exponentially decaying inventory system," *International Journal of Production Research*, vol. 21, pp. 449–460, 1963.

[3] R. P. Covert and G. C. Philip, "An EOQ model for items with Weibull distribution deterioration," *AIIE Transactions*, vol. 5, no. 4, pp. 323–326, 1973.

[4] S. K. Goyal and B. C. Giri, "Recent trends in modeling of deteriorating inventory," *European Journal of Operational Research*, vol. 134, no. 1, pp. 1–16, 2001.

[5] P. L. Abad, "Optimal pricing and lot-sizing under conditions of perishability and partial backordering," *Management Science*, vol. 42, no. 8, pp. 1093–1104, 1996.

[6] P. L. Abad, "Optimal price and order size for a reseller under partial backordering," *Computers and Operations Research*, vol. 28, no. 1, pp. 53–65, 2001.

[7] C. Y. Dye, "Joint pricing and ordering policy for a deteriorating inventory with partial backlogging," *Omega*, vol. 35, no. 2, pp. 184–189, 2007.

[8] C. Y. Dye, L. Y. Ouyang, and T. P. Hsieh, "Inventory and pricing strategies for deteriorating items with shortages: a discounted cash flow approach," *Computers and Industrial Engineering*, vol. 52, no. 1, pp. 29–40, 2007.

[9] H. J. Chang, J. T. Teng, L. Y. Ouyang, and C. Y. Dye, "Retailer's optimal pricing and lot-sizing policies for deteriorating items with partial backlogging," *European Journal of Operational Research*, vol. 168, no. 1, pp. 51–64, 2006.

[10] K. I. Nakhai and R. Maihami, "Joint pricing and inventory control for deteriorating items with partial backlogging," *International Journal of Industrial Engineering and Production Management*, vol. 21, pp. 167–177, 2011.

[11] Y. C. Tsao and G. J. Sheen, "Dynamic pricing, promotion and replenishment policies for a deteriorating item under permissible delay in payments," *Computers and Operations Research*, vol. 35, no. 11, pp. 3562–3580, 2008.

[12] B. Sarkar, "An EOQ model with delay in payments and stock dependent demand in the presence of imperfect production," *Applied Mathematics and Computation*, vol. 218, pp. 8295–8308, 2012.

[13] B. Sarkar, "An EOQ model with delay in payments and time varying deterioration rate," *Mathematical and Computer Modelling*, vol. 55, pp. 367–377, 2012.

[14] B. K. Sett, B. Sarkar, and A. Goswami, "A two-warehouse inventory model with increasing demand and time varying deterioration," *Scientia Iranica E*, vol. 19, no. 6, pp. 1969–1977, 2012.

[15] B. Sarkar, S. S. Sana, and K. Chaudhuri, "An economic production quantity model with stochastic demand in an imperfect production system," *International Journal of Services and Operations Management*, vol. 9, no. 3, pp. 259–283, 2011.

[16] J. A. Buzacott, "Economic order quantity with inflation," *Operational Research Quarterly*, vol. 26, no. 3, pp. 553–558, 1975.

[17] R. B. Misra, "A note on optimal inventory management under inflation," *Naval Research Logistics Quarterly*, vol. 26, no. 1, pp. 161–165, 1979.

[18] F. Jolai, R. Tavakkoli-Moghaddam, M. Rabbani, and M. R. Sadoughian, "An economic production lot size model with deteriorating items, stock-dependent demand, inflation, and partial backlogging," *Applied Mathematics and Computation*, vol. 181, no. 1, pp. 380–389, 2006.

[19] K. S. Park, "Inflationary effect on EOQ under trade-credit financing," *International Journal on Policy and Information*, vol. 10, no. 2, pp. 65–69, 1986.

[20] T. K. Datta and A. K. Pal, "Effects of inflation and time-value of money on an inventory model with linear time-dependent demand rate and shortages," *European Journal of Operational Research*, vol. 52, no. 3, pp. 326–333, 1991.

[21] S. Goel, Y. P. Gupta, and C. R. Bector, "Impact of inflation on economic quantity discount schedules to increase vendor profits," *International Journal of Systems Science*, vol. 22, no. 1, pp. 197–207, 1991.

[22] R. W. Hall, "Price changes and order quantities: impacts of discount rate and storage costs," *IIE Transactions*, vol. 24, no. 2, pp. 104–110, 1992.

[23] B. R. Sarker and H. Pan, "Effects of inflation and the time value of money on order quantity and allowable shortage," *International Journal of Production Economics*, vol. 34, no. 1, pp. 65–72, 1994.

[24] M. A. Hariga and M. Ben-Daya, "Optimal time varying lot-sizing models under inflationary conditions," *European Journal of Operational Research*, vol. 89, no. 2, pp. 313–325, 1996.

[25] I. Horowitz, "EOQ and inflation uncertainty," *International Journal of Production Economics*, vol. 65, no. 2, pp. 217–224, 2000.

[26] I. Moon and S. Lee, "Effects of inflation and time-value of money on an economic order quantity model with a random product life cycle," *European Journal of Operational Research*, vol. 125, no. 3, pp. 558–601, 2000.

[27] A. Mirzazadeh and A. R. Sarfaraz, "Constrained multiple items optimal order policy under stochastic inflationary conditions," in *Proceedings of the 2nd Annual International Conference on Industrial Engineering Application and Practice*, pp. 725–730, San Diego, Calif, USA, November 1997.

[28] J. K. Dey, S. K. Mondal, and M. Maiti, "Two storage inventory problem with dynamic demand and interval valued lead-time over finite time horizon under inflation and time-value of money," *European Journal of Operational Research*, vol. 185, no. 1, pp. 170–194, 2008.

[29] A. Mirzazadeh, M. M. S. Esfahani, and S. M. T. F. Ghomi, "An inventory model under uncertain inflationary conditions, finite production rate and inflation-dependent demand rate for deteriorating items with shortages," *International Journal of Systems Science*, vol. 40, no. 1, pp. 21–31, 2009.

[30] H. M. Wee and S. T. Law, "Replenishment and pricing policy for deteriorating items taking into account the time-value of money," *International Journal of Production Economics*, vol. 71, no. 1–3, pp. 213–220, 2001.

[31] T. P. Hsieh and C. Y. Dye, "Pricing and lot-sizing policies for deteriorating items with partial backlogging under inflation," *Expert Systems With Applications*, vol. 37, pp. 7234–7242, 2010.

[32] B. Sarkar and I. Moon, "An EPQ model with inflation in an imperfect production system," *Applied Mathematics and Computation*, vol. 217, no. 13, pp. 6159–6167, 2011.

[33] B. Sarkar, S. S. Sana, and K. Chaudhuri, "An imperfect production process for time varying demand with inflation and time value of money—an EMQ model," *Expert Systems with Applications*, vol. 38, no. 11, pp. 13543–13548, 2011.

[34] B. Sarkar, S. S. Sana, and K. Chaudhuri, "A finite replenishment model with increasing demand under inflation," *International Journal of Mathematics and Operational Research*, vol. 4, no. 2, pp. 347–385, 2010.

[35] J. D. Hess and G. E. Mayhew, "Modeling merchandise returns in direct marketing," *Journal of Interactive Marketing*, vol. 11, no. 2, pp. 20–35, 1997.

[36] E. T. Anderson, K. Hansen, D. Simister, and L. K. Wang, "How are demand and returns related? Theory and empirical evidence," Working Paper, Kellogg School of Management, Northwestern University, 2006.

[37] J. Chen and P. C. Bell, "The impact of customer returns on pricing and order decisions," *European Journal of Operational Research*, vol. 195, no. 1, pp. 280–295, 2009.

[38] B. A. Pasternack, "Optimal pricing and returns policies for perishable commodities," *Marketing Science*, vol. 4, no. 2, pp. 166–176, 1985.

[39] S. X. Zhu, "Joint pricing and inventory replenishment decisions with returns and expediting," *European Journal of Operational Research*, vol. 216, pp. 105–112, 2012.

15

Sequential Failure Analysis Using Novel Algorithms in Sequence Determination of Petri Nets Firing

Abolfazl Doostparast Torshizi, Jamshid Parvizian, and Farshad Tooyserkani

Department of Industrial Engineering, Isfahan University of Technology, Isfahan 84156, Iran

Correspondence should be addressed to Abolfazl Doostparast Torshizi; a.doostparast@aut.ac.ir

Academic Editor: Josefa Mula

Failure occurrence in industrial systems can be a result of a sequence of failures leading to a total system failure. Up to now, several methods to determine failure sequences and to calculate probability of such failures have been proposed. These methods primarily focus on modeling aspects of the problem and do not present a certain framework to determine potential failure sequences. In this paper, a novel approach based on Petri net modeling of the systems is proposed and several heuristic algorithms are developed. Determination of potential failures in sample industrial problems and comparing the results with other existing methods demonstrates that the presented algorithms are much more efficient in dealing with complex Petri net models while existing methods are not capable of handling such complicated models.

1. Introduction

Risk analysis of complicated systems, such as flexible manufacturing cells, is a challenging task. There are diverse approaches aiming in describing different risky behaviors of the systems. One of the most applicable tools in this field is the Fault Tree Analysis (FTA) method. This method, presented in early 1960s, is only a static graphical technique to find correlations among principal reasons of a system failure [1] which makes it difficult in dealing with complicated systems. Other methods, including Failure Mode and Effect Analysis (FMEA), suffer from a similar deficiency [2, 3].

Failures occurring in systems are not confined to failures of each independent sub-system. Sequential failures of sub-systems may also lead to the failure of the entire system. Sequential Failure Logic (SFL) was presented by Fussell et al. [4]. In this research, the focus is on analyzing non-repairable electric supply systems with main and standby power units and switch controls. Exact and approximate methods are used to calculate the probability of occurrence of the output event from priority-AND SFL. It is assumed that elementary events are independent and stochastic [4].

The approach proposed in [4] is then adopted by some researchers, for example, in risk analysis of a human-robot

system [5], in the field of product liability prevention [6], and quantitative analysis of dynamic systems like space satellites [7].

The concept of sequential failure analysis [1] has been further developed by introducing counters of transitions in stochastic Petri nets (SPNs) located in various network connections [8]. The probabilities of sequential failures are calculated based on the obtained counters of failure transitions in the net.

A fuzzy approach to the problem of sequential failure is presented in [9]. Here, the authors combined adaptability of fuzzy logic with accuracy and modeling power of Petri nets to perform an efficient failure analysis.

Stochastic Petri nets have also been under attention during last years. For example in [10] Wang et al. have used stochastic Petri nets to assess reliability of systems based on non-homogenous Markov isomorphism. Useless service failures are a serious issue in real world problems so Zhao et al. [11] have used stochastic Petri nets models to detect useless service failures.

Uncertainty is an inherent characteristic of industrial systems. Such uncertainties can be handled by stochastic Petri nets. Garg and Sharma [12] have utilized stochastic Petri nets to model the behavior of complex industrial systems and then on its basis they try to find some of the reliability measure such as mean time between failures (MTBF) using Lambda Tau methodology. Another important property of industrial machines is their availability. Availability is a crucial topic especially in heavy industries since idle times of machines can impose thousands of dollars to the company. According to this, Beirong et al. [13] have used Generalized stochastic Petri nets (GSPN) to model complex industrial systems to maximize the machine availability. In another research, stochastic failure sequence has been investigated by Su and Wang [14] in order to simulate the dynamic reliability of manufacturing systems using stochastic Petri nets.

Although SFL provides an appropriate tool for evaluating systems, it has some drawbacks. For instance, in SFL it is assumed that failure sequences are known. Of course, this cannot be true in real world problems where there may be many unknown sequences of failures. In order to overcome such deficiencies, a novel approach for calculating probabilities of occurrences of sequential failures is presented in [15]. This research adopts the concept of reachability trees in Petri Nets, and then determines different failure sequences by drawing reachability tree of the Petri nets model of the system. Although this approach seems to be suitable for small systems with limited number of states, it is not beneficial for complicated systems with several states since to draw the reachability tree for such systems is nearly impossible. Hence, our goal in this paper is to enhance the method introduced in [15] and develop new algorithms to determine failure sequences of large systems automatically.

The reminder of the paper is as follows. In Section 2, basic concepts of Petri nets and their application in failure analysis are discussed. The framework of sequential failure analysis is presented in Section 3. In Section 4, the developed method is discussed and the performance of the developed algorithms is analyzed. Section 5 is devoted to an illustrative example in order to demonstrate capacities of the developed method. The paper concludes in Section 6.

2. Petri Nets and Their Application in Failure Analysis

Petri Nets are graphical and mathematical modeling tools applicable to many systems. They offer formal graphical description possibilities for modeling systems consisting of concurrent processes. Petri Nets have been used extensively for modeling and analyzing of discrete event systems. As a graphical tool, Petri nets can be used for visual communication aims similar to flow charts, block diagrams, and networks. In addition, *tokens* are used in these nets to simulate the dynamic and concurrent activities of systems.

For more details about evolution of Petri nets, reader is referred to [16, 17]. A Petri net is a 5-tuple, $PN = (P, T, F, W, M_0)$, where

$$P = \{p_1, p_2, \ldots, p_m\},$$

$$T = \{t_1, t_2, \ldots, t_n\},$$

$F \subseteq (P \times T) \cup (T \times P)$ is a set of arcs (flow relations),

$W : F \rightarrow \{1, 2, 3, \ldots\}$ is a weight function,

$M_0 : P \rightarrow \{0, 1, 2, \ldots\}$ is the initial marking,

$P \cap T = \emptyset$ and $P \cup T \neq \emptyset$.

The dynamic behavior of a system is modeled by changing state or marking in Petri nets according to the following (firing) rules.

(1) A transition t is said to be enabled if each input place p of t is marked with at least $w(p, t)$ tokens, where $w(p, t)$ is the weight of the arc from p to t.

(2) An enabled transition may or may not fire depending on whether or not the event actually takes place (firing conditions are ok).

(3) Firing of an enabled transition t removes $w(p, t)$ tokens from each input place p to t and adds $w(t, p)$ tokens to each output place p of t, where $w(p, t)$ and $w(t, p)$ are the weights of the arcs from p to t or t to p, respectively.

In graphical representation of a Petri net, places are represented by circles and transitions are shown by hollow bars. The relationships between places and transitions are represented by direct arcs. For example, the Petri net of Figure 1 depicts the firing of a transition.

In un-timed Petri net one can prohibit controlled transitions from firing but cannot force the firing of a transition at a particular time. In timed Petri nets controlled transitions are forced to fire by observing the time dependent firing functions. In timed Petri nets, each transition has its specific time which determines the transition's holding time. When a transition is fired during its holding time, markings of networks are not changed. By elapsing holding time, the markings will change according to the firing rules.

Application of Petri nets in failure analysis is an emerging active field of research. The application of PNs is similar to the application of "Fault (Event) Tree Analysis (FTA and ETA)" which are two strong graphical tools for pre (post) event reliability and risk analysis. As this is a rather new field, the literature is not yet rich; however researches on safety analysis and reliability growth [18, 19], reliability evaluation [20–22], and reliability of manufacturing systems [23–25] have already been presented.

Some researchers believe that PNs can be an appropriate alternative for FTA [19, 20], since it not only graphically symbolizes the cause and effect relationships among the events, but also represent dynamic behavior of the system. Fault trees, which are basic graphical risk analysis tools, can be transformed to Petri Nets. For more details, readers are referred to [19].

3. Framework of Sequential Failure Analysis

General framework of Sequential Failure Analysis (SFA) in the literature of reliability and risk analysis is shown in Figure 2 [1].

This methodology utilizes Fault Tree Analysis (FTA) and FMEA and dynamic Petri net modeling for identifying all possible failures and their sequences of occurrence. As shown in Figure 2, the framework of sequential failure analysis

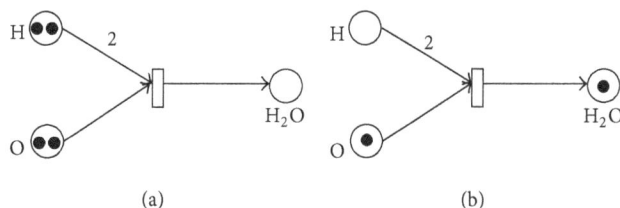

FIGURE 1: Transition firing: (a) marking before firing, (b) marking after firing.

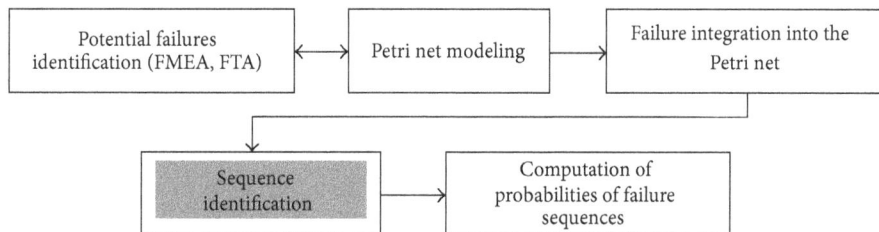

FIGURE 2: Framework of sequential failure analysis [1].

consists of five steps. Since there is no specific algorithm for sequence identification in the literature, the main goal of this paper is to develop new algorithms in the fourth step, highlighted in Figure 2.

Sequential failure analysis steps start with using FMEA or FTA techniques in order to predict all potential failures. Although FMEA is a general term, it is divided to different branches such as Quality FMEA (QFMEA), Design FMEA (DFMEA), Process FMEA (PFMEA), and so forth. The second step, Petri net modeling, includes modeling the system in a dynamic manner so that all tasks and activities taking place in the system can be seen. The third step is similar to the second one except that it considers system failures and merges such possible failures with the main body of the system Petri model.

Step 4 is our main focus and we will discuss it in next sections. The last step has been considered by many other researchers [1, 15] and we do not discuss it anymore.

4. Methodology

As noted earlier, failures of a system are not limited to failures in sub-systems but they also include a hierarchy of failures in relevant sub-systems. On the other hand, various analyzing methods of Petri nets fail in determining failure sequences leading to total system failure, despite making a schematic view of system behaviors during time. In spite of the capacity of reachability trees in showing sequences of events, they are not efficient in analyzing complicated nets. On the other hand, to the knowledge of authors, no specific algorithm capable of constructing reachability trees, combined with determination of sequences of events, can be found in the literature.

One of the key factors in calculating sequential failures of a system is to determine behavioral sequences of the net leading to the failure. Hence, the proposed algorithm must be able to construct different behavioral states (markings of the net) and the entire sequences of events in a combinatorial manner. In the following, we describe symbols utilized in our methodology and then present our approach.

4.1. Variables and Symbols Definitions

P: Number of all timed and untimed places existing in the Petri net.

T: Number of all timed and untimed transitions existing in the Petri net.

External: An external $T \times P$ matrix. The entries of this matrix are the weights of all arcs connecting each transition (in rows) to each place (in column).

Internal: An internal $P \times T$ matrix. The entries of this matrix are the weights of all arcs connecting each place (in rows) to each transition (in column).

Status: State $T \times P$ matrix. The entries of this matrix are 0 and 1. In fact, this matrix shows how a place (in row) is connected to a transition (in column). If the arc connecting place i to transition j is ordinary, then entry (i, j) of the status matrix is 1; in case of inhibitor arc this component is 0. If there are no arcs between a place and a transition, then the corresponding entry in the status matrix will be again 1.

Info: Evolutionary behavioral matrix of the net. This matrix plays the main role in our heuristic algorithm and it becomes more complete during each step. The entries include markings (behavioral states of the net) and existing firing sequences of the net. We will discuss the structure of this matrix in more details in the following sections.

Level: The last level among different levels of the net being considered.

$M(0)$: Initial marking of the Petri net.

$M(i)$: Marking i of the Petri net.

```
MAIN Algorithm
Input: Internal = []_{P×T}, External = []_{T×P}, M(0), Status = []_{P×T}
Output: Info matrix              // Info is defined in Section 4.3.2
External = External*; Info = [0]_{n×n};
Info{1, 1} = M(0); Info{2, 1} = Enabling(M(0), Internal, External);
Stop = 0; level = 1;
while Stop = 0 do
    for g = 2 → 3 →Linefinder(Info)
        Info= Copier(Info, g, level, T);
    end for
    Info = Filler(Info, level, T, Internal, External, Status);
    for s = 1 → Linefinder(Info)
        if Info{s, level} = 0
            Stop = Stop + 1;
        end if
    end for
    if Stop = Linefinder(Info)
        end while
    else Stop = 0; level = level + 1;
    end if
end while
return Info
```

ALGORITHM 1: Main algorithm.

4.2. Assumptions. Petri nets are:

(i) pure. Purity means that a place cannot be at the same time the input and output of a specific transition,

(ii) live, and

(iii) bounded.

4.3. The Heuristic Algorithm.

In this section we present our heuristic algorithm. The following sub-sections describe the *main* algorithm and relevant *functions*.

4.3.1. Main Algorithm.

Here we present the main body of the proposed algorithm in Algorithm 1.

4.3.2. The Performance of Algorithm.

Algorithm 1 is able to construct reachability tree, entire markings of the systems, and all the firing sequences occurring in the Petri net model, simultaneously. The significance of this algorithm is in analyzing sequential failures where identification of sequential failures is vital in calculating sequences of event leading to total failure of the system.

In this algorithm, firstly a square matrix **Info** with zero elements is constructed. The size of the matrix depends on the size of the Petri net model being considered. In order to avoid confusion when using the matrix **External**, the algorithm uses the transpose of the input matrix **External**. Hence, when speaking about matrix **External**, we mean the transpose of the input matrix **External**. Then, the algorithm substitutes the initial marking $M(0)$ in the cell (1, 1) of **Info** and also substitutes the result of an internal function "Enabling" in cell (2, 1). This internal function gets the marking of a Petri net and gives a row matrix as its output. This row matrix consists of 0 and 1 and demonstrates enabled transitions of the corresponding marking. It is apparent that the number of columns of the output of the function "Enabling" equals T.

As shown in Figure 3, the algorithm is designed to operate level by level. In this paper, levels in reachability tree mean sets of markings in which there are equal numbers of firings from the initial marking to reach such markings.

Then during the next step, the internal algorithm Copier, which will be discussed later, operates on elements of the second row of the matrix **Info** and advances with triple steps. Cells considered by this algorithm are the row matrices demonstrating the enabled transitions of each marking.

Each sequence in matrix **Info** is made of three rows. The algorithm Copier copies three rows of each firing sequence n times. Here n means the number of enabled transitions in each considered sequence.

According to above notations, our proposed algorithm considers three rows for each sequence of firing. The first row shows the markings of each sequence, the second row shows enabled transitions of its corresponding marking, and the third row demonstrates the number of the fired transition in each marking.

Then firing process of the existing transitions in each feasible sequence of the considered level is performed, and new markings, resulted from firing of these transitions, are transported to the next level (column) of the matrix **Info**. This process is performed by internal algorithm Filler which will be explained in the next sections. In order to explain the proposed algorithm, consider the Petri net of Figure 4. Reachability tree and the output matrix **Info** from operation of the algorithm are displayed in Figure 3 and Figure 5, respectively.

In fact, in this algorithm, elements of the matrix **Info** are completed via a wave process (Figure 5). This means

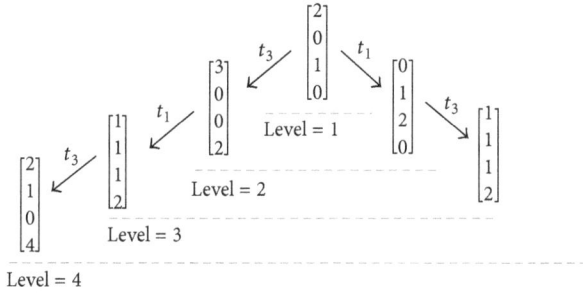

FIGURE 3: Firing levels form an initial marking in a Petri net.

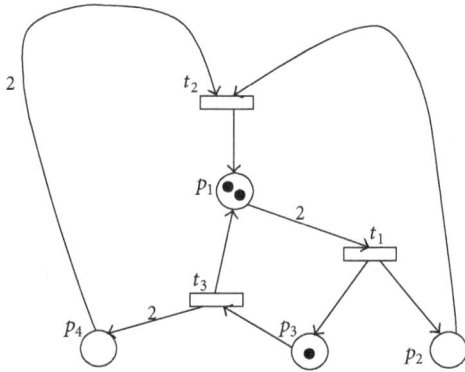

FIGURE 4: An ordinary Petri net [17].

Level = 1	Level = 2	Level = 3	Level = 4
[2; 0; 1; 0]	[3; 0; 0; 2]	[1; 1; 1; 2]	[2; 1; 0; 4]
[0, 0, 0]	[0, 0, 0]	[0, 0, 0]	
3	1	3	
[2; 0; 1; 0]	[0; 1; 2; 0]		
[0, 0, 0]	[0, 0, 0]		
1	3		

FIGURE 5: Output of the algorithm with the process of wave operation.

that completion of each level of the reachability tree and evolution of the constructed sequences in the next level are performed simultaneously. It is noteworthy to mention that during evolution of matrix **Info,** elements of rows 2, 5, 8, . . . transform to zero through a multilevel process. The aim is to prevent the algorithm from making repeated markings in the sequence. For example in Figure 5, in the marking of the cell (4, 2) of the matrix **Info,** transition 3 is enabled. If this transition is fired the resulting marking in the fourth level will be equal to the marking of the cell (4, 1) which is iterated. Hence, the sequence terminates at this level.

Evolution of the matrix **Info** terminates only if the new generated marking exists in the existing set of the markings of the sequence. This is to prevent the algorithm from generating repeated sequences. If the new sequence being generated by the algorithm already exists, then the algorithm skips this sequence. This is to prevent the reachability tree of the net from having state space explosion.

We should note that the function Linefinder in Algorithm 1 is an internal function which returns the number of the last row of the matrix **Info** in which there is a positive value. Here we consider the main condition of stop in the main body function and prove its termination.

Termination Condition. According to the assumption of Section 4.2, the proposed algorithm operates only on bounded Petri nets. Internal construction of the function Filler prevents extension of each sequence in case of iteration. Therefore, the possibility of constructing unbounded sequences is zero. On the other hand, since the Petri nets are bounded and algorithm has a performance similar to complete counting process then the entire markings will be definitely counted. This means that according to the principle of the boundedness of the net, there will be one iteration in any of the sequences and all the sequences will be terminated at a level, and the condition for termination condition will be satisfied.

4.3.3. Function Filler. This function is one of the main operators in the main algorithm. This algorithm operates on levels (columns of the matrix **Info**) and fires enabled transitions in such a way that no iteration happens. It also constructs next level in matrix **Info** gradually by a wave shaped motion. Algorithm 2 presents this function.

This function returns a two-dimensional row matrix called "*a*" and a one-dimensional row matrix, "*b*". In fact, this function operates on the considered level with triple steps and gives the kinds and numbers of different markings in that level in matrixes "*a*" and "*b*", respectively.

Algorithm "Filler" has no termination condition and operates according to the dimension of its input data. This algorithm performs on the first row of each sequence and then analyzes each sequence. In case of existence of enabled transition in the last marking of the sequence, it generates an experimental marking. If this new marking is not iterated then the function adds this new marking to the end of the sequence in the next level and also adds the enabled transitions of the new marking to the next level. In Algorithm 2, we have used a function called "Newmarking" which will be discussed in detail in the coming sections.

4.3.4. Function Copier. This algorithm plays an important role in generating different sequences, by operating on the second row of each existing sequence. This row demonstrates enabled transitions of its corresponding marking. According to the number of enabled transitions in each marking, function "Copier" firstly checks whether firing a transition leads to a new sequence or not. If this is true then the relevant

Filler Algorithm
Input: Info, level, T, **Internal, External, Status**
Ouput: Cons $= []_{n \times n}$
Cons = **Info**;
$[ab]$ = Finder(**Info**, level); $\backslash \backslash a = []_{1 \times k}, b = []_{1 \times k}$
for $i = 1 \ \rightarrow \ 3 \ \rightarrow$ Linefinder(**Info**) + 1
　　if Cons$\{i - 2,$ level$\} \neq 0$ **then**
　　　　for $m = 1 \ \rightarrow \ k$
　　　　　　if Info$\{i - 2,$ level$\} - a\{1, m\} = 0$ **then**
　　　　　　　　for $j = 1 \ \rightarrow \ T$
　　　　　　　　　　if Cons$\{i - 1,$ level$\}(j) = 1$ **then**
　　　　　　　　　　　　Cons$\{i - 1,$ level$\}(j) = 0$; Cons$\{i,$ level$\} = j$;
　　　　　　　　　　end if
　　　　　　　　　　for $z = 2 \ \rightarrow \ 3 \ \rightarrow$ Linefinder(**Info**)
　　　　　　　　　　　　Update **Info**;
　　　　　　　　　　end for
　　　　　　　　　　for $l = 1 \ \rightarrow$ level
　　　　　　　　　　　　If Newmarking(Cons$\{i - 2,$ level$\}$) is not iterated **then**
　　　　　　　　　　　　　　Add Newmarking and enabling matrices to the sequence;
　　　　　　　　　　　　end if
　　　　　　　　　　end for
　　　　　　　　end for
　　　　　　end if
　　　　end for
　　end if
end for
return Cons

ALGORITHM 2: Function Filler of the proposed algorithm.

Copier Algorithm
Input: Info, row, level, T
Output: the same **Info** matrix which is modified
ifInfo$\{$row$-1,$ level$\} \neq 0$ **then**
dd = number of ones in **Info**$\{$row, level$\}$;
　　if level = 1 **then**
　　　　for $k = 1 \ \rightarrow \ dd - 1$
　　　　　　for $l = 1 \ \rightarrow$ level
　　　　　　　　Copy the sequence to **Info** with one row spacing;
　　　　　　end for
　　　　end for
　　else
　　　　　　for $k = 1 \ \rightarrow \ dd - 1$
　　　　　　　　for $l = 1 \ \rightarrow$ level
　　　　　　　　　　Copy the sequence to **Info**;
　　　　　　　　end for
　　　　　　end for
　　end if
end if
return Info

ALGORITHM 3: Function copier.

sequence will be added to the end of the evolutionary matrix "Info". Output of this function is constantly processed by the function "Filler". The corresponding algorithm of this function is presented in Algorithm 3.

4.3.5. Function Newmarking. This algorithm is designed based on dominating concepts in the field of Petri nets. This function constructs and solves linear systems of the Petri net using basic concepts of token transfer and so forth. For more

NEWMARKING Algorithm
Input: Marking $=[]_{P \times 1}$, **Internal**, **External**, **Status**, Tranum
Output: Outmarking (a new marking resulting from firing transition Tranum)
Difference $= [0]_{P \times T}$;
For $k = 1 \rightarrow T$
   for $i = 1 \rightarrow P$
     if Status$(i, k) = 1$ **then**
       Difference = **External** − **Internal**;
     else
       Difference = **External**;
     end if
   end for
end for
Outmarking = Marking + Difference $\times [0]_{T \times 1}$ (with element 1 in rownumber Tranum);
return Outmarking

ALGORITHM 4: Function newmarking.

details, the reader is referred to [17]. The algorithm of this function is demonstrated in Algorithm 4.

At the end of this section, the proposed method is entirely represented in Figure 6.

5. An Illustrative Example

In this section we solve a sequential failure problem and demonstrate the capabilities of the proposed method. This example is adopted from [10]. We have coded the proposed algorithms in MATLAB. Consider the Petri net of a machining cell in Figure 7. Input matrices of the proposed heuristic algorithm are

$$
\text{initial marking} = \begin{bmatrix} 1 \\ 0 \\ 0 \\ 1 \\ 1 \\ 0 \\ 0 \\ 1 \\ 1 \\ 0 \\ 0 \end{bmatrix},
$$

$$
\text{internal matrix} = \begin{bmatrix} 1 & 0 & 0 & 0 & 0 & 0 & 0 \\ 0 & 1 & 0 & 0 & 0 & 0 & 0 \\ 0 & 0 & 1 & 0 & 0 & 0 & 0 \\ 0 & 1 & 0 & 0 & 0 & 0 & 0 \\ 0 & 1 & 0 & 0 & 0 & 0 & 0 \\ 0 & 0 & 0 & 1 & 0 & 0 & 0 \\ 0 & 0 & 0 & 0 & 0 & 0 & 1 \\ 0 & 0 & 0 & 0 & 1 & 0 & 0 \\ 0 & 0 & 0 & 0 & 0 & 1 & 0 \\ 0 & 0 & 0 & 0 & 0 & 0 & 1 \\ 0 & 0 & 1 & 0 & 0 & 0 & 0 \end{bmatrix},
$$

$$
\text{external matrix} = \begin{bmatrix} 0 & 1 & 0 & 0 & 0 & 0 & 0 & 0 & 0 & 0 & 0 \\ 1 & 0 & 1 & 0 & 0 & 1 & 0 & 0 & 0 & 0 & 0 \\ 0 & 0 & 0 & 1 & 1 & 0 & 0 & 0 & 0 & 0 & 0 \\ 0 & 0 & 0 & 0 & 0 & 0 & 1 & 0 & 0 & 0 & 0 \\ 0 & 0 & 0 & 0 & 0 & 0 & 0 & 0 & 0 & 1 & 0 \\ 0 & 0 & 0 & 0 & 0 & 0 & 0 & 0 & 0 & 1 & 0 \\ 0 & 0 & 0 & 0 & 0 & 0 & 0 & 0 & 0 & 0 & 1 \end{bmatrix}.
$$

$$(1)$$

According to the proposed main algorithm represented in Figure 6, the wavy procedure is applied to the manufacturing cell represented in Figure 7. This manufacturing cell consists of one robotic arm, and a single machine to process the incoming parts. The robotic arm is responsible for loading and unloading the parts to and from the machine. In cases when the robotic arm drops a part then an operator should enter the hazardous zone to solve the problem and load or unload the machine manually. In such situation the operator is in danger of having accident with the robotic arm. This process is totally depicted in Figure 7.

Computational results of executing the heuristic algorithm are presented in Table 1. As it can be seen from Figure 7, the aim is to determine all the firing sequences of transitions leading to firing of transition t_7.

Here we will describe the solution procedure of this problem step by step. According to the flowchart represented in Figure 6 firstly the internal, external, and status matrices should be determined according to Petri net model of the system. Then in the next step the first two elements of the matrix **Info** should be filled using the Enabling function. Then for each firing level, enabled transition are discovered and firing sequences for the entire enabled transitions are performed until reaching a similar firing sequence which has been obtained before.

The functions Linefinder, Filler, and Copier are intermediate algorithms which are responsible for checking uniqueness of a firing sequence, and implementing the wavy procedure to fill the **Info** matrix. As noted before, the wavy procedure tries to find the possible unique firing sequences and find firing levels, simultaneously.

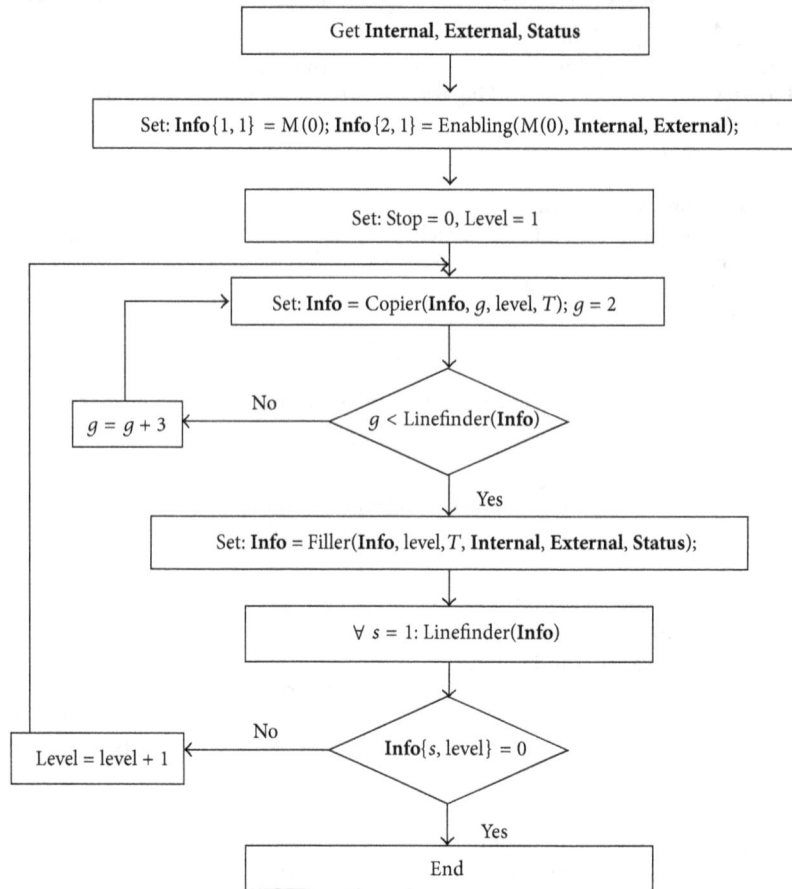

FIGURE 6: Flowchart of the main algorithm.

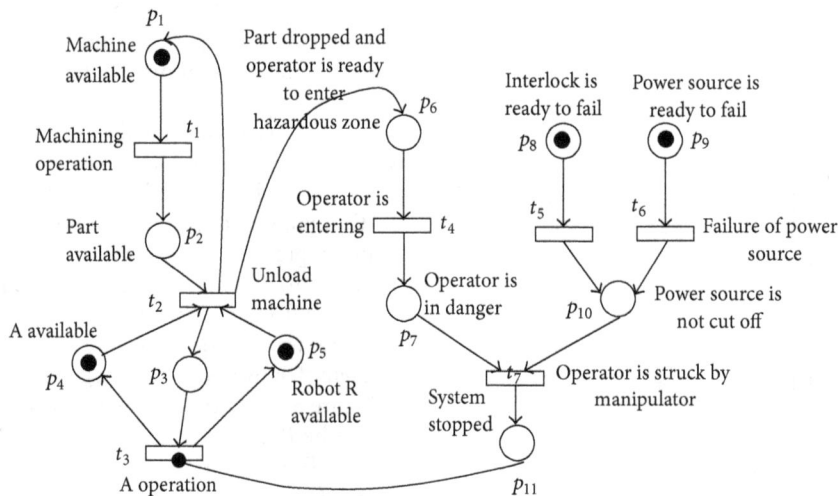

FIGURE 7: Petri net model of a manufacturing cell.

The main body algorithm in this paper then terminates until the entire possible firing sequences for the entire firing levels have been investigated and then they are reported.

Transition 7 plays the role of hitting an operator by robotic manipulator. Since the number of firing sequences leading to firing t_7 is large we present only some of firing sequences leading to firing failure transition (t_7).

TABLE 1: Firing sequences leading to firing transition t_7 of Figure 4.

1	$t_5 t_1 t_2 t_1 t_3 t_2 t_4 t_7$
2	$t_5 t_1 t_2 t_4 t_7$
3	$t_5 t_1 t_2 t_4 t_1 t_7$
4	$t_1 t_2 t_5 t_4 t_7$
5	$t_1 t_2 t_5 t_1 t_4 t_7$
6	$t_1 t_2 t_1 t_4 t_5 t_7$
7	$t_1 t_2 t_4 t_5 t_7$
8	$t_1 t_2 t_4 t_1 t_5 t_7$
9	$t_1 t_2 t_3 t_4 t_1 t_6 t_7$
10	$t_1 t_2 t_1 t_4 t_5 t_3 t_7$
11	$t_1 t_2 t_1 t_5 t_4 t_6 t_3 t_7$
12	$t_6 t_1 t_2 t_4 t_1 t_5 t_7$
13	$t_1 t_2 t_3 t_1 t_4 t_2 t_1 t_5 t_6 t_7$
14	$t_1 t_2 t_6 t_3 t_1 t_4 t_2 t_5 t_7$
15	$t_1 t_2 t_1 t_3 t_2 t_1 t_5 t_4 t_4 t_3 t_7$
16	$t_5 t_1 t_2 t_1 t_3 t_2 t_4 t_3 t_1 t_7$
17	$t_5 t_1 t_2 t_1 t_3 t_2 t_6 t_3 t_1 t_4 t_7$
18	$t_1 t_2 t_1 t_3 t_2 t_3 t_2 t_3 t_4 t_4 t_1 t_6 t_7$

TABLE 2: Firing sequences leading to firing transition t_7 of Figure 4 by the method presented in [15].

1	$t_5 t_1 t_2 t_4 t_7$
2	$t_5 t_1 t_2 t_4 t_1 t_7$
3	$t_1 t_2 t_5 t_4 t_7$
4	$t_1 t_2 t_5 t_1 t_4 t_7$
5	$t_5 t_1 t_2 t_1 t_3 t_2 t_4 t_7$
6	$t_1 t_2 t_1 t_4 t_5 t_7$
7	$t_1 t_2 t_4 t_5 t_7$
8	$t_1 t_2 t_4 t_1 t_5 t_7$

with one of the main existing methods in the literature. This comparison demonstrated precision and accuracy of the proposed method.

Our proposed algorithm is general and can handle different firing sequences in addition to failure analysis. Table 2 represents computational results of the technique presented in [15]. According to Table 2, this method which is based upon drawing reachability tree of the Petri nets can detect only 8 failure sequences while the proposed method in this paper has detected 18 failure sequences led to firing of t_7. This proves that the older technique can detect only 40% of potential sequential failures but the proposed heuristic algorithms in this paper are capable to approximately detect the whole sequences. Hence, performance of the method [15] cannot be trusted in complex systems.

By analyzing the results of the two tables above, it can be concluded that the maximum number of transition firings detected by the method [15] is 7 firings but according to Table 1 maximum number of firings is 12 which is considerably greater than of [15]. This is because the older method is a graphical-based method and cannot handle complex nets. On the other hand, the technique presented in this paper represents a systematic approach and does not need to draw reachability graphs of the net and has omitted some time consuming parts of the older method.

Computational results of the above example shows that the method used in [15] just considers some of firing sequences leading to failure while method adopted in this paper is much stronger and can determine all the firing sequences.

6. Conclusions

In this paper, some novel algorithms in order to determine firing sequences leading to failures in systems were developed. The proposed method not only can present entire firing sequences in a Petri net but also it can draw reachability tree of that Petri net, simultaneously. We coded these algorithms MATLAB programming language and compared the results

References

[1] A. Adamyan and D. He, "Sequential failure analysis using counters of Petri net models," *IEEE Transactions on Systems, Man, and Cybernetics Part A:Systems and Humans*, vol. 33, no. 1, pp. 1–11, 2003.

[2] M. Braglia, M. Frosolini, and R. Montanari, "Fuzzy critically assessment for failure modes and effect analysis," *International Journal of Quality & Reliability Management*, vol. 20, no. 4, pp. 503–524, 2003.

[3] K. Xu, L. C. Tang, M. Xie, and M. L. Zhu, "Fuzzy assessment of FMEA for engine systems," *Reliability Engineering & System Safety*, vol. 75, no. 1, pp. 19–27, 2002.

[4] J. B. Fussell, E. F. Aber, and R. G. Rahl, "On the quantity analysis of priority-AND failure logic," *IEEE Transactions on Reliability*, vol. R-25, no. 5, pp. 324–326, 1976.

[5] Y. Sato, E. J. Henley, and K. Inoue, "Action-chain model for the design of hazard-control systems for robots," *IEEE Transactions on Reliability*, vol. 39, no. 2, pp. 151–157, 1990.

[6] Y. Shibata and Y. Sato, "Case study of risk assessment for product liability prevention," in *Proceedings of the PSAM-4*, vol. 2, pp. 1215–1220, 1998.

[7] L. Ngom, A. Cabarbaye, and C. Barpm, "Taking into account of dependency relations in the Monte Carlo simulation of non-coherent fault trees," in *Proceedings of the PSAM-4*, vol. 3, pp. 2067–2072, 1998.

[8] F. Baccelli, G. Cohen, G. J. Olsder, and J. P. Quadrat, *Synchronization and Linearity*, John Wiley, New York, NY, USA, 1992.

[9] D. Torshizi A and S. R. Hejazi, "A fuzzy approach to sequential failure analysis using Petri nets," *International Journal of Industrial Engineering and Production Research*, vol. 21, no. 2, pp. 53–60, 2010.

[10] Q. Wang, J. Gao, K. Chen, and P. Yang, "Reliability assessment of Manufacturing system based on HSPN models and non-homogeneous isomorphism Markov," in *Proceedings of the International Conference on Quality, Reliability, Risk, Maintenance, and Safety Engineering*, pp. 182–186, 2011.

[11] M. Zhao, Y. Zhou, Y. Yang, W. Song, and Y. Du, "A new method to detect useless service failure model in SPN," *Journal of Convergence Information Technology*, vol. 5, no. 3, pp. 129–134, 2010.

[12] H. Garg and S. P. Sharma, "Stochastic behavior analysis of complex repairable industrial systems utilizing uncertain data," *ISA Transactions*, vol. 51, no. 6, pp. 752–762, 2012.

[13] Z. Beirong, X. Xiaowen, and X. Wei, "Availability modeling and analysis of equipment based on generalized stochastic petri nets,," *Research Journal of Applied Sciences, Engineering and Technology*, vol. 4, no. 21, pp. 4362–4366, 2012.

[14] C. Su and S. Wang, "Dynamic reliability simulation for manufacturing system based on stochastic failure sequence analysis," *Journal of Mechanical Engineering*, vol. 47, no. 24, pp. 165–170, 2011.

[15] A. Adamyan and D. He, "Analysis of sequential failures for assessment of reliability and safety of manufacturing systems," *Reliability Engineering and System Safety*, vol. 76, no. 3, pp. 227–236, 2002.

[16] C. A. Petri, "Kommunikation mit automaten," Schriften Des IIM no. 3, Institut für Instrumentelle Mathematik, Bonn, Germany, 1962.

[17] T. Murata, "Petri nets: properties, analysis and applications," *Proceedings of the IEEE*, vol. 77, no. 4, pp. 541–580, 1989.

[18] N. G. Leveson and J. L. Stolzy, "Safety analysis using Petri nets," *IEEE Transactions on Software Engineering*, vol. SE-13, no. 3, pp. 386–397, 1987.

[19] S. K. Yang and T. S. Liu, "Failure analysis for an airbag inflator by petri nets," *Quality and Reliability Engineering International*, vol. 13, no. 3, pp. 139–151, 1997.

[20] T. S. Liu and S. B. Chiou, "The application of Petri nets to failure analysis," *Reliability Engineering and System Safety*, vol. 57, no. 2, pp. 129–142, 1997.

[21] G. S. Hura and J. W. Atwood, "Use of Petri nets to analyze coherent fault trees," *IEEE Transactions on Reliability*, vol. 37, no. 5, pp. 469–474, 1988.

[22] V. Kumar and K. K. Aggarwal, "Petri Net modelling and reliability evaluation of distributed processing systems," *Reliability Engineering and System Safety*, vol. 41, no. 2, pp. 167–176, 1993.

[23] J. Changjun, D. Baiqing, and W. Feng, "Study on reliability of manufacturing system based on petri net," *High Technology Letters*, vol. 1, no. 2, pp. 25–30, 1995.

[24] H. Xiong and Y. He, "GSPN based reliability modeling and analysis of CIMS," *Mechanical Science Technology*, vol. 16, pp. 1103–1106, 1997.

[25] C. H. Kuo and H. P. Huang, "Failure modeling and process monitoring for flexible manufacturing systems using colored timed petri nets," *IEEE Transactions on Robotics and Automation*, vol. 16, no. 3, pp. 301–312, 2000.

16

Optimizing Inventory and Pricing Policy for Seasonal Deteriorating Products with Preservation Technology Investment

Yong He and Hongfu Huang

Institute of Systems Engineering, School of Economics and Management, Southeast University, Nanjing 210096, China

Correspondence should be addressed to Yong He; heyong@126.com

Academic Editor: Kung-Jeng Wang

The paper studies a kind of deteriorating seasonal product whose deterioration rate can be controlled by investing on the preservation efforts. In contrast to previous studies, the paper considers the seasonal and deteriorating properties simultaneously. A deteriorating inventory model is developed for this problem. We also provide a solution procedure to find the optimal decisions about the preservation technology investment, the market price, and the ordering frequency. Then a case study is used to illustrate the model and the solution procedure. Finally, sensitive analysis of the optimal solution with respect to major parameters is carried out.

1. Introduction

The research on deteriorating items has begun from 1963. A model with exponentially decaying inventory was initially proposed by Ghare [1]. In recent years, many researchers have done a lot of work on inventory problems about deteriorating products. Deterioration is defined as decay, change, or spoilage such that the items are not in a condition of being used for their original purpose [2]. Electronic goods, radioactive substances, grains, alcohol, and gasoline are examples of deteriorating products. Also, for some products, the demand may exist for just a limited time horizon. We call such products as seasonal products, for example, Christmas trees or fireworks. Now, more and more products become deteriorating and seasonal simultaneously because of the competition and technology development, such as seasonal fashion goods (clothes, sweaters, shoes, etc.), high-tech electronics products (e.g., laptops, computers, mobiles, and cameras), and some seasonal food products (such as Chinese moon cake).

Hence, this will become a very difficult problem to decide the inventory if the product is both deteriorating and seasonal. In this paper, we mainly study the optimal inventory decision of the seasonal deteriorating products.

Some researchers have studied such deteriorating inventory model, but they do not consider that the deterioration rate can be controlled.

In reality, the deterioration rate can be controlled through preservation technology investment. For example, the fruit retailer can reduce the rate of product deterioration by adopting the cool supply chain. But the preservation technology investment will lead to additional cost. Hence, a key inventory problem is to find the optimal replenishment and preservation technology investment policy which maximizes the unit time profit.

This paper is the first paper to study both the preservation technology investment and pricing strategies of deteriorating seasonal products. In this paper, a model for deteriorating seasonal products is built, in which deterioration rate can be controlled by preservation technology investment. The decision variables are the market demand, the preservation technology investment parameter, and the ordering frequency. To get the optimal solution, an algorithm is designed. To foster additional managerial insights, we perform extensive sensitivity analyses and illustrate our results with a case study.

The rest of the paper is organized as follows. Section 2 is the review of the related papers. Section 3 is the notations and assumptions. Section 4 is the description of model. Section 5 is the algorithm and numerical examples. The last section provides concluding remarks and describes future research.

2. Literature Review

The property of the deterioration rate is very important in the research of deteriorating inventory. In most literatures till now, it is assumed that deterioration rate is a constant [3–8] or an exogenous variable [2, 9]. But in many practical situations, the deterioration rate can be controlled and reduced through various efforts such as procedural changes and specialized equipment acquisition. Especially for these products with high deterioration rate such as refrigerated food, fruit and vegetable, and fresh seafood, the firm has strong willingness to adopt the preservation technology to decrease the deterioration rate. Recently, some papers started to study the deteriorating inventory with preservation technology investment. Blackburn and Scudder [10] studied the optimal control of warehouse temperature under warehouse capacity constraints. They also proved that it is beneficial to share the inventory between supply chain members. Kouki et al. [11] found that when warehouse temperature can be controlled, a continuous temperature control policy can be very efficient. Musa and Sani [12] studied the model when the deterioration rate is noninstantaneous and deterioration rate can be controlled by preservation technology investment. Hsu et al. [13] proposed that the cost of preservation and the deterioration rate satisfy the equation $\lambda = k - M(\varepsilon)$. $M(\varepsilon)$ is the reduced deterioration rate after investing on preservation technology. The first derivative of $M(\varepsilon)$ is positive, while the second derivative is negative. Similar to Hsu et al. [13], Dye and Hsieh [14] proposed that the cost of preservation and the deterioration rate satisfy the equation $\lambda(t) = \lambda_0(t)(1 - M(\varepsilon))$.

In addition to the deterioration rate, market demand is another very important factor considered in this paper. In some situations, the demand rate is assumed to be a constant (see [15]). But in real life, demand can hardly be a constant. It may change with time (see [16–18]), or it can be influenced by inventory level [19, 20] and marketing efforts [2]. At the same time, market price is highly related to demand. It is an important decision variable in many literatures. For example, in the papers of Shah et al. [2], Dye et al. [17], and Liang and Zhou [7], they also regarded price as a decision variable.

For some seasonal deteriorating products, the demand can only exist for a limited time horizon. Since the time horizon is fixed, it is necessary to decide the ordering frequency in a limited time horizon instead of the ordering period length. Some people have considered such situation, such as Sana et al. [3] and Yang et al. [21]. But they did not consider that the deterioration rate can be controlled.

3. Notation and Assumptions

3.1. Notation. The notation in this paper is listed below.

Decision Variables
  n : Ordering frequency
  α: Cost of preservation technology investment per unit time

p: Market price
q: Ordering quantity.

Constant Parameters
  c: Buying cost per unit
  h: Inventory holding cost per unit per time
  $I(t)$: Inventory level of a time point
  A: Ordering cost per order
  $D(p)$: Market demand, $D(p) = b - ap$
  b: Demand scale
  a: Price sensitive parameter
  TP: Total profit of the selling season.

3.2. Assumptions. The model in this paper is built on the base of the following assumptions.

(1) Market demand is linear related to market price.

(2) Market demand only exists in a limited time horizon T.

(3) Demand cannot be backlogged.

(4) Ordering lead time is zero.

(5) Deteriorated products have no value, and there is no cost to dispose or store them.

(6) The relationship of deterioration rate and the preservation technology investment parameter satisfies $\partial\lambda(\alpha)/\partial\alpha < 0, \partial^2\lambda(\alpha)/\partial\alpha^2 > 0$. Hence, in this paper we assume that $\lambda(\alpha) = \lambda_0 e^{-\delta\alpha}$. Here, $\lambda(\alpha)$ is the deterioration rate after investing on preservation technology, λ_0 is the deterioration rate without preservation technology investment, and δ is the sensitive parameter of investment to the deterioration rate.

(7) The cost of preservation technology investment per unit time is restricted to $\alpha \in [0, \overline{\alpha}]$.

4. Model

This study considers a single retailer's inventory policy in which the deterioration rate is affected by the preservation technology investment. For seasonal products, the decision variables are the market price, the ordering frequency, and the preservation technology investment parameter.

In this model, there are two tradeoffs. The first one is the tradeoff between the ordering frequency and the ordering cost per order. By increasing ordering frequency, we can decrease the deteriorating cost. But the ordering cost increases. The second tradeoff is the preservation technology investment and the deteriorating cost. By increasing the preservation technology investment, deteriorating cost decreases.

According to the assumption, the time length is equal in all the ordering periods. So, we only study the first period. In the first period, according to the modeling of exponential

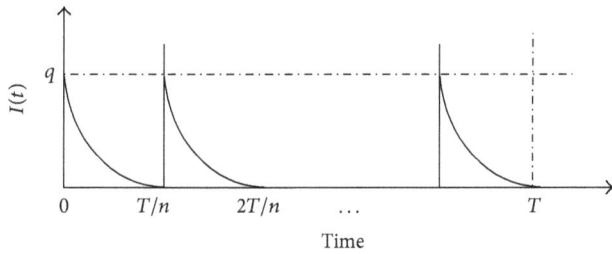

FIGURE 1: The inventory system.

deteriorating inventory in Ghare [1], the inventory level $I(t)$ can be depicted as Figure 1 and formulated as

$$\frac{\partial I(t)}{\partial t} = -\lambda(\alpha) I(t) - D(p) \quad \left(t \in \left[0, \frac{T}{n}\right]\right). \quad (1)$$

The boundary condition is

$$I\left(\frac{T}{n}\right) = 0. \quad (2)$$

By solving (1), we have

$$I(t) = \frac{D(p)}{\lambda(\alpha)}\left(e^{\lambda(\alpha)(T/n-t)} - 1\right). \quad (3)$$

The total profit of the season can be formulated as

TP = Sales revenue − Buying cost − Ordering cost

    − Inventory cost − Preservation cost. $\quad (4)$

(1) Sales revenue: The total revenue in time T can be formulated as

$$R = p \cdot D(p) \cdot T. \quad (5)$$

(2) Buying cost: According to (3), we can know the ordering quantity. The total buying cost can be formulated as

$$C_p = n \cdot c \cdot q = n \cdot c \cdot \frac{D(p)}{\lambda(\alpha)}\left(e^{\lambda(\alpha)T/n} - 1\right). \quad (6)$$

(3) Inventory cost: The total inventory quantity

$$\text{Inv} = n \cdot D(p) \cdot \int_0^{T/n} I(t)\,dt$$

$$= n \cdot \frac{D(p)\left(e^{\lambda(\alpha)(T/n)} - \lambda(\alpha)(T/n) - 1\right)}{\lambda^2(\alpha)}. \quad (7)$$

The formulation of the total inventory cost is

$$C_h = h \cdot \text{Inv}. \quad (8)$$

(4) Ordering cost: Ordering cost is a constant in every period. The total cost can be formulated as

$$C_o = n \cdot A. \quad (9)$$

(5) Preservation cost:

$$I_0 = \alpha T. \quad (10)$$

Hence, the profit function is

$$\text{TP}(n, \alpha, p) = p \cdot D(p) \cdot T - n \cdot c \cdot q$$
$$- h \cdot \text{Inv} - n \cdot A - I_0. \quad (11)$$

The problem is to solve the next program

$$\begin{aligned}
\min \quad & \text{TP}(n, \alpha, p) \\
\text{s.t.} \quad & D(p) > 0 \\
& p > 0 \\
& 0 \le \alpha \le \overline{\alpha}.
\end{aligned} \quad (12)$$

According to the Taylor series theory, for small λ and T/n values, the exponential function can have an approximation of $e^{\lambda T/n} \approx 1 + \lambda T/n + (\lambda T/n)^2/2$. This assumption is very common in many other papers, such as Lo et al. [22] and Wee et al. [23]. Substituting the equation into the target function, we have

$$\text{TP}(n, \alpha, p) = pD(p)T - cD(p)\left(T + \frac{T^2}{2n}\lambda(\alpha)\right)$$
$$- hD(p) \cdot \frac{T^2}{2n} - \alpha T - nA. \quad (13)$$

Proposition 1. *When market price p and preservation cost α are fixed, the profit function $TP(n, \alpha, p)$ is concave in ordering frequency n.*

Proof. The first and second partial derivatives of the target function $\text{TP}(n, \alpha, p)$ with respect to n are as follows:

$$\frac{\partial \text{TP}(n, \alpha, p)}{\partial n}$$
$$= \frac{\left(cD(p)\lambda(\alpha)T^2/2 + hD(p)T^2/2\right)}{n^2} - A, \quad (14)$$

$$\frac{\partial^2 \text{TP}(n, \alpha, p)}{\partial n^2} = -\frac{\left(cD(p)\lambda(\alpha)T^2 + hD(p)T^2\right)}{n^3} < 0. \quad (15)$$

According to (15), we can know that the profit function is concave in n. The ordering frequency is an integer. So, the search for the optimal ordering frequency is reduced to find a local optimal solution. $\qquad\square$

Proposition 2. *For known n and fixed p, we have the following.*

(1) *If $\Delta_1(n, p) \le 0$, $TP(n, \alpha, p)$ has a maximum value at $\alpha^* = 0$.*

(2) *If $\Delta_2(n, p) \ge 0$, $TP(n, \alpha, p)$ has a maximum value at $\alpha^* = \overline{\alpha}$.*

(3) If $\Delta_1(n, p) > 0$ and $\Delta_2(n, p) < 0$, $TP(n, \alpha, p)$ is concave and reaches its global maximum at point $\alpha^* \in (0, \overline{\alpha})$ to set $\partial TP(n, \alpha, p)/\partial \alpha = 0$.

($\Delta_1(n, p)$ and $\Delta_2(n, p)$ are defined in the following proof.)

Proof. The first and second partial derivatives of the target function $TP(n, \alpha, p)$ with respect to α give

$$\frac{\partial TP(n, \alpha, p)}{\partial \alpha} = \frac{cD(p)T^2\delta\lambda(\alpha)}{2n} - T, \quad (16)$$

$$\frac{\partial^2 TP(n, \alpha, p)}{\partial \alpha^2} = -\frac{cD(p)T^2\delta^2\lambda(\alpha)}{2n} < 0. \quad (17)$$

For simplicity, we set $G(\alpha) = cD(p)T^2\delta\lambda(\alpha)/2n - T$.

We define $\Delta_1(n, p) = G(\alpha)|_{\alpha=0} = cD(p)T^2\delta\lambda_0/2n - T$, $\Delta_2(n, p) = G(\alpha)|_{\alpha=\overline{\alpha}} = cD(p)T^2\delta\lambda(\overline{\alpha})/2n - T$.

It is obvious that $G'(\alpha) < 0$. So, $G(\alpha)$ is strictly decreasing in α.

(1) If $\Delta_1(n, p) \le 0$, $G(\alpha) \le 0$, and $\forall \alpha \in [0, \overline{\alpha}]$, $TP(n, \alpha, p)$ is decreasing in $\alpha \in [0, \overline{\alpha}]$. So, the optimal preservation cost is $\alpha^* = 0$.

(2) If $\Delta_2(n, p) \ge 0$, $G(\alpha) \ge 0$, and $\forall \alpha \in [0, \overline{\alpha}]$, $TP(n, \alpha, p)$ is increasing in $\alpha \in [0, \overline{\alpha}]$. So, the optimal preservation cost is $\alpha^* = \overline{\alpha}$.

(3) If $\Delta_1(n, p) > 0$ and $\Delta_2(n, p) < 0$, according to the intermediate value theorem, there exists unique value $\alpha^* \in (0, \overline{\alpha})$ to satisfy $G(\alpha^*) = 0$, that is,

$$\frac{cD(p)T^2\delta\lambda(\alpha^*)}{2n} - T = 0. \quad (18)$$

□

Proposition 3 indicates that when the initial deterioration rate is sufficiently small or the efficiency of the invested capital is low, there is no need to invest in preservation technology, for it is unbeneficial. Besides, if there is a constraint of the investment capital, there may be a potential for the firm to get more profit. Conclusions are proved in the case study.

Proposition 3. *There exists unique p^* which maximizes profit function $TP(n, \alpha, p)$ for fixed n and α.*

Proof. The first and second partial derivatives of the target function $TP(n, \alpha, p)$ with respect to n are as follows:

$$\frac{\partial TP(n, \alpha, p)}{\partial p} = (b - 2ap)T + ac\left[T + \frac{T^2\lambda(\alpha^*)}{2n}\right]$$

$$+ ah\frac{T^2}{2n}. \quad (19)$$

Let $\partial TP(n, \alpha, p)/\partial p$ be zero and solve for the optimal p^*, we have

$$p^* = \frac{b}{2a} + \frac{c}{2}\left[1 + \frac{T\lambda(\alpha)}{2n}\right] + \frac{hT}{4n}. \quad (20)$$

At point $p = p^*$, we have $\partial^2 TP(n, \alpha, p)/\partial p^2|_{p=p^*} = -2aT < 0$.

Thus, p^* is the global optimal which maximizes the profit function $TP(n, \alpha, p)$ for fixed n and α. □

Combining Propositions 1, 2, and 3, we have Proposition 4.

Proposition 4. *For fixed n, the optimal solution (α^*, p^*) that maximizes profit function $TP(n, \alpha, p)$ exists and is unique. The optimal solution can be obtained through some interaction algorithms.*

In the subsection, we use an interaction algorithm to solve numerical examples.

5. Algorithm

Step 1. Set $n = 1$.

Step 2. Set $k = 1$ and initialize the value of $p^k = p_0$.

Step 3. Calculate $\Delta_1(n, p)$, $\Delta_2(n, p)$ and execute any one of the following three cases 1, 2, or 3.

(1) If $\Delta_1(n, p) \le 0$, then $\alpha_1^k = 0$. Obtain p_1^k from (20).

(2) If $\Delta_2(n, p) \ge 0$, then $\alpha_1^k = \overline{\alpha}$. Obtain p_1^k from (20).

(3) If $\Delta_1(n, p) > 0$ and $\Delta_2(n, p) < 0$, obtain the value of α_1^k by solving (18). Substitute the value of α_1^k into (20) and to obtain the corresponding value of p_1^k.

Set $p^{k+1} = p_1^k$ and $\alpha^k = \alpha_1^k$.

Step 4. If $|p^{k+1} - p^k| \le 10^{-4}$, then $(\alpha^*, p^*) = (\alpha^k, p^{k+1})$ and go to Step 5. Otherwise, set $k = k + 1$ and go to Step 3.

Step 5. Calculate $TP(n, \alpha^*, p^*)$. It is the maximum of profit function for fixed n.

Step 6. Set $n' = n + 1$, repeat Step 2 to 5 and find $TP(n', \alpha^*, p^*)$. Go to Step 7.

Step 7. If $TP(n', \alpha^*, p^*) \ge TP(n, \alpha^*, p^*)$, set $n = n'$. Go to Step 6. Otherwise go to Step 8.

Step 8. Set $(n^*, \alpha^*, p^*) = (n, \alpha^*, p^*)$, then (n^*, α^*, p^*) is the optimal solution.

Step 9. Calculate corresponding Q according to (3).

6. Case Study

To better illustrate our conclusions, we proposed four cases. The first one is a normal case, which is a benchmark for the other two. In the second case, there is a constraint for the investment capital. In the third and fourth case, the value of initial deterioration rate and the efficiency parameter is relatively small. Here, we apply the above algorithm to solve the problem.

TABLE 1: Initial values of the parameters.

A	b	a	c	h	λ_0	$\bar{\alpha}$	T	δ
500	10	0.2	10	1	0.02	10	100	0.5

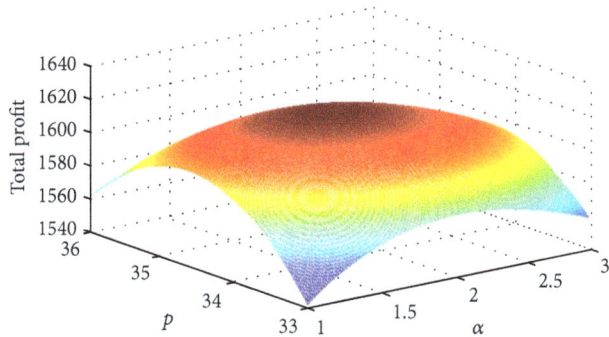

FIGURE 2: The function with respect to p and α for fixed $n = 6$.

Example 1. The initial values of the parameters are listed in Table 1.

By calculating with Matlab 7.1, when $n = 6$, we can plot the relation of total profit to market price and preservation cost as Figure 2. From Figure 2, we can see that the profit function is jointly concave in market price and preservation cost. For different n, the optimal profit always exists. As is shown in Table 2, the profit function is concave in ordering frequency. And when $n = 6$, the profit function reaches its maximum. So, the maximum of profit is TP = \$1621.9.

In this example, the upper bound of investment on preservation cost is sufficiently large. The optimal solution is not on the boundary.

Example 2. In this example, we set $\bar{\alpha} = 1.7500$ without change of other parameters of Example 1. From Table 3, we can see that the optimal solution is $\alpha^* = 1.7500$ which is on the boundary and the profit is smaller than that of the first example. In practice, the maximum capital on preservation technology investment can have significant influence on the benefit of retailer.

Example 3. The initial values of the parameters are shown in Table 4. When the initial value of the deterioration rate is relatively low, that is, $\lambda_0 = 0.001$ in Table 4, according to the algorithm, $\Delta_1(n, p) < 0$, the optimal investment capital is $\alpha^* = 0$. This indicates that invest in preservation technology is not beneficial when the initial value of the deterioration rate is low.

Example 4. The initial values of the parameters are shown in Table 5. When the sensitive parameter $\delta = 0.01$, according to algorithm, $\Delta_1(n, p) < 0$, so the optimal solution for the investment is $\alpha^* = 0$. This also indicates that when the efficiency of the parameter is low, it is not beneficial to invest in preservation technology.

TABLE 2: The search process of the problem.

n	α	p	λ	q	TP
3	2.6383	38.78	0.0053	81.47	754.4
4	2.4144	36.62	0.0060	71.88	1337.0
5	2.1512	35.34	0.0068	62.64	1582.6
6	*1.8996*	*34.49*	*0.0077*	*55.04*	*1621.9*
7	1.6681	33.88	0.0087	48.91	1529.2
8	1.4567	33.43	0.0097	43.93	1347.8
9	1.2634	33.07	0.0106	39.84	1104.0

TABLE 3: The search process of the problem.

n	α	p	λ	q	TP
3	1.7500	39.03	0.0083	83.31	707.65
4	1.7500	36.78	0.0083	73.04	1300.1
5	1.7500	35.42	0.0083	63.20	1553.4
6	*1.7500*	*34.51*	*0.0083*	*55.21*	*1596.3*
7	1.6681	33.88	0.0087	48.91	1529.2
8	1.4567	33.43	0.0097	43.93	1347.8
9	1.2634	33.07	0.0106	39.84	1104.0

TABLE 4: Initial values of the parameters for Example 3.

A	b	a	c	h	λ_0	$\bar{\alpha}$	T	δ
500	10	0.2	10	1	0.001	10	100	0.5

TABLE 5: Initial values of the parameters for Example 4.

A	b	a	c	h	λ_0	$\bar{\alpha}$	T	δ
500	10	0.2	10	1	0.02	10	100	0.01

7. Sensitive Analysis

In this part, we performed the sensitivity analysis on the optimal solution of the model with respect to parameters $(A, \lambda_0, c, h, \delta)$ by changing each of the parameters by -50%, -40%, -30%, -20%, -10%, 10%, 20%, 30%, 40%, and 50%, taking one parameter at a time and keeping the remaining parameters unchanged. Table 6 is the sensitive analysis results with respect to Example 1. Figure 3 is the percent changes of parameter on total profit for Example 1.

From Table 6, we can conclude the following.

(1) The retailer's ordering frequency is insensitive to the change of λ_0 and δ. While, the retailer's ordering frequency is decreasing in A and c, it is increasing in h. It means that when the ordering cost is high, retailer will order less frequently to reduce the cost. When buying cost is high, the increasing ordering frequency leads to a lower deterioration cost, which is beneficial for the retailer. But when the inventory holding cost rate is low, the less ordering frequency is much beneficial for the retailer.

(2) The retailer's total ordering quantity $n \cdot q$ and profit TP are both decreasing in A and c, and insensitive on the change of λ_0. While the retailer's total ordering

TABLE 6: Sensitive analysis results for Example 1 ($\overline{\alpha} = 5.0000$).

		−50%	−40%	−30%	−20%	−10%	0	+10%	+20%	+30%	+40%	+50%
	n	9	8	7	7	6	6	5	5	5	5	4
	α	1.2634	1.4567	1.6681	1.6681	1.8996	1.8996	2.1512	2.1512	2.1512	2.1512	2.4144
A	p	33.07	33.43	33.88	33.88	34.49	34.49	35.34	35.34	35.34	35.34	36.62
	$n \cdot q$	358.56	351.44	342.37	342.37	330.24	330.24	313.20	313.20	313.20	313.20	287.52
	TP	3354.0	2947.8	2579.2	2229.2	1921.9	1621.9	1332.6	1082.6	832.6	582.6	337.0
	n	6	6	6	6	6	6	6	6	6	6	6
	α	0.5133	0.8779	1.1862	1.4533	1.6889	1.8996	2.0902	2.2642	2.4243	2.5725	2.7105
λ_0	p	34.49	34.49	34.49	34.49	34.49	34.49	34.49	34.49	34.49	34.49	34.49
	$n \cdot q$	330.24	330.24	330.24	330.24	330.24	330.24	330.24	330.24	330.24	330.24	330.24
	TP	1760.5	1724.0	1693.2	1666.5	1642.9	1621.9	1602.8	1585.4	1569.4	1554.6	1540.8
	n	6	6	6	6	6	6	6	6	5	5	5
	α	0.8172	1.1247	1.3743	1.5806	1.7536	1.8996	2.0232	2.1278	2.4540	2.5220	2.5762
c	p	31.94	32.45	32.96	33.47	33.98	34.49	35.00	35.51	36.88	37.40	37.91
	$n \cdot q$	401.10	384.30	369.36	355.62	342.66	330.24	318.18	306.42	277.75	266.35	255.05
	TP	3439.0	3046.5	2669.7	2307.3	1958.2	1621.9	1297.7	985.4	696.7	424.6	164.0
	n	4	5	5	5	6	6	6	6	6	7	7
	α	2.8430	2.4120	2.3501	2.2860	1.9537	1.8996	1.8439	1.7866	1.7275	1.4783	1.4277
h	p	33.43	33.30	33.81	34.32	34.06	34.49	34.91	35.34	35.77	35.34	35.71
	$n \cdot q$	351.48	354.00	343.80	333.60	338.70	330.24	321.72	313.20	304.62	313.18	305.83
	TP	3209.2	2837.0	2508.1	2189.4	1883.9	1621.9	1366.9	1119.0	878.26	649.85	443.05
	n	6	6	6	6	6	6	6	6	6	6	6
	α	0.9388	1.4150	1.6682	1.8032	1.8712	1.8996	1.9036	1.8927	1.8723	1.8461	1.8163
δ	p	34.83	34.71	34.63	34.57	34.53	34.49	34.46	34.43	34.41	34.40	34.38
	$n \cdot q$	343.50	339.12	335.94	333.54	331.68	330.24	328.98	327.96	327.12	326.40	325.74
	TP	1511.3	1533.1	1557.0	1580.3	1602.0	1621.9	1640.0	1656.5	1671.6	1685.4	1699.8

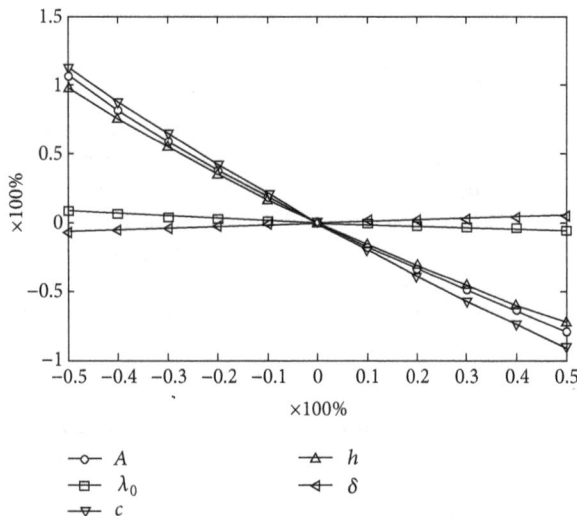

FIGURE 3: Percent changes of parameter on total profit for Example 1.

quantity is decreasing in δ, the profit is increasing in δ.

(3) The market price is insensitive to λ_0, increasing in A and c and decreasing in δ. For the same ordering frequency, market price is increasing in h.

(4) The preservation cost is increasing in A, c, and λ_0, while decreasing in h. It means that when the ordering cost, the buying cost, and initial deterioration rate are high, the retailer will invest more to reduce the deteriorating cost. But when the holding cost rate is high, the retailer can reduce the cost by ordering more frequently instead of investing more on preservation cost.

8. Conclusions and Future Research

In this paper, we study a kind of deteriorating seasonal products whose deterioration rate can be controlled by investing on the preservation efforts. Then, we propose an algorithm to solve the nonlinear program problem. By analysis, we can find some properties when parameters changed. Smaller buying cost per unit, holding cost per unit time, and ordering cost can all benefit the company. Besides, when deterioration rate is relatively small or the sensitivity parameter of the investment (δ) is small, there is no need to invest the preservation technology. Also, the profit can be influenced by the constraint of the investment capital.

For future research, we can take the backlogged demand into our model. Furthermore, we can assume that the ordering lead time exists and can be controlled by extra investment.

Also, we can extend the model to the deteriorating problems in multiechelon supply chains.

Acknowledgments

The authors thank the valuable comments of the anonymous referees for an earlier version of this paper. Their comments have significantly improved the paper. This work is supported by the National Natural Science Foundation of China (no. 71001025). Also, this research is partly supported by the Program for New Century Excellent Talents in University (no. NCET-10-0327) and the Ministry of Education of China. Grant-in-Aid for Humanity and Social Science Research (no. 11YJCZH139).

References

[1] P. M. Ghare, "A model for exponentially decaying inventory," *The Journal of Industrial Engineering*, vol. 5, no. 14, pp. 238–243, 1963.

[2] N. H. Shah, H. N. Soni, and K. A. Patel, "Optimization inventory and marketing policy for non-instantaneous deteriorating items with generalized type deterioration and holding cost rates," *Omega*, vol. 41, no. 2, pp. 421–430, 2012.

[3] S. Sana, S. K. Goyal, and K. S. Chaudhuri, "A production-inventory model for a deteriorating item with trended demand and shortages," *European Journal of Operational Research*, vol. 157, no. 2, pp. 357–371, 2004.

[4] A. Thangam and R. Uthayakumar, "Two-echelon trade credit financing for perishable items in a supply chain when demand depends on both selling price and credit period," *Computers and Industrial Engineering*, vol. 57, no. 3, pp. 773–786, 2009.

[5] Y. He and J. He, "A production model for deteriorating inventory items with production disruptions," *Discrete Dynamics in Nature and Society*, vol. 2010, Article ID 189017, 14 pages, 2010.

[6] Y. He, S. Y. Wang, and K. K. Lai, "An optimal production-inventory model for deteriorating items with multiple-market demand," *European Journal of Operational Research*, vol. 203, no. 3, pp. 593–600, 2010.

[7] Y. Liang and F. Zhou, "A two-warehouse inventory model for deteriorating items under conditionally permissible delay in payment," *Applied Mathematical Modelling*, vol. 35, no. 5, pp. 2221–2231, 2011.

[8] Y. He and S. Y. Wang, "Analysis of production-inventory system for deteriorating items with demand disruption," *International Journal of Production Research*, vol. 50, no. 16, pp. 4580–4592, 2012.

[9] K. Skouri, I. Konstantaras, S. Papachristos, and I. Ganas, "Inventory models with ramp type demand rate, partial backlogging and Weibull deterioration rate," *European Journal of Operational Research*, vol. 192, no. 1, pp. 79–92, 2009.

[10] J. Blackburn and G. Scudder, "Supply chain strategies for perishable products: the case of fresh produce," *Production and Operations Management*, vol. 18, no. 2, pp. 129–137, 2009.

[11] C. Kouki, E. Sahin, Z. Jemaï, and Y. Dallery, "Assessing the impact of perishability and the use of time temperature technologies on inventory management," *International Journal of Production Economics*, vol. 2010, 2010.

[12] A. Musa and B. Sani, "Inventory ordering policies of delayed deteriorating items under permissible delay in payments,"
International Journal of Production Economics, vol. 136, no. 1, pp. 75–83, 2012.

[13] P. H. Hsu, H. M. Wee, and H. M. Teng, "Preservation technology investment for deteriorating inventory," *International Journal of Production Economics*, vol. 124, no. 2, pp. 388–394, 2010.

[14] C. Y. Dye and T. P. Hsieh, "An optimal replenishment policy for deteriorating items with effective investment in preservation technology," *European Journal of Operational Research*, vol. 218, no. 1, pp. 106–112, 2012.

[15] G. C. Mahata, "An EPQ-based inventory model for exponentially deteriorating items under retailer partial trade credit policy in supply chain," *Expert Systems With Applications*, vol. 39, no. 3, pp. 3537–3550, 2012.

[16] M. Cheng and G. Wang, "A note on the inventory model for deteriorating items with trapezoidal type demand rate," *Computers and Industrial Engineering*, vol. 56, no. 4, pp. 1296–1300, 2009.

[17] C. Y. Dye, H. J. Chang, and J. T. Teng, "A deteriorating inventory model with time-varying demand and shortage-dependent partial backlogging," *European Journal of Operational Research*, vol. 172, no. 2, pp. 417–429, 2006.

[18] B. C. Giri, A. K. Jalan, and K. S. Chaudhuri, "Economic order quantity model with Weibull deterioration distribution, shortage and ramp type demand," *International Journal of Systems Science*, vol. 34, no. 4, pp. 237–243, 2003.

[19] T. H. Burwell, D. S. Dave, K. E. Fitzpatrick, and M. R. Roy, "Economic lot size model for price-dependent demand under quantity and freight discounts," *International Journal of Production Economics*, vol. 48, no. 2, pp. 141–155, 1997.

[20] B. R. Sarker, S. Mukherjee, and C. V. Balan, "An order-level lot size inventory model with inventory-level dependent demand and deterioration," *International Journal of Production Economics*, vol. 48, no. 3, pp. 227–236, 1997.

[21] H. L. Yang, J. T. Teng, and M. S. Chern, "An inventory model under inflation for deteriorating items with stock-dependent consumption rate and partial backlogging shortages," *International Journal of Production Economics*, vol. 123, no. 1, pp. 8–19, 2010.

[22] S. T. Lo, H. M. Wee, and W. C. Huang, "An integrated production-inventory model with imperfect production processes and Weibull distribution deterioration under inflation," *International Journal of Production Economics*, vol. 106, no. 1, pp. 248–260, 2007.

[23] H. M. Wee, J. F. Jong, and J. C. Jiang, "A note on a single-vendor and multiple-buyers production-inventory policy for a deteriorating item," *European Journal of Operational Research*, vol. 180, no. 3, pp. 1130–1134, 2007.

17

A Study of Multicriteria Decision Making for Supplier Selection in Automotive Industry

Nadia Jamil, Rosli Besar, and H. K. Sim

Faculty of Engineering and Technology, Multimedia University, Jalan Ayer Keroh Lama, 75450 Bukit Beruang, Melaka, Malaysia

Correspondence should be addressed to Nadia Jamil; diajh288@yahoo.com.my

Academic Editor: C. K. Kwong

This paper is designed to present the effectiveness of group multicriteria decision making in automotive manufacturing company focusing on the selection of suppliers in Malaysia. The process of selecting suppliers is one of the most critical and challenging endeavor in any supply chain management. There are five decision making tools being analyzed in this study, namely, analytical hierarchy process (AHP), fuzzy analytical hierarchy process (FAHP), technique for order performance by similarity to ideal solution (TOPSIS), fuzzy technique for order performance by similarity to ideal solution (FTOPSIS), and fuzzy analytical hierarchy process integrated with fuzzy technique for order performance by similarity to ideal solution (FAHPiFTOPSIS). The scores of ranking among the suppliers in each MCDM tools (AHP, FAHP, TOPSIS, FTOPSIS, and FAHPiFTOPSIS) show significantly comparable variation. Scores of the best supplier is then compared to the lowest supplier for all MCDM tools whereby this reflects that the highest percentage goes to TOPSIS with scoring of 79.37%. On the contrary, FAHPiFTOPSIS demonstrated the lowest score variation of 22.42% which indicates that FAHPiFTOPSIS is able to eliminate biasness in supplier selection process.

1. Introduction

A supply chain is a system which connects several departments from procurement of raw materials, to manufacturing, warehousing, and distribution of the products to the customers. Part of the contribution to supply chain complexity is the geographical outsourcing for cheaper supply and new market penetration. The complexity of supply chain is aggravated further when industry rely too much on multirange products and frequent introduction of new products as a strategy to meet different segmented market demands.

In automotive industry, such situation is rampant. The frequent introduction of new models and shorter product lifecycles compounded by fast order-delivery require high level of agility and flexibility of the suppliers, thereby, exacerbating the supply chain complexity. Hence, the right selection of supplier becomes more complicated. With the mounting complexity of supply chain, the selection of the suppliers becomes very challenging. The recent incident in Fukushima, Japan, devastated by massive earthquake and

nuclear disaster, and major floods in Thailand, had affected severely many Malaysian industries as well as industries in other parts of the world [1]. Those Malaysian companies which have their suppliers associated with these companies have suffered critical production problems, in particular, automotive industry. Thus, purchasing function for each organization is increasingly seen as the most important role in supplier selection [2].

In the previous years the automotive industry has witnessed an unprecedented turmoil. Such crises had affected the European and Asian automotive industry and had gravely stricken the American automobile industry. The first half of 2009 had indeed been a very faltering year due to economy recession. Based on press report released by MAA [3] dated 19 January 2011 (Figure 1), sales of passenger cars in Malaysia had dropped by 7.5% from year-to-date September 2008 to September 2009 (see Table 1) affected by the bad economic slowdown.

The economy is currently in transition from recession to recovery in tandem with financial markets improvement.

TABLE 1: Vehicles Sales in 2008 and 2009.

Segment	September			Year-to-date-September		
	2009	2008	Variance	2009	2008	Variance
PV (passenger vehicles)	42,039	46,476	0.90	361,463	392,393	0.92
CV (commercial vehicles)	4,030	4,253	0.95	36,156	37,520	0.96
Total	46,069	50,729	0.91	397,619	429,913	0.92

TABLE 2: Summary of passenger and commercial vehicles produced and assembled in Malaysia from 1980 to 2010.

Year	Passenger cars	Commercial vehicles	4×4 vehicles	Total vehicles
1980	80,422	23,805	—	104,227
1985	69,769	37,261	—	107,030
1990	116,526	63,181	11,873	191,580
1995	231,280	45,805	11,253	288,338
2000	295,318	36,642	27,235	359,195
2005	422,225	95,662	45,623	563,510
2006	377,952	96,545	28,551	503,048
2007	403,245	38,433	—	441,678
2008	484,512	46,298	—	530,810
2009	447,002	42,267	—	489,269
2010	522,568	45,147	—	567,715

(Source: [3]).
Note:
(i) Passenger vehicle industry reclassified in January 2007 and includes all passenger carrying vehicles, that is, passenger cars, 4WD/SUV, window van, and MPV models.
(ii) Commercial vehicles also reclassified on 1 January 2007 and include trucks, prime movers, pick-up, panel vans, buses and others.

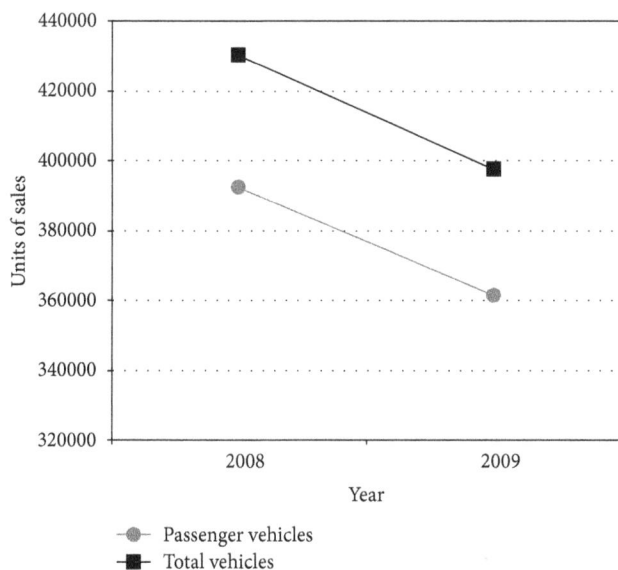

FIGURE 1: Vehicles sales in 2008 and 2009.

Year 2010 shows an increase trend in automotive industry. In 2010, the total industry volume of vehicles has picked up and increased to 605,156 units and, in fact, surpassed the highest recorded volume of 552,316 units in 2005.

Table 1 shows total vehicles produced and assembled from 1980 to 2010. From Table 2, total vehicles produced and assembled in 2010 were 605,156 units, increased by 12.7% from 536,905 units produced the year before. In 1980 the production statistics shows only two types of vehicles: passenger cars and commercial vehicles. In 1990, the statistics had included production figures for 4WD vehicles. In 2007, there was a reclassification where passenger cars, 4WD/SUV, window van, and MPV models are all categorized as passenger vehicles. Commercial vehicles had been reclassified to include trucks, prime movers, pick-up, panel vans, bus, and others. The data in Table 1 is summarized graphically in Figure 2 which illustrates the bar graph for passenger and commercial vehicles produced and assembled in Malaysia from year 1980 to 2010.

The data in Table 2 is summarized in bar graph shown in Figure 3 to highlight the significant increase in the number of passengers cars produced and assembled from 1980 to 2010. The total number of vehicles produced and assembled also showed a good progress of increment from year to year. To see the production trend of commercial vehicles from 2005 to 2010, Figure 3 is presented. From Figure 3, it is noted that the production of commercial vehicles did not exceed more than 100,000 units.

In Figure 3, the line graph representing the commercial vehicles illustrates a slight increment of production from 2009 to 2010 which is an increase of 6.8% of 45,147 units. In terms of percentage contribution of commercial vehicles to the market, it gradually increases from 8.8% in 2005 to 10.2% in 2010 [4].

In May 2011, based on a press release by Malaysian Automotive Association (MAA), automotive industry has

TABLE 3: Summary of passenger and commercial vehicles produced and assembled in Malaysia from May 2010 to May 2011.

Segment	May			Year-to-date May		
	2011	2010	Variance	2011	2010	Variance
PV (passenger vehicles)	40,936	46,259	0.88	228,816	222,977	1.03
CV (commercial vehicles)	5,109	4,624	1.1	26,597	24,133	1.1
Total	46,045	50,883	0.9	255,413	247,110	1.03

(Source: [3]).

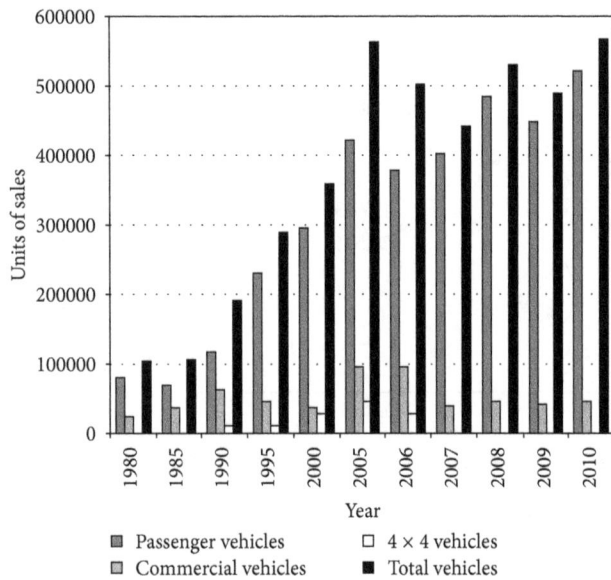

FIGURE 2: Bar graph of passenger and commercial vehicles produced and assembled in Malaysia from 1980 to 2010 [3].

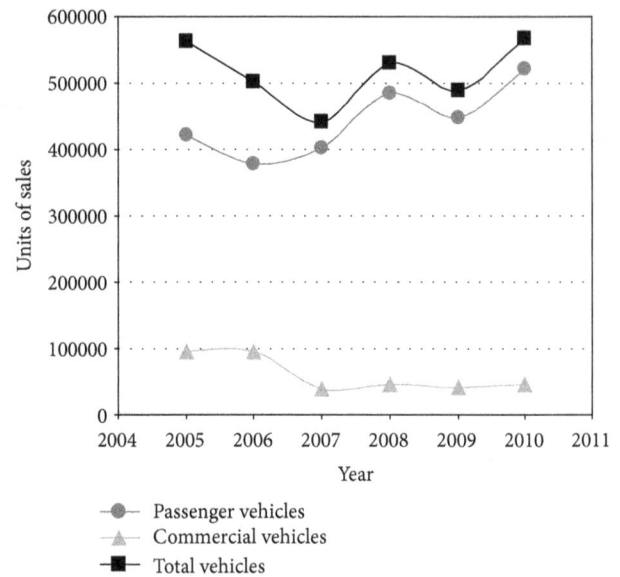

FIGURE 3: Line graph of passenger and commercial vehicles produced and assembled in Malaysia from 2005 to 2010 [3].

shown a healthy increase of sales compared to the same period the year before (see Table 3). However, in comparison to sales in May 2010 and May 2011, it showed that there was a reduction of 11.5%. This impact is due to the earthquake and tsunami which happened in Japan. More than 80% of the vehicles produced were dominated by passenger vehicles.

Currently, automotive industry is slowly and gradually shifting towards Asian countries, mainly due to high cost and saturation of automobile industry in the west and the increase in demand in Asia. The principal driving markets for Asian automobile industry are China, India, and ASEAN nations. The future of automotive industry in the Asian countries in particular Thailand, Philippines, Malaysia, and Indonesia is bright and promising because of the (ASEAN Free Trade Area) AFTA with tariffs currently at 0 to 5% [5].

Malaysia is a country that has a long history of making cars. There are many local and international cars assemblers and manufacturers in the country. The employees of the industry are widely regarded as skilled, well educated, and trainable. Located strategically in the ASEAN region which has a population of more than 500 million people, Malaysia offers vast opportunities for global automotive and component manufacturers and suppliers to set up their manufacturing and distribution operations in the country. Mercedes-Benz assembly plant has proven its great prospect

where its plant located in Pekan, Pahang, initially assembled only 4 units per day for one model and today assembles the S-Class, E-Class, and C-Class Mercedes with annual volume reaching 5,000 units [6].

Supplier selection decision process considers qualitative and quantitative criteria [2, 7]. It is a complex process whereby it involves many criteria, not necessarily quality, cost, and delivery. Such decision making process that involves multiple criteria is classified as a multicriteria decision making (MCDM) process.

One of the MCDM is the analytical hierarchy process (AHP) whereby it is a theory of mathematical for decision making and measurement introduced by Saaty [8]. It assists decision makers to make effective decision based on its goals, criteria, and alternatives. AHP can be applied in making decisions which are unstructured and complex and consists of multiple criteria [9]. This method has been used in various areas including performance evaluation, supplier selection, credit scoring, project management, resource allocation, distribution channel management, inventory management, promotion and recruitment decisions, portfolio management, energy resources planning, technology management, financial planning, budgeting decisions, socioeconomic planning, common vote prediction, and conflict resolution [10]. The Technique for Order Preference by Similarity to the

Ideal Solution (TOPSIS) is another mathematical model for MCDM. TOPSIS advocates two artificial alternatives, that is, the ideal alternative, the one which has the best level for all attributes considered, and the negative ideal alternative, the one which has the worst attribute values. TOPSIS selects the alternative that is the closest to the ideal solution and farthest from negative ideal alternative [11]. For the purpose of this research and without any prejudice to the others, two MCDM mathematical models have been selected, that is, AHP and TOPSIS.

In making decision for supplier selection, it is agreed by [2, 12] that the decisions made are very often involved by several decision makers. The purpose of having more people in decision making is to avoid any weaknesses or prejudice in the selection process. In order to make a decision which reflects human thinking, a system needs to be realistic. In making decision, decision makers prefer to evaluate a criterion with a certain level of tolerance rather than a fixed value judgment [13]. Due to this, one system which implies a human-like thinking model is introduced which is known as fuzzy logic [14].

This system describes a matter with a certain degree of characteristic which is also known as membership function. Membership function is a graphical representation which associates with the magnitude of input and ultimately determines an output response. There are different membership functions associated with each input and output response. Details about membership function are explained in Section 2.

The remainder of this paper is organized as follows: in Section 2, fuzzy concepts and the integration between fuzzy AHP and TOPSIS are explained; Section 3 shows the methodology used to conduct this study; Section 4 explained the results and discussion, finally in Section 5, the conclusions are presented.

2. Background of Fuzzy Concepts

Fuzzy sets can be simply defined as a set with fuzzy boundaries whereby the values of boundaries is multivalued unlike the two-valued Boolean logic. In the fuzzy theory, fuzzy set X of universe Y is defined by function $\mu_X(y)$ which is called the membership functions of set X. This notation can be expressed as follows [15]:

$$\mu_X(y) : Y \longrightarrow [0,1], \tag{1}$$

where

$$\mu_X(y) = 1 \quad \text{if } y \text{ is totally in } X,$$

$$\mu_X(y) = 0 \quad \text{if } y \text{ is not in } X, \tag{2}$$

$$0 < \mu_X(y) < 1 \quad \text{if } y \text{ is partly in } X.$$

The above set explains the membership (characteristic) functions of X which has the value ranges from 0 to 1. It allows a wide range of possible values. The value from 0 to 1 in this set represents the degree of membership of element y in set X. The membership function is commonly illustrated in terms of membership curve.

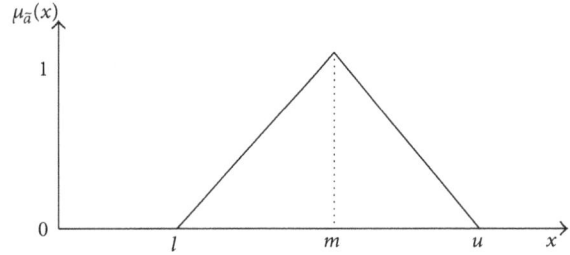

FIGURE 4: Symmetry TFN.

2.1. Triangular Fuzzy Number (TFN). There are 2 types of TFN which are symmetry and unsymmetry. Symmetry TFN is used in this paper since it enables users to easily calculate, understand, and capture the vagueness in people's verbal assessments [16].

TFN can be defined in three numbers, (l, m, u) which represented the smallest possible value, the most promising value, and the largest value which describes the fuzzy event. This representation can be observed from Figure 4 which illustrates symmetry TFN [17].

Throughout this study, the commonly used algebraic operations for fuzzy numbers are addition and multiplication. The fuzzy operators shown below were adapted from [18]. Let A and B be two triangular fuzzy numbers with their parameters shown as follows:

$$A = (a_1, a_2, a_3), \qquad B = (b_1, b_2, b_3). \tag{3}$$

Fuzzy numbers addition is defined by

$$A + B = (a_1 + b_1, a_2 + b_2, a_3 + b_3). \tag{4}$$

On the other hand, fuzzy numbers multiplication is calculated as shown below:

$$A \times B \approx (a_1 b_1, a_2 b_2, a_3 b_3). \tag{5}$$

2.2. AHP and Fuzzy AHP. The analytical hierarchy process (AHP) was first developed by Saaty, in mid of 1970 [19]. It is a decision analysis method that considers both qualitative and quantitative attributes in decision making. It decomposes complex problem into tangible and intangible entities, transforms them into systematic hierarchies of criteria and alternatives, and ranks them in order from most to least importance or desirable. As such the complex problem can easily be comprehended and each level can be analyzed independently. Each criterion from the AHP structure will be compared against a given alternatives, thus, allowing it to judge the intensity of the importance of one criterion over the other based on the given alternatives [19].

AHP is simple, systematic, and a very useful approach which integrates the matrix theory. In defining the weights of criteria and comparing the alternatives, a set of pair-wise comparison has been developed by Saaty [8] as part of the steps in AHP. Since the concept of AHP considers only real exact number, the set of pair-wise comparison by Saaty has

TABLE 4: Linguistic variables for importance of the criteria.

Linguistic variables	Scale of fuzzy number
Very low (VL)	(0, 0, 0.1, 0.2)
Low (L)	(0.1, 0.2, 0.2, 0.3)
Medium low (ML)	(0.2, 0.3, 0.4, 0.5)
Medium (M)	(0.4, 0.5, 0.5, 0.6)
Medium high (MH)	(0.5, 0.6, 0.7, 0.8)
High (H)	(0.7, 0.8, 0.8, 0.9)
Very high (VH)	(0.8, 0.9, 1.0, 1.0)

TABLE 5: Linguistic variables for performance of the alternatives.

Linguistic variables	Scale of fuzzy number
Very poor (VP)	(0, 0, 1, 2)
Poor (P)	(1, 2, 2, 3)
Medium poor (MP)	(2, 3, 4, 5)
Fair (F)	(4, 5, 5, 6)
Medium good (MG)	(5, 6, 7, 8)
Good (G)	(7, 8, 8, 9)
Very good (VG)	(8, 9, 10, 10)

been modified to suit human's judgement. This is the main reason, and the scale of weight and criteria preferences are defined in terms of linguistic variables in fuzzy numbers as presented in Tables 4 and 5 [2]. Linguistic variables are used to evaluate and assess the ratings of suppliers and weight of criteria. It is defined as variables with values expressed by words in a natural language [30].

There are three principles used in solving problems with AHP [9] in supplier selection which are as follows.

(i) AHP establishes the priorities based on sets of pairwise comparisons.

(ii) AHP score is built on human attributes and judgements where the intensity of each attribute or judgment is set according to its hierarchy over the other.

(iii) AHP synthesizes these judgments by using the hierarchy framework to obtain the overall priority of the elements or factors.

The concept of AHP and FAHP is mainly the same, and the difference of FAHP is that it analyses the numbers in terms of fuzzy numbers but as for AHP the numbers analyzed are crisp numbers. FAHP was firstly introduced by Van Laarhoven and Pedrycz in 1983 [31] whereby the fuzzy judgment is represented by triangular fuzzy number.

There are 6 steps in the process of decision making using FAHP. These steps are as follows.

(1) Firstly, form a decision matrix of the importance of each criterion with respect to each other.

(2) The membership function of triangular fuzzy number is defined by three real numbers (l, m, u) which is

mathematically described as $[\mu A(x)]$. The comparison matrix is derived from (6) as

$$\mu_A(x) = \begin{cases} \dfrac{(x-l)}{m-l}, & l \leq x \leq m \\ \dfrac{u-x}{u-m}, & m \leq x \leq u \\ 0, & \text{otherwise,} \end{cases} \tag{6}$$

$$a_{ij} = \frac{1}{a_{ij}} \begin{cases} a_{ij} = \left(l_{ij}, m_{ij}, u_{ij}\right) \\ a_{ij} = \left(\dfrac{1}{u_{ij}}, \dfrac{1}{m_{ij}}, \dfrac{1}{l_{ij}}\right). \end{cases}$$

(3) Calculate the $\widehat{G}_i$ (objective) which is defined by Chang in 1992 [32] in his study of fuzzy extent analysis. The computations of $\widehat{G}_i$ are explained in (7) as

$$\widehat{G}_i = \left(l_i, m_i, u_i\right),$$

$$l_i = \left(l_{i1} \otimes l_{i2} \otimes \cdots \otimes l_{ik}\right)^{1/k} \quad i = 1, 2, \ldots, k,$$

$$m_i = \left(m_{i1} \otimes m_{i2} \otimes \cdots \otimes m_{ik}\right)^{1/k} \quad i = 1, 2, \ldots, k, \tag{7}$$

$$u_i = \left(u_{i1} \otimes u_{i2} \otimes \cdots \otimes u_{ik}\right)^{1/k} \quad i = 1, 2, \ldots, k.$$

(4) The next step is the calculation of $\widehat{G}_T$ (total objective) whereby the equation used is displayed in (8) as

$$\widehat{G}_T = \left(\sum_{i=1}^{k} l_i, \sum_{i=1}^{k} m_i, \sum_{i=1}^{k} u_i\right). \tag{8}$$

(5) Upon getting the result for $\widehat{G}_T$, weight of each criterion with respect to the objective is calculated. The equations used are shown in (9) as

$$\widehat{w} = \frac{\widehat{G}_i}{\widehat{G}_T} = \frac{(l_i, m_i, u_i)}{\left(\sum_{i=1}^{k} l_i, \sum_{i=1}^{k} m_i, \sum_{i=1}^{k} u_i\right)}$$

$$= \left[\frac{l_i}{\sum_{i=1}^{k} u_i}, \frac{m_i}{\sum_{i=1}^{k} m_i}, \frac{u_i}{\sum_{i=1}^{k} l_i}\right]. \tag{9}$$

(6) The final stage is defuzzification whereby triangular fuzzy numbers are transformed into real numbers which is defined as weights (W_{in}) and can be computed in (10) as

$$W_{in} = \frac{w_{id}}{\sum_{i=1}^{k} w_{id}}, \quad i = 1, 2, \ldots k. \tag{10}$$

The steps of 1 to 6 will be repeated for each alternative in terms of the criterion which will be called as the weight of factors in terms of alternatives.

2.3. TOPSIS and Fuzzy TOPSIS. TOPSIS has been used in various fields and a number of applications such as outsourcing of logistics service, weapon selection, supplier selection analyzing business competition, and many other applications [24, 25, 33]. This method considers three types of criteria which are cost, qualitative, and quantitative. Generally, there are five steps in solving a problem using TOPSIS [34]. These steps are as follows:

(1) Obtain normalized decision matrix for m alternatives over n criteria. Let x_{ij} score of option i with respect to criterion j. Data can be normalized as below:

$$r_{ij} = \frac{x_{ij}}{\left(\Sigma x^2_{ij}\right)}, \quad (11)$$

for $i = 1, \ldots, m$ and $j = 1, \ldots, n$.

(2) Construct weighted normalized decision matrix, v_{ij},

$$v_{ij} = w_j \times r_{ij}, \quad (12)$$

whereby w_j is a set of weights for each criteria and r_{ij} is the normalized decision matrix.

(3) Identify the ideal alternative (extreme performance on each criterion, A^+) and negative ideal alternative (reverse extreme performance on each criterion, A^-) as

$$A^+ = \{v_1^+, \ldots, v_n^+\}, \quad (13)$$

whereby $v_1^+ = \{\max_i(v_{ij})$ if $j \in J^+$; $\min_i(v_{ij})$ if $j \in J^-\}$ and

$$A^- = \{v_1^-, \ldots, v_n^-\}, \quad (14)$$

whereby $v_1^- = \{\min_i(v_{ij})$ if $j \in J^-$; $\min_i(v_{ij})$ if $j \in J^-\}$.

(4) Develop a distance measure over criterion to both ideal (D^+) and negative ideal alternative (D^-) as

$$D_i^+ = \left[\Sigma\left(v_j^+ - v_{ij}\right)^2\right]^{1/2}, \quad D^- = \left(\Sigma\left(v_j^- - v_{ij}\right)^2\right)^{1/2}. \quad (15)$$

(5) For each alternative, determine the relative closeness to the ideal solution, C_i^+. C_i^+ is equal to the distance of the negative ideal solution divided by the sum of the distance of the negative ideal and ideal solution given by

$$C_i^+ = \frac{D^-}{D^- + D^+}. \quad (16)$$

TOPSIS only considers crisp values, whereas human judgments are usually uncertain and could not be evaluated using fix numbers. In spite of this, fuzzy numbers are used to replace all the crisp values in TOPSIS. In decision making, it is difficult to give a certain judgement, hence by integrating fuzzy logic and TOPSIS it will eliminate the uncertainty of the decision made [35]. Fuzzy TOPSIS has been successfully applied in several MCDM problems [36–39].

In performing FTOPSIS, Chen in 2000 [36] underlined 9 steps to be followed. The steps of the algorithm in this multicriteria decision making tool are as follows.

TABLE 6: Linguistic variables for the importance weight of each criterion.

Linguistic variables	Fuzzy value
Very low (VL)	(0, 0, 0.1)
Low (L)	(0, 0.1, 0.3)
Medium low (ML)	(0.1, 0.3, 0.5)
Medium high (MH)	(0.5, 0.7, 0.9)
High (H)	(0.7, 0.9, 1.0)
Very high (VH)	(0.9, 1.0, 1.0)

TABLE 7: Linguistic variables for the ratings.

Linguistic variables	Fuzzy value
Very poor (VP)	(0, 0, 1)
Poor (P)	(0, 1, 3)
Medium poor (MP)	(1, 3, 5)
Fair (F)	(3, 5, 7)
Medium good (MG)	(5, 7, 9)
Good (G)	(7, 9, 10)
Very good (VG)	(9, 10, 10)

Step 1. A committee of decision makers were formed to evaluate the alternatives based on the goal defined using the linguistic variables in Table 6.

Step 2. Appropriate linguistic variables for the ratings of the alternative with respect to criteria were chosen and these linguistic ratings used are taken from Table 7 [36].

Step 3. The importance weights of criteria and the ratings of three candidates are converted into fuzzy values. In order to obtain the fuzzy decision matrix and fuzzy weights of three alternatives, average value of each criteria for 3 decision makers was taken.

Step 4. The next step of the FTOPSIS analysis is the normalization of fuzzy decision matrix. This step is performed by taking each respective value and divide it with the maximum number of the particular criteria for all the 3 candidates.

Step 5. Construction of fuzzy weighted normalized decision matrix was then generated. In order to generate these values, values in Step 4 were divided by the weight of the respective column of fuzzy weighted normalized decision matrix.

Step 6. Fuzzy positive ideal solution (FPIS) and fuzzy negative ideal solution (FNIS) were then defined based on the values of normalized positive and negative triangular fuzzy numbers.

Step 7. Distance of each alternative from FPIS and FNIS was then calculated. The distance of two fuzzy numbers is calculated in (17) as

$$\sqrt{\frac{1}{3}\left[(m_1 - n_1)^2 + (m_2 - n_2)^2 + (m_3 - n_3)^2\right]}. \quad (17)$$

TABLE 8: Summary of advantages of MCDM models.

Differences	AHP	FAHP	TOPSIS	FTOPSIS	FAHPiFTOPSIS
Evaluators are able to represent the relative importance and interaction of multiple criteria in the supplier selection process [20]	Y				
Bias in decision making can be reduced by the flexibility and ability to check on inconsistency and able to decompose and problems into hierarchies of criteria. [21]	Y	Y			
Accurate, effective, and systematic decision support tool [22]					Y
Effectively handle both qualitative and quantitative data and easy to implement and understand [23, 24]	Y	Y	Y	Y	Y
No tedious pairwise comparison and weights can be directly assigned by decision makers which makes the practical application of the methodology very straightforward [22, 25]			Y	Y	
TOPSIS has been proved to be one of the best methods addressing rank reversal issue, that is, the change in the ranking of the alternatives when a nonoptimal alternative is introduced [22]			Y	Y	Y
Fuzzy AHP is preferable for widely spread hierarchies, where few importance/rating pair-wise comparisons are required at lower level trees [22]		Y			
Can adopt linguistic variables [22]		Y		Y	Y
By using fuzzy AHP and fuzzy TOPSIS, uncertainty and vagueness from subjective perception and the experiences of decision maker can be effectively represented and reached to a more effective decision [22]					Y
Ranking results for both methods are similar which shows that when decision makers are consistent in determining the data, two methods independently, and the ranking results will be the same and will handle fuzziness of data involved in decision making effectively [22]		Y		Y	

Note: Y means the differences are applicable for the respective MCDM.

TABLE 9: Summary of disadvantages of MCDM models.

Differences	AHP	FAHP	TOPSIS	FTOPSIS	FAHPiFTOPSIS
When a problem is decomposed into subsystems, the decision problem might become very large and lengthy [7]	Y				
AHP's using crisp number, hence not able to reflect human thinking style [7]	Y				
When a number of alternatives and criteria increased, pair-wise comparison becomes cumbersome and risk of inconsistencies grows [22, 25–27]		Y			
Problem is not decomposed into hierarchy hence decision maker might encounter difficulty to simplify the problem			Y	Y	
Integration with FAHP resulted in a number of extra steps to be followed					Y
Does not take into account the uncertainty associated with the mapping of one's judgment to a number [23]	Y				
FAHP requires more complex computations than FTOPSIS which includes pair wise comparison [22, 28]		Y			
In the extent analysis of FAHP, the priority weights of criterion or alternative can be equal to zero [22]		Y			

Note: Y means the differences are applicable for the respective MCDM.

Step 8. After complete computing the positive and negative distance, these values were then summed up to obtain the total distance measurement for positive values, A^*, and negative values, A^-.

Step 9. Finally, obtain the final ranking of the alternative closeness coefficient, CC. In order to achieve the CC, (18) is used as

$$\frac{d_i^-}{d_i^- + d_i^+}. \tag{18}$$

As per quoted by Krohling and Campanharo in 2011 [35], the triangular fuzzy number used is very effective for solving decision making problem which involved uncertainty and imprecise judgement.

2.4. Fuzzy AHP Integrated with Fuzzy TOPSIS (FAHPiFTOP-SIS). Integration between fuzzy AHP and fuzzy TOPSIS is believed to be able to make decision making more practical and reliable for decision makers due to its human-like thinking capability. For FAHPiFTOPSIS, the process of decision making comprises of 9 steps [2, 30].

TABLE 10: Weight and level of importance of criteria 1 to 9 of 12. Major automotive manufacturers.

Criteria	Level of importance	Respondent	Mean
(a) Delivery/lead time (DT)			
(1) On time delivery	58	12	4.83
(2) No shipping error (incorrect shipment)	53	12	4.42
(3) Able to deliver supplies within short lead time	50	12	4.17
(4) Products delivered in good conditions	56	12	4.67
(5) Proper delivery record and followup	58	12	4.83
(6) Applies JIT (just in time) concept	50	12	4.17
Total	325	72	
Weightage (mean)	4.5139	σ	0.31
Weightage (mean) (%)	11.7247	Var	0.10
Importance weight	0.1175		
(b) Support service (SS)			
(1) Handled by supplier technical experts	56	12	4.67
(2) Supplier support readily available	57	12	4.75
(3) Promptness of response	56	12	4.67
Total	169	36	
Weightage (mean)	4.6944	σ	0.05
Weightage (mean) (%)	12.1937	Var	0.002
Importance weight	0.12194		
(c) Quality factor (QF)			
(1) Meeting customer requirement and expectations	55	12	4.58
(2) Reliability in supply quality	58	12	4.83
(3) Product certification	55	12	4.58
(4) Provide sample of supply before first ordering	53	12	4.42
(5) Proper record on complaints and followup	55	12	4.58
(6) Raw material quality and supply reliability	57	12	4.75
(7) Special packaging and shipping requirement	51	12	4.25
(8) Proper marking and labeling of materials	54	12	4.50
(9) Conformance to environmental standard specifications	53	12	4.42
(10) Safety and health record performance	53	12	4.42
Total	544	120	
Weightage (mean)	4.5333	σ	0.17
Weightage (mean) (%)	11.7753	Var	0.03
Importance weight	0.11775		
(d) Technology (TE)			
(1) Availability of production facilities and capacity	56	12	4.67
(2) Technological capabilities for future improvement	48	12	4.00
(3) Innovativeness in product design and development	47	12	3.92
(4) Technical support on product development	51	12	4.25
(5) Intellectual property	45	12	3.75
(6) Information system capability (ICT, EDI, ERP, barcode)	46	12	3.83
(7) Shorter product development lead time	50	12	4.17
Total	343	84	
Weightage (mean)	4.0833	σ	0.31
Weightage (mean) (%)	10.6064	Var	0.10
Importance weight	0.10606		

TABLE 10: Continued.

Criteria	Level of importance	Respondent	Mean
(e) Price/cost (PR)			
(1) Competitive pricing	56	12	4.67
(2) Additional discount for large volume purchase	52	12	4.33
(3) No charges to distribution/logistic costs	44	12	3.67
(4) No charges after sales services costs	46	12	3.83
(5) No additional costs for small volume order	46	12	3.83
Total	244	60	
Weightage (mean)	4.0667	σ	0.42
Weightage (mean) (%)	10.5631	Var	0.18
Importance weight	0.10563		
(f) Factory capacity and capability (FC)			
(1) High production capacity	51	12	4.25
(2) Capability to cope with any order changes	49	12	4.08
(3) Maintain workforce competency	49	12	4.08
(4) Reliable production facilities	54	12	4.50
(5) Reliable maintenance programme	52	12	4.33
(6) Apply OEE concept	51	12	4.25
Total	306	72	
Weightage (mean)	4.2500	σ	0.16
Weightage (mean) (%)	11.0393	Var	0.03
Importance weight	0.11039		
(g) Supplier background (SB)			
(1) Industry and technological knowledge on product	54	12	4.50
(2) Having own transportation	48	12	4.00
(3) Strong financial management and support	52	12	4.33
(4) Approved supplier to other established Tier 1 or 2 company	50	12	4.17
(5) Available supply of skilled workforce	49	12	4.08
(6) Strategic geographical location of the supplier	50	12	4.17
Total	303	72	
Weightage (mean)	4.2083	σ	0.18
Weightage (mean) (%)	10.9311	Var	0.03
Importance weight	0.10931		
(h) Flexibility (FL)			
(1) Flexible manufacturing system	47	12	3.92
(2) Flexibility in operation and production	50	12	4.17
(3) Flexibility in product design changes	45	12	3.75
(4) Flexibility to respond on changes in requirement	48	12	4.00
Total	190	48	
Weightage (mean)	3.9583	σ	0.17
Weightage (mean) (%)	10.2817	Var	0.30
Importance weight	0.10282		
(i) Other management system requirements (MS)			
(1) Kanban	52	12	4.33
(2) Kaizen	53	12	4.42
(3) TPS (Toyota production system)	50	12	4.17
(4) Six sigma	42	12	3.50
(5) 5S implementation	53	12	4.42
(6) Failure modes and effect analysis (FMEA)	50	12	4.17

TABLE 10: Continued.

Criteria	Level of importance	Respondent	Mean
(7) Statistical process control (SPC)	46	12	3.83
(8) Production part approval process (PPAP)	50	12	4.17
(9) Advanced product quality planning (APQP)	45	12	3.75
(10) Measurement systems analysis (MSA)	44	12	3.67
(11) TS16949 technical specification	56	12	4.67
(12) ISO 9001 quality management system	56	12	4.67
(13) ISO 14001 environmental management system	54	12	4.50
(14) OHSAS 18001 safety and health management system	53	12	4.42
Total	704	168	
Weightage (mean)	4.1905	σ	0.32
Weightage (mean) (%)	10.8847	Var	0.56
Importance weight	0.10885		
Grand total weightage (mean)	38.50		

(1) Construct pair-wise comparison matrices among all the elements or criteria in the dimensions of the hierarchy system and convert it into fuzzy.

(2) Define the fuzzy geometric mean as

$$\widetilde{r}_i = \left(\widetilde{a}_{i1} \otimes \cdots \otimes \widetilde{a}_{ij} \otimes \cdots \otimes \widetilde{a}_{in} \right)^{1/n}. \tag{19}$$

(3) Define the fuzzy weights as

$$\widetilde{w}_i = \widetilde{r} \otimes \left[\widetilde{r}_i \oplus \cdots \oplus \widetilde{r}_i \oplus \cdots \oplus \widetilde{r}_n \right]^{-1}. \tag{20}$$

(4) Aggregating the ranking of criteria and suppliers' performance criteria ranking as

$$\widetilde{x}_{ijk} = \left(a_{ijk}, b_{ijk}, c_{ijk} \right), \tag{21}$$

where

$$a_{ij} = \min_{k} \left\{ a_{ijk} \right\},$$

$$b_{ij} = \frac{1}{k} \sum_{k=1}^{k} b_{ijk}, \tag{22}$$

$$c_{ij} = \max_{k} \left\{ c_{ijk} \right\}.$$

(5) Suppliers' performance as

$$\widetilde{w}_j = \left(w_{j1}, w_{j2}, w_{j3} \right), \tag{23}$$

where

$$w_{j1} = \min_{k} \left\{ w_{jk1} \right\},$$

$$w_{j2} = \frac{1}{k} \sum_{k=1}^{k} w_{jk2}, \tag{24}$$

$$w_{j3} = \max_{k} \left\{ w_{jk3} \right\}.$$

(6) Normalizing the fuzzy decision matrix as

$$\widetilde{R} = \left[\widetilde{r}_{ij} \right]_{m \times n}, \quad i = 1, 2, \ldots m, \ j = 1, 2, \ldots n, \tag{25}$$

where

$$\widetilde{r}_{ij} = \left(\frac{a_{ij}}{c_j^*}, \frac{b_{ij}}{c_j^*}, \frac{c_{ij}}{c_j^*} \right),$$

$$c_j^* = \max_{i} c_{ij}. \tag{26}$$

(7) Weighted normalization of the fuzzy decision matrix as

$$\widetilde{V} = \left[\widetilde{v}_{ij} \right]_{m \times n}, \quad i = 1, 2, \ldots m, \ j = 1, 2, \ldots n, \tag{27}$$

where

$$\widetilde{v}_{ij} = \widetilde{r}_{ij} (\bullet) \widetilde{w}_j. \tag{28}$$

(8) Distance to positive and negative ideal solution using vertex method [36]. Let $\widetilde{m} = (m_1, m_2, m_3)$ and $\widetilde{n} = (n_1, n_2, n_3)$ between two triangular fuzzy numbers is as

$$d_v (\widetilde{m}, \widetilde{n}) = \sqrt{\frac{1}{3} [(m_1 - n_1)^2 + (m_2 - n_2)^2 + (m_3 - n_3)^2}. \tag{29}$$

The best level of solution denoted by A^+ and the worst level of solution denoted by A^- are defined in (30) as

$$A^+ = \left(\widetilde{v}_1^*, \widetilde{v}_2^*, \ldots, \widetilde{v}_n^* \right),$$

$$A^- = \left(\widetilde{v}_1^-, \widetilde{v}_2^-, \ldots, \widetilde{v}_n^- \right), \tag{30}$$

TABLE 11: Weight and level of importance of nine (9) criteria of 4 major tyre suppliers.

Criteria	Level of importance	Respondent	Mean
(a) Delivery/lead time (DT)			
(1) On time delivery	19	4	4.75
(2) No shipping error (incorrect shipment)	17	4	4.25
(3) Able to deliver supplies within short lead time	15	4	3.75
(4) Products delivered in good conditions	19	4	4.75
(5) Proper delivery record and followup	19	4	4.75
(6) Applies JIT (just in time) concept	16	4	4
Total	105	24	
Weightage (mean)	4.3750	σ	0.44
Weightage (mean) (%)	11.6600	Var	0.19
Importance weight	0.11660		
(b) Support service (SS)			
(1) Handled by supplier technical experts	19	4	4.75
(2) Supplier support readily available	17	4	4.25
(3) Promptness of response	18	4	4.5
Total	54	12	
Weightage (mean)	4.5000	σ	0.25
Weightage (mean) (%)	11.9931	Var	0.06
Importance weight	0.11993		
(c) Quality factor (QF)			
(1) Meeting customer requirement and expectations	19	4	4.75
(2) Reliability in supply quality	19	4	4.75
(3) Product certification	15	4	3.75
(4) Provide sample of supply before first ordering	15	4	3.75
(5) Proper record on complaints and followup	15	4	3.75
(6) Raw material quality and supply reliability	19	4	4.75
(7) Special packaging and shipping requirement	14	4	3.5
(8) Proper marking and labeling of materials	15	4	3.75
(9) Conformance to environmental standard specifications	15	4	3.75
(10) Safety and health record performance	17	4	4.25
Total	163	40	
Weightage (mean)	4.0750	σ	0.50
Weightage (mean) (%)	10.8605	Var	0.25
Importance weight	0.10860		
(d) Technology (TE)			
(1) Availability of production facilities and capacity	16	4	4
(2) Technological capabilities for future improvement	16	4	4
(3) Innovativeness in product design and development	16	4	4
(4) Technical support on product development	19	4	4.75
(5) Intellectual property	12	4	3
(6) Information system capability (ICT, EDI, ERP, barcode)	15	4	3.75
(7) Shorter product development lead time	16	4	4
Total	110	28	
Weightage (mean)	3.9286	σ	0.51
Weightage (mean) (%)	10.4702	Var	0.26
Importance weight	0.10470		

TABLE 11: Continued.

Criteria	Level of importance	Respondent	Mean
(e) Price/cost (PR)			
(1) Competitive pricing	18	4	4.5
(2) Additional discount for large volume purchase	15	4	3.75
(3) No charges to distribution/logistic costs	13	4	3.25
(4) No charges after sales services costs	13	4	3.25
(5) No additional costs for small volume order	16	4	4
Total	75	20	
Weightage (mean)	3.7500	σ	0.53
Weightage (mean) (%)	9.9943	Var	0.28
Importance weight	0.09994		
(f) Factory capacity and capability (FC)			
(1) High production capacity	16	4	4.00
(2) Capability to cope with any order changes	16	4	4.00
(3) Maintain workforce competency	17	4	4.25
(4) Reliable production facilities	19	4	4.75
(5) Reliable maintenance programme	17	4	4.25
(6) Apply OEE concept	15	4	3.75
Total	100	24	
Weightage (mean)	4.1667	σ	0.34
Weightage (mean) (%)	11.1048	Var	0.12
Importance weight	0.11105		
(g) Supplier background (SB)			
(1) Industry and technological knowledge on product	16	4	4
(2) Having own transportation	15	4	3.75
(3) Strong financial management and support	19	4	4.75
(4) Approved supplier to other established Tier 1 or 2 company	16	4	4
(5) Available supply of skilled workforce	16	4	4
(6) Strategic geographical location of the supplier	16	4	4
Total	98	24	
Weightage (mean)	4.0833	σ	0.34
Weightage (mean) (%)	10.8827	Var	0.12
Importance weight	0.10883		
(h) Flexibility (FL)			
(1) Flexible manufacturing system	17	4	4.25
(2) Flexibility in operation and production	18	4	4.50
(3) Flexibility in product design changes	17	4	4.25
(4) Flexibility to respond on changes in requirement	18	4	4.50
Total	70	16	
Weightage (mean)	4.3750	σ	0.14
Weightage (mean) (%)	11.6600	Var	0.02
Importance weight	0.11660		
(i) Other management system requirements (MS)			
(1) Kanban	17	4	4.25
(2) Kaizen	18	4	4.50
(3) TPS (Toyota production system)	17	4	4.25
(4) Six sigma	16	4	4.00
(5) 5S implementation	19	4	4.75
(6) Failure modes and effect analysis (FMEA)	16	4	4.00
(7) Statistical process control (SPC)	15	4	3.75

TABLE 11: Continued.

Criteria	Level of importance	Respondent	Mean
(8) Production part approval process (PPAP)	17	4	4.25
(9) Advanced product quality planning (APQP)	18	4	4.50
(10) Measurement systems analysis (MSA)	18	4	4.50
(11) TS16949 technical specification	17	4	4.25
(12) ISO 9001 quality management system	17	4	4.25
(13) ISO 14001 environmental management system	17	4	4.25
(14) OHSAS 18001 safety and health management system	17	4	4.25
Total	239	56	
Weightage (mean)	4.2679		0.25
Weightage (mean) (%)	11.3745		0.06
Importance weight	0.11374		
Grand total weightage (mean)	37.5214		

where

$$\widetilde{v}_j^* = (1,1,1) \otimes \widetilde{w}_j = \left(lw_j; mw_j; uw_j \right),$$
$$\widetilde{v}_j^- = (0,0,0), \quad j = 1,2,\dots,n. \tag{31}$$

The distances (d_i^+ and d_i^-) between suppliers from A^+ and A^- can be achieved by the area compensation method which is shown in the following equations:

$$\widetilde{d}_i^+ = \sum_{j=1}^{n} d\left(\widetilde{v}_{ij}, \widetilde{v}_j^*\right), \quad i = 1,2,\dots,m; \ j = 1,2,\dots,n,$$
$$\widetilde{d}_i^- = \sum_{j=1}^{n} d_v\left(\widetilde{v}_{ij}, \widetilde{v}_j^-\right), \quad i = 1,2,\dots,m; \ j = 1,2,\dots,n. \tag{32}$$

(9) Closeness coefficient as

$$\widetilde{CC}_i = \frac{\widetilde{d}_i^-}{\widetilde{d}_i^+ + \widetilde{d}_i^-}, \quad i = 1,2,\dots,m. \tag{33}$$

2.5. Summary of Advantages and Disadvantages of MCDM Models. All the 5 models mentioned above are basically focusing on the theoretical part of the decision making. In this subsection the models are compared with respect to the differences. Based on Table 8, every model has its advantages. However, there is one common advantage which caters for all 5 models which is the effectiveness to handle qualitative and quantitative data.

Table 8 summarized the advantages of 5 models of focus in this study. For FAHP, AHP, and FAHPiFTOPSIS, these models are able to decompose complex problems into hierarchy which simplify the process of decision making and reduce the biasness by checking the inconsistency of rating. This is different for TOPSIS and FTOPSIS whereby the weights defined by decision makers are directly inserted into the models without any pairwise comparison. Models incorporated with fuzzy logic such as FAHP, FTOPSIS, and

FAHPiFTOPSIS can adopt linguistic variables which convert expression of words to fuzzy numbers. FAHPiFTOPSIS is a versatile MCDM which adopted the decomposed hierarchy system of FAHP and the best method of ranking issue by FTOPSIS.

Referring to Table 9, it analysed the disadvantages of every model. These disadvantages can be compensated by combining the model with another MCDM model, for instance FAHPiFTOPSIS whereby FAHP is integrated with FTOPSIS.

Since FAHPiFTOPSIS integrated with FAHP and FTOPSIS, there are more steps that need to be performed. When comparing FAHP to FTOPSIS, FAHP requires more complex computation which includes pairwise comparison. When the number of alternatives and criteria increased, pairwise comparison becomes cumbersome. For AHP and FAHP besides the advantage of decomposing the problems, this might also lead to a large and lengthy process. AHP and TOPSIS are not integrated; hence, the models will not be able to reflect human thinking style.

3. Research Methodology

Companies involved in this study are automotive supply companies, Malaysian automotive manufacturers, and foreign automotive manufacturing plants operating in Malaysia together with the cooperation of four major tyre manufacturers in Malaysia. The process of the study started with the determination of the important criteria and their ranking which was made through surveys conducted on supply companies for automotive manufacturers. The data and information collected were analysed in depth and compared against traditional criteria.

Five methods of MCDM which are AHP, FAHP, TOPSIS, FTOPSIS, and an integrated FAHPiFTOPSIS model will be investigated. The effectiveness of all the five MCDM models will be compared. This study will be conducted through questionnaires distributed to Malaysian automotive manufacturers and foreign companies which supply preassembled parts to automotive manufacturers. The respondents of this study comprise 4 major tyre manufacturers and 12 automotive manufacturing plants.

3.1. Phase 1: Data Collection.

3.1. Phase 1: Data Collection. This phase starts with the design of the data collection protocol, which concerns data to be collected and how to collect the data. The data used in this paper are collected and obtained from several sources such as questionnaires, face-to-face interviews, and phone interviews. The questionnaire aimed at identifying the priorities of various criteria highlighted in selecting suppliers as attached in Appendix C. In this phase, the interviews have been conducted by visiting the selected automotive industry. Phone interviews were also conducted whenever necessary. The purpose of this interview is to have a clearer picture on how supplier selection process is performed and at the same time to clarify any doubts regarding the questionnaires answered earlier. Nine main criteria were identified by the author for supplier selection process in automotive industry based on initial interview conducted with one of the major local car manufacturers. These criteria are shown in Appendix A.

3.2. Phase 2: Analyzing and Comparing Supplier Selection Models Using AHP, FAHP, TOPSIS, FTOPSIS, and FAHPiFTOPSIS. In this phase, real quantitative data are used. After identifying the criteria which are critical and important in selecting suppliers, the next stage of the project will be structuring the existing AHP, FAHP, TOPSIS, FTOPSIS, and FAHPiFTOPSIS models for supplier selection using data obtained from the questionnaire. Results obtained from the survey which consists of goal, criteria, subcriteria, and alternatives are used in structuring the models.

The models used for AHP and TOPSIS are from the founders themselves which are Saaty [8] and Hwang and Yoon 1981 [11], whereas for FAHP, the calculation method applied in this research is based on Chang's approach [32]. Chang introduced a new approach for handling the fuzzy AHP—the use of TFNs for a pairwise comparison scale of the fuzzy AHP and the use of the extent analysis method for the synthetic extent values of the pairwise comparisons. Chang's approach is one of the most popular approaches in the fuzzy AHP field. In this work, Chang's extent analysis method is preferred whereby the steps of this approach are relatively easier than the other fuzzy AHP approaches and are similar to the conventional AHP. Considering Buckley [40], the model developed utilizes trapezoidal calculation and is tremendous. For FTOPSIS, a general extension of TOPSIS in fuzzy environment has been developed by Chen 2000 [36] which is the model utilized in this study, whereas for FAHPiFTOPSIS the model used is by Sun 2010 [30].

Once these five models are structured, the results produced by the models are then analysed. The purpose of this analysis is to determine the qualities, shortcomings, and biasness of the models. For AHP, FAHP, and FAHPiFTOPSIS, bias in decision making can be eliminated due to its flexibility to check the inconsistency and ability to decompose problems into hierarchies [21]. To determine which model has the less

variation which means less bias in selection process, computation of variation percentage between models is performed in the analysis section [41, 42].

3.3. Phase 3: Testing and Validation of the Models. The final step of this paper is to validate the models using existing data collected from the questionnaire. This is to prove the consistency of the results for the five models with the results of the supplier selection of in-house procedures of the companies under study. The results of the models are then analyzed with the current practice of the selected companies. The main purpose of this step is to test if the results are consistent with the models and also to observe the best method relevant to the companies in obtaining the best supplier for automotive industry.

The system architecture example of the models is illustrated in Appendix B which involves one decision maker, decision maker 1 (DM1), and three potential alternatives, supplier 1 (S1), supplier 2 (S2), and supplier 3 (S3) with 9 criteria adopting the triangular values of fuzzy. From the figure, it can be seen that the system has three major components which are

> "**A**": selection of criteria ranking and supplier's performance,
>
> "**B**": interpretation of data,
>
> "**C**": results.

In the first component (**A**), decision maker is required to pairwise compare the suppliers' performance with respect to the criteria. Further ranking of the criteria based on its importance towards the goal is then performed. Accordingly, performance of the suppliers based on the criteria is then selected. In the second component (**B**), the interpretation of data takes place where all the data collected from the survey are inserted as input for all the models. All the calculation of analysis is done using Microsoft Excel 2010. The pairwise comparison and ranking utilized the linguistic variable method which are variables with values expressed by words in a natural language [30]. All the data inputs are selected based on the linguistic variable table provided. Component "**C**" is the final major component of the example in obtaining the best supplier according to the desired criteria.

3.4. Application in Automotive Industries. In Tables 10 and 11, the data tabulated are the number of respondents and level of importance for criteria 1 to 9 responded by 12 major automotive manufacturers and 4 major tyre suppliers in Malaysia.

4. Results and Discussion

Upon analyzing the various decision making tools described in Section 3, Figure 5 is presented. The figure illustrates the overall ranking for 5 different decision making tools used for supplier selection process in automotive industries using the results tabulated in Table 12.

Referring to Table 12, these are the values of actual scores for Supplier 1, 2, 3, and 4 according to the decision making

TABLE 12: Actual scores of Supplier 1, 2, 3, and 4 according to 5 decision making tools.

	AHP	FAHP	TOPSIS	FTOPSIS	FAHPiFTOPSIS
Supplier 1	0.16451	0.18905	0.23400	0.31199	0.25827
Supplier 2	0.16414	0.11949	0.18632	0.29032	0.25693
Supplier 3	0.43261	0.41440	0.98365	0.92561	0.48116
Supplier 4	0.23874	0.25259	0.46652	0.49101	0.37255

TABLE 13: Normalised scores of Supplier 1, 2, 3, and 4 according to 5 decision making tools.

	AHP	FAHP	TOPSIS	FTOPSIS	FAHPiFTOPSIS
Supplier 1	0.16451	0.14389	0.12510	0.15453	0.18867
Supplier 2	0.16414	0.16597	0.09961	0.14380	0.18769
Supplier 3	0.43261	0.49307	0.52588	0.45847	0.35149
Supplier 4	0.23874	0.19707	0.24941	0.24320	0.27215

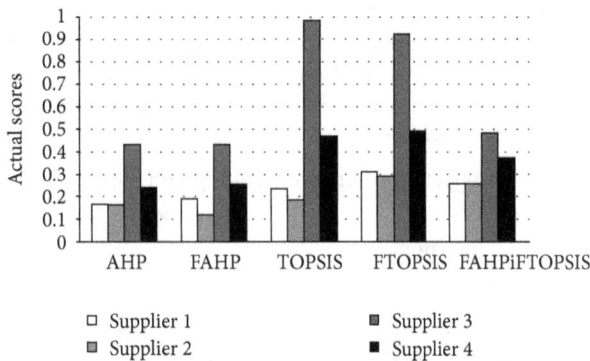

FIGURE 5: Summary of various decision making tools based on actual score.

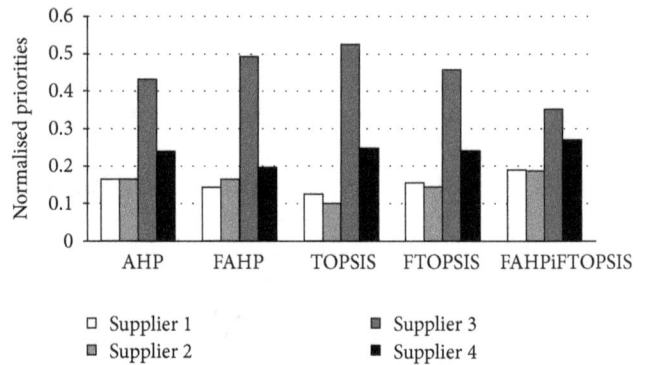

FIGURE 6: Summary of various decision making tools based on normalised priority.

tools applied. From this table, the values are then normalized as in Table 13. The equation used for normalization is presented in (34) below. The normalised scores in Table 13 were then used to plot the bar graph of Figure 6. Normalized values are used to make values comparable with each other. It is a process of reducing the values into a standard scale [43].

Normalized value:

$$\frac{\text{Actual score}}{\text{Total actual score of the related decision making tool}}. \quad (34)$$

Generally, it can be observed from Figure 6 that the ranking of supplier selection process for each different decision making tool followed the order of S3 > S4 > S2 > S1 where Supplier 3 exhibits the highest priority level while Supplier 1 exhibits the lowest priority level. The normalized score for Supplier 3 hits above average of 0.3, whereas for Supplier 1 and 2 the difference in normalized scoring is very small which indicates that both suppliers are competitively comparable.

The scores for the ranking among the four suppliers in each MCDM tool (AHP, FAHP, TOPSIS, and FTOPSIS) show significantly comparable variation. The percentage of variation scores for 4 suppliers is −36.9%, −40.33%, 26.8%, and 32.2% when comparing AHP, FAHP, TOPSIS, and

FTOPSIS with FAHPiFTOPSIS, respectively. These variations are obtained from the average of the original scoring and then compared with respect to FAHPiFTOPSIS. Table 14 showed the variations of models with respect to each other. The computation can be performed as in (34) whereby the calculation for AHP percentage of variation is taken here, for example. The reliability of the model is based on the less variation of a model. Based on Table 14, when the variations are summed up for each MCDM model, FAHPiFTOPSIS achieved less variation, whereas the high variation model is FAHP for 3 computations with respect to FAHPiFTOPSIS, FTOPSIS, and TOPSIS

Percentage of variation:

$$\Big(\big(\text{Average of AHP actual score} - \text{average of FAHPiFTOPSIS} $$
$$\text{actual score} \big) \times \big(\text{Average of AHP actual score} \big)^{-1} \Big) \times 100. \quad (35)$$

The following evaluation to determine the variation is based on the difference of scores achieved by the best supplier with the less preferred supplier, and Table 15 is generated. The data for Table 15 can be obtained from Table 13. Referring to Table 15 and Figure 7, the highest percentage goes to

TABLE 14: Variations of models with respect to each other.

	AHP	FAHP	TOPSIS	FTOPSIS	FAHPiFTOPSIS
With respect to AHP	—	−2.5%	46.5%	50.5%	26.9%
With respect to FAHP	2.4%	—	47.8%	51.7%	28.7%
With respect to TOPSIS	−87%	−91.7%	—	7.4%	−36.6%
With respect to FTOPSIS	−101.9%	−107%	−7.9%	—	−47.5%
With respect to FAHPiFTOPSIS	−36.9%	−40.33%	26.8%	32.2%	—
Total percentage of variations	−223.5%	241.53%	113.2%	141.8%	−28.5%

TABLE 15: Difference of normalized scores for the best supplier (Supplier 3) and the less preferred supplier (Supplier 2).

	AHP	FAHP	TOPSIS	FTOPSIS	FAHPiFTOPSIS
Supplier 2	0.16414	0.16597	0.09961	0.14380	0.18769
Supplier 3	0.43261	0.49307	0.52588	0.45847	0.35149
Score difference	0.26847	0.29491	0.79734	0.63529	0.22422
Percentage	26.84	29.49	79.73	63.53	22.42

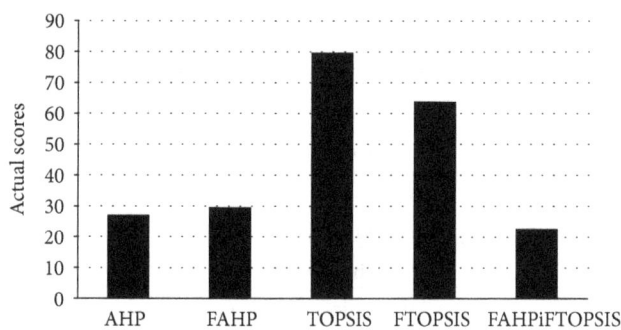

FIGURE 7: Difference in percentage comparison between the best supplier and the less preferred supplier.

TOPSIS with scoring of 79.73% followed by FTOPSIS with percentage of 63.53 and FAHP with 29.49%. On the contrary, FAHPiFTOPSIS decision making tool demonstrated the lowest score variation of 22.42% followed by 26.84% of AHP. This indicates that FAHPiFTOPSIS is able to eliminate the "biasness" and ambiguity in supplier selection process.

5. Conclusions

This research has successfully demonstrated the applicability of the MCDM tools (AHP, FAHP, TOPSIS, FTOPSIS, and FAHPiFTOPSIS) by automotive industry in the selection of suppliers. As a result, few points can be concluded as follows.

(1) Six additional parameters were observed to have equal important criteria in automotive industry besides three classic criteria, namely, price, delivery, and quality which are support service, technology, factory capacity and capability, supplier background, flexibility, and other management system requirements.

(2) AHP, FAHP, TOPSIS, FTOPSIS, and FAHPiFTOPSIS are all applicable and accurate for supplier selection in automotive industry.

(a) For AHP and FAHP, when the number of suppliers and alternatives becomes big, more criteria will need to be considered and decision maker will face a challenge to perform pairwise comparison in a big matrix.

(b) For TOPSIS and FTOPSIS, it is a simpler method whereby weights are needed as the input of this decision making tool irrespective with the number of alternatives.

(c) The results from AHP, FAHP, TOPSIS, and FTOPSIS show great variation in the final ranking scores. However, FAHPiFTOPSIS decision making tool demonstrated the lowest score variation of 22.42%. This indicates that FAHPiFTOPSIS application is able to effectively eradicate any ambiguity or fuzziness in the process.

Appendices

A. Supplier Selection Criteria

See Table 16.

B. Generic System Architecture of FAHP, FTOPSIS, and FAHPiFTOPSIS for Decision Makers in Selecting the Best Supplier

See Figure 8.

C. Questionnaire Developed for the Study

C.1. Developed Questionnaire. The developed questionnaire is called:

Supplier selection process for automotive industry.

C.1.1. Supplier Selection Criteria. Supplier selection evaluations are normally being conducted by companies in

TABLE 16: Supplier selection criteria.

Category	Criteria
(1) Delivery	(i) Applies JIT concept
	(ii) Delivery lead time
	(iii) Delivery quality
	(iv) Packaging quality
(2) Support service	(i) Handled by expertise of the related field
	(ii) Availability
	(iii) Promptness
	(iv) Applies OEE concept
(3) Quality	(i) Quality performance
	(ii) Durability
	(iii) Ergonomic qualities
	(iv) Reliability
(4) Technology	(i) Collaboration with established R&D organization on referral designs
	(ii) Product certification
	(iii) High technology machines and processes
	(iv) Align with current technology (product/process/design)
	(v) Suppliers capable of modifying product/design/process
	(vi) Material availability
	(vii) Alternative material and technologies
	(viii) EDI capability
	(a) inventory
	(b) logistics
	(c) production
	(d) transaction
	(ix) Design capability
	(a) Flexibility to respond to design changes
	(b) Product innovativeness
	(c) Product performance
	(d) Ability to modify product/process
	(x) R&D
	(a) Availability of testing laboratory
	(b) Reliability of testing laboratory
	(c) Calibration program
(5) Price	(i) Flexibility in price reduction
	(ii) Competitive operating costs
	(iii) Flexibility in payment
(6) Factory capacity and capability	(i) Level of capability to cope with rush orders
	(ii) Sufficient product capacity
	(iii) Sufficient product facilities

TABLE 16: Continued.

Category	Criteria
(7) Supplier background	(i) Industry knowledge (specifically on the related process or parts)
	(ii) Geographical location of the supplier
	(iii) Geographical condition such as labor cost and traffic congestion
	(a) suppliers have dedicated supply point
	(iv) Flexibility in freight
	(a) suppliers have their own transportation
	(v) Position in the industry and reputation
	(a) suppliers supply to other established car manufacturers
	(vi) Performance history
	(a) suppliers have been blacklisted or not
	(vii) Financial
	(a) Table financial management system
	(b) bankruptcy
	(viii) Manning
	(a) sufficient workers
	(b) low turnover rate
(8) Flexibility	(i) Practices flexible manufacturing system in terms of design and process
	(ii) Flexibility of operation
	(iii) Flexibility in production
	(iv) Flexibility in order frequency and amount
	(v) Flexibility to respond requirement volume changes
(9) Management tool system	(i) Kanban
	(ii) Kaizen
	(iii) TPS (Toyota production system)
	(iv) ISO 9000
	(v) Six sigma
	(vi) 5S
	(vii) FMEA
	(viii) SPC
	(ix) Strong safety processes and culture
	(x) Environmental performance

order to evaluate their suppliers' performance and reliability. Traditionally, supplier selection evaluations are based on price/cost, quality, and delivery. However, in today's management, most companies do not only consider the three primary criteria, but they also take into account technology, support

TABLE 17

Criteria
(1) On time delivery (minimize lateness and earliness)
(2) Proper delivery record and followup
(3) Minimum shipping error (number of incorrect shipment)
(4) Able to deliver supplies within short lead time
(5) Products are checked thoroughly prior delivery
(6) Applies JIT (just in time) concept

TABLE 18

Criteria
(1) Provide supporting staff to overcome any technical and supply problems
(2) Supporting staff and technical experts are readily available
(3) Promptness to response
(4) Professional attitude in dealing with problems

TABLE 19

Criteria
(1) Meeting customer requirement and expectations
(2) Zero reject rate
(3) Reliability in raw material supplies
(4) Conformance to environmental standard specifications
(5) Proper record on complaints and followup
(6) Provide sample of components or raw material before delivery
(7) Ergonomic quality in product/raw material design
(8) Special packaging and shipping requirement
(9) Proper marking and labeling of materials (legible, durable, and follow specification)
(10) Safety record of suppliers

TABLE 20

Criteria
(1) Technological capabilities for future improvement
(2) Collaboration with established R&D organization or institution
(3) Information system capabilities (electronic data interchanges, ERP, barcode)
(4) Innovativeness in product development and improvement
(5) Have intellectual property (trademarks, patents) and unique technologies capability
(6) Product development facility
(7) Efficient production facilities and capacity
(8) Complying to international certifications

TABLE 21

Criteria
(1) Low or reasonable price of supplies
(2) Minimize distribution/logistic costs (transportation, administrative, customs, risk, and handling)
(3) Avoid additional cost incurred in manufacturing, labor, maintenance, inspection, rework, or due to raw materials
(4) Eliminate after sales service costs
(5) Additional discount for large volume purchase
(6) Additional cost required due to low volume ordered

TABLE 22

Criteria
(1) Manufacturing capability to cope with rush orders
(2) Sufficient product capacity
(3) Sufficient product facilities

TABLE 23

Criteria
(1) Experts in the industry
(2) Suppliers having own transportation
(3) Suppliers having strong financial standing
(4) Suppliers having sufficient workers
(5) Strategic geographical location

TABLE 24

Criteria
(1) Flexible in design and process
(2) Flexibility in manufacturing operations
(3) Flexibility in production planning
(4) Flexibility in order frequency and amount
(5) Flexibility in response to changes on requirement
(6) Speedy product conception to product realization

service, flexibility, supplier background, management tool systems, and factory capacity and capability.

As part of my Master research in manufacturing engineering, I would like to seek you a great assistance to provide me your critical input on supplier selection. Below you will find the criteria listed. What you have to do is to fill in your inputs in each question by writing the numbers according to the ranking in the right box.

(1) *Delivery/Lead Time.* Delivery of supply: whichever is applied to your company, please indicate the degree of importance on the following delivery criteria as expected from your suppliers. See Table 17.

(2) *Support Service.* See Table 18.

(3) *Quality Factor.* Supplier selection is not only limited to cost, but also to the performance of other competitive dimensions of quality. Meeting customer requirements is one example, but there are other important factors involved. Please indicate the degree of importance on the following quality criteria as expected from your suppliers. See Table 19.

TABLE 25

Criteria
(1) Kanban
(2) Kaizen
(3) TPS (Toyota production system)
(4) ISO 9000
(5) Six sigma
(6) 5S
(7) FMEA
(8) SPC
(9) TS16949
(10) Safety management and culture
(11) Takt time
(12) Overall equipment effectiveness

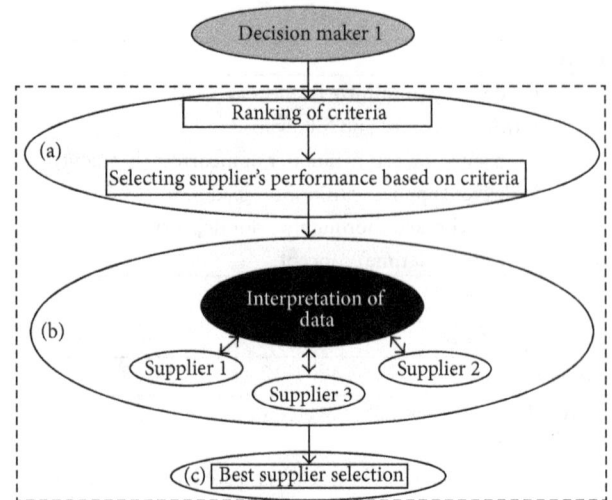

FIGURE 8

TABLE 26

Criteria	Supplier 1	Supplier 2	Supplier 3	Supplier 4
(1) Delivery/lead time				
(2) Support service				
(3) Quality factor				
(4) Technology				
(5) Price/cost				
(6) Factory capacity and capability				
(7) Supplier background				
(8) Flexibility				
(9) Management system				

Criteria rating: (1) Very poor → (9) very good.

TABLE 27: Saaty's pair-wise comparison table.

Numerical rating	Verbal comparison of preference
1	Equal importance
3	Moderate importance of one over another
5	Strong or essential importance
7	Very strong or demonstrated importance
9	Extreme importance
2, 4, 6, 8	Intermediate values

(Source: [29]).

(4) *Technology*. Suppliers have to cope with technology to be more competitive. Suppliers which are responsive to new technology are more favorable to any company. Please indicate the degree of importance on the following criteria. See Table 20.

(5) *Price/Cost*. Traditionally, any company will go for the lowest cost from their suppliers. Consequently, the cost control system administered by any company may have some effect on the suppliers. The criteria below show some of the important issues related to cost and price. Please indicate their level of importance as applicable to your company. See Table 21.

(6) *Factory Capacity and Capability*. See Table 22.

(7) *Supplier Background*. The additional requirements criteria involve supplier's background. Below are some of the important factors for suppliers. Please indicate their level of importance. See Table 23.

(8) *Flexibility*. See Table 24.

(9) *Management System*. See Table 25.

C.1.2. Rating of Suppliers under Various Criteria. Please state your level of importance of the supplier with respect to the criteria on the right boxes regarding the numbers below. See Table 26.

C.1.3. Pairwise Comparison of 9 Criteria. According to Saaty's table, compare the 9 criteria accordingly and fill in the table. Criteria on the left side of the table need to be compared to the criteria at the horizontal columns. See Tables 27 and 28.

C.1.4. Pairwise Comparison of Suppliers with Respect to 9 Criteria. According to Saaty's table, compare the suppliers (S1, S2, S3, and S4) with respect to the criteria. Suppliers on the left side columns are to be compared with the suppliers on the right columns. Kindly mark (/) at the preferred value when pairwise comparison is performed. See Tables 29, 30, 31, 32, 33, 34, 35, 36, 37, and 38.

TABLE 28

	Criteria 1	Criteria 2	Criteria 3	Criteria 4	Criteria 5	Criteria 6	Criteria 7	Criteria 8	Criteria 9
Criteria 1									
Criteria 2									
Criteria 3									
Criteria 4									
Criteria 5									
Criteria 6									
Criteria 7									
Criteria 8									
Criteria 9									

TABLE 29: Saaty's pair-wise comparison table.

Numerical rating	Verbal comparison of preference
1	Equal importance
3	Moderate importance of one over another
5	Strong or essential importance
7	Very strong or demonstrated importance
9	Extreme importance
2, 4, 6, 8	Intermediate values

(Source: [29]).

TABLE 30: Criterion 1: delivery/lead time.

	9 8 7 6 5 4 3 2 1 2 3 4 5 6 7 8 9	
S1		S2
S1		S3
S1		S4
S2		S3
S2		S4
S3		S4

TABLE 31: Criterion 2: support service.

	9 8 7 6 5 4 3 2 1 2 3 4 5 6 7 8 9	
S1		S2
S1		S3
S1		S4
S2		S3
S2		S4
S3		S4

TABLE 32: Criterion 3: quality.

	9 8 7 6 5 4 3 2 1 2 3 4 5 6 7 8 9	
S1		S2
S1		S3
S1		S4
S2		S3
S2		S4
S3		S4

TABLE 33: Criterion 4: technology.

	9 8 7 6 5 4 3 2 1 2 3 4 5 6 7 8 9	
S1		S2
S1		S3
S1		S4
S2		S3
S2		S4
S3		S4

TABLE 34: Criterion 5: price/cost.

	9 8 7 6 5 4 3 2 1 2 3 4 5 6 7 8 9	
S1		S2
S1		S3
S1		S4
S2		S3
S2		S4
S3		S4

TABLE 35: Criterion 6: factory capacity and capability.

	9 8 7 6 5 4 3 2 1 2 3 4 5 6 7 8 9	
S1		S2
S1		S3
S1		S4
S2		S3
S2		S4
S3		S4

TABLE 36: Criterion 7: supplier background.

	9 8 7 6 5 4 3 2 1 2 3 4 5 6 7 8 9	
S1		S2
S1		S3
S1		S4
S2		S3
S2		S4
S3		S4

TABLE 37: Criterion 8: flexibility.

	9	8	7	6	5	4	3	2	1	2	3	4	5	6	7	8	9	
S1																		S2
S1																		S3
S1																		S4
S2																		S3
S2																		S4
S3																		S4

TABLE 38: Criterion 9: management tool system.

	9	8	7	6	5	4	3	2	1	2	3	4	5	6	7	8	9	
S1																		S2
S1																		S3
S1																		S4
S2																		S3
S2																		S4
S3																		S4

Acknowledgments

The authors would like to thank Associate Professor Dr. Mohd. Khaled Omar and all the survey respondents for their kind support and help contributed towards the implementation of this paper.

References

[1] V. Vaidya, *Impact of Thailand Floods on Automotive Industry and Supply Chain*, Frost and Sullivan, 2011.

[2] C. T. Chen, C. T. Lin, and S. F. Huang, "A fuzzy approach for supplier evaluation and selection in supply chain management," *International Journal of Production Economics*, vol. 102, no. 2, pp. 289–301, 2006.

[3] Malaysia Automotive Association, MAA, *Summary of Sales & Production Data*, 2011.

[4] A. Ahmad, *Market Review of 2010 and Outlook for 2011*, Malaysia Automotive Association, 2011.

[5] Secretariat, *ASEAN Free Trade Area (AFTA) : An update*, Association of Southeast Nation, 1999.

[6] Bernama, "DRB-Hicom's Pekan Plant to See Higher Production," Bernama, 2009.

[7] C. Kahraman, U. Cebeci, and Z. Ulukan, "Multi-criteria supplier selection using fuzzy AHP," *Logistics Information Management*, vol. 16, pp. 382–394, 2003.

[8] T. L. Saaty, *The Analytical Hierarchy Process*, McGrawHill, New York, NY, USA, 1980.

[9] T. L. Saaty, "Axiomatic foundation of the analytical hierarchy process," *Management Science*, vol. 32, no. 7, pp. 841–855, 1986.

[10] O. Cakir and M. S. Canbolat, "A web-based decision support system for multi-criteria inventory classification using fuzzy AHP methodology," *Expert Systems with Applications*, vol. 35, no. 3, pp. 1367–1378, 2008.

[11] C. L. Hwang and K. Yoon, *Multiple Attribute Decision Making-Methods and Applications*, Springer, Heidelberg, Germany, 1981.

[12] L. De Boer, L. Van Der Wegen, and J. Telgen, "Outranking methods in support of supplier selection," *European Journal of Purchasing and Supply Management*, vol. 4, no. 2-3, pp. 109–118, 1998.

[13] J. Wang, K. Fan, and W. Wang, "Integration of fuzzy AHP and FPP with TOPSIS methodology for aeroengine health assessment," *Expert Systems with Applications*, vol. 37, no. 12, pp. 8516–8526, 2010.

[14] L. A. Zadeh, "Fuzzy sets," *Information and Control*, vol. 8, no. 3, pp. 338–353, 1965.

[15] M. Negnevitsky, *Artificial Intelligence: A Guide To Intelligent Systems*, Addison-Wesley: Pearson Education Limited, Essex, UK, 2002.

[16] M. Bevilacquaa, F. E Ciarapicab, and G. Giacchettab, "A fuzzy-QFD approach to supplier selection," *Journal of Purchasing & Supply Management*, vol. 12, pp. 14–27, 2006.

[17] H. Deng, "Multicriteria analysis with fuzzy pair-wise comparison," *International Journal of Approximate Reasoning*, vol. 21, pp. 215–231, 1999.

[18] S. M. Ordoobadi, "Development of a supplier selection model using fuzzy logic," *Supply Chain Management*, vol. 14, no. 4, pp. 314–327, 2009.

[19] T. L. Saaty, "A scaling method for priorities in hierarchical structures," *Journal of Mathematical Psychology*, vol. 15, no. 3, pp. 234–281, 1977.

[20] C. C. Yang and B. S. Chen, "Supplier selection using combined analytical hierarchy process and grey relational analysis," *Journal of Manufacturing Technology Management*, vol. 17, no. 7, pp. 926–941, 2006.

[21] R. L. Nydick and R. P. Hill, "Using the analytic hierarchy process to structure the supplier selection procedure," *International Journal of Purchasing and Materials Management*, pp. 31–36, 1992.

[22] I. Ertuğrul and N. Karakaşoğlu, "Comparison of fuzzy AHP and fuzzy TOPSIS methods for facility location selection," *International Journal of Advance Manufacturing Technology*, vol. 39, pp. 783–795, 2008.

[23] F. Tüysüz and C. Kahraman, "Project risk evaluation using a fuzzy analytic hierarchy process: an application to information technology projects," *International Journal of intelligent Systems*, vol. 21, pp. 559–584, 2006.

[24] A. Kelemenis, K. Ergazakis, and D. Askounis, "Support managers' selection using an extension of fuzzy TOPSIS," *Expert Systems with Applications*, vol. 38, no. 3, pp. 2774–2782, 2011.

[25] E. Bottani and A. Rizzi, "A fuzzy TOPSIS methodology to support outsourcing of logistics services," *Supply Chain Management*, vol. 11, no. 4, pp. 294–308, 2006.

[26] H. J. Shyur, "COTS evaluation using modified TOPSIS and ANP," *Applied Mathematics and Computation*, vol. 177, no. 1, pp. 251–259, 2006.

[27] M. F. Shipley, A. de Korvin, and R. Obid, "A decision making model for multi-attribute problems incorporating uncertainty and bias measures," *Computers and Operations Research*, vol. 18, no. 4, pp. 335–342, 1991.

[28] O. Durán and J. Aguilo, "Computer-aided machine-tool selection based on a Fuzzy-AHP approach," *Expert Systems with Applications*, vol. 34, no. 3, pp. 1787–1794, 2008.

[29] T. L. Saaty, *Multi-Criteria Decision Making: the Analytic Hierarchy Process-Planning, Priority Setting, Resource Allocation*, R W S Publications, Pittsburgh, Pa, USA, 1996.

[30] C. C. Sun, "A performance evaluation model by integrating fuzzy AHP and fuzzy TOPSIS methods," *Expert Systems with Applications*, vol. 37, no. 12, pp. 7745–7754, 2010.

[31] P. J. M. van Laarhoven and W. Pedrycz, "A fuzzy extension of Saaty's priority theory," *Fuzzy Sets and Systems*, vol. 11, no. 3, pp. 229–241, 1983.

[32] D. Y. Chang, *Extent Analysis and Synthetic Decision, Optimization Techniques and Applications*, vol. 1, World Scientific, Singapore, 1992.

[33] A. T. Gumus, "Evaluation of hazardous waste transportation firms by using a two step fuzzy-AHP and TOPSIS methodology," *Expert Systems with Applications*, vol. 36, no. 2, pp. 4067–4074, 2009.

[34] D. L. Olson, "Comparison of weights in TOPSIS models," *Mathematical and Computer Modelling*, vol. 40, no. 7-8, pp. 721–727, 2004.

[35] R. A. Krohling and V. C. Campanharo, "Fuzzy TOPSIS for group decision making: a case study for accidents with oil spill in the sea," *Expert Systems with Applications*, vol. 38, no. 4, pp. 4190–4197, 2011.

[36] C. T. Chen, "Extensions of the TOPSIS for group decision-making under fuzzy environment," *Fuzzy Sets and Systems*, vol. 114, no. 1, pp. 1–9, 2000.

[37] T. Y. Chen and C. Y. Tsao, "The interval-valued fuzzy TOPSIS method and experimental analysis," *Fuzzy Sets and Systems*, vol. 159, no. 11, pp. 1410–1428, 2008.

[38] Y. M. Wang and T. M. S. Elhag, "Fuzzy TOPSIS method based on alpha level sets with an application to bridge risk assessment," *Expert Systems with Applications*, vol. 31, no. 2, pp. 309–319, 2006.

[39] D. Yong, "Plant location selection based on fuzzy TOPSIS," *International Journal of Advanced Manufacturing Technology*, vol. 28, no. 7-8, pp. 839–844, 2006.

[40] J.-J. Buckley, "Fuzzy hierarchical analysis," *Fuzzy Sets and Systems*, vol. 17, no. 1, pp. 233–247, 1985.

[41] U. Nirmal, B. F. Yousif, D. Rilling, P. V. Brevern, and N. Jamil, "The potential of using treated betelnut fibres as reinforcement for tribo-bio polymeric composites subjected to dry/wet contact conditions," in *Proceedings of the International Conference on Natural Polymers (ICNP '10)*, Kottayam, Kerala, India, September 2010.

[42] U. Nirmal, N. Singh, J. Hashim, S. T. W. Lau, and N. Jamil, "On the effect of different polymer matrix and fibre treatment on single fibre pullout test using betelnut fibres," *Materials and Design*, vol. 32, no. 5, pp. 2717–2726, 2011.

[43] S. Borgatti, Research Method, BA 762, May 2012.

18

PI Controller Design for Time Delay Systems Using an Extension of the Hermite-Biehler Theorem

Sami Elmadssia, Karim Saadaoui, and Mohamed Benrejeb

Unité de Recherche LARA-Automatique, Ecole Nationale d'Ingénieurs de Tunis, BP 37, le Belvédère, 1002 Tunis, Tunisia

Correspondence should be addressed to Sami Elmadssia; sami_elmadssia@yahoo.fr

Academic Editor: Alan Chan

We consider stabilizing first-order systems with time delay. The set of all stabilizing proportional-integral PI controllers are determined using an extension of the Hermite-Biehler theorem. The time delay is approximated by a second-order Padé approximation. For uncertain plants, with interval type uncertainty, robust stabilizing PI controllers are determined.

1. Introduction

In process control, many systems are represented by first-order plants with time delay. Although several tuning rules are reported in the literature [1, 2], the problem of determining the entire set of stabilizing controllers of a given order, being PI or PID, for such systems is recently addressed in [3, 4]. In [5], the Hermite-Biehler theorem is used to determine analytically the set of stabilizing gains k_p, k_i, and k_d of a PID controller by replacing the time delay by a first-order Padé approximation. In fact, extensions of the Hermite-Biehler theorem were effectively used to determine the set of all stabilizing controllers of a given order and a given structure for systems without delay, see [1, 6, 7].

In this paper, we use an extension of the Hermite-Biehler theorem to determine the set of all stabilizing PI controllers for a first-order system with time delay, where the time delay is replaced by a second-order Padé approximation. We show that for a fixed value of the proportional gain k_p, the set of stabilizing k_i gains is a single interval. This conclusion still holds for higher-order Padé approximations. Next, we consider uncertain second-order systems with time delay and robust stabilizing PI controllers are determined.

The paper is organized as follows. In Section 2 we present some preliminary results which can be used to determine stabilizing proportional gains for systems without delay. These results are used in Section 3 to determine stabilizing PI controllers for first-order systems with time delay. In Section 4 robust stabilizing PI controllers are determined for

uncertain systems. Illustrative examples are given in Section 5. Finally Section 6 contains some concluding remarks.

2. Proportional Controllers

In this section, an algorithm that determines the set of all proportional controllers [6] is reviewed. Let us first fix the notation used in this paper. Let **R** denote the set of real numbers and **C** denote the set of complex numbers and let $\mathbf{C}_-$, $\mathbf{C}_0$, $\mathbf{C}_+$ denote the points in the open left-half, $j\omega$-axis, and the open right-half of the complex plane, respectively. Given a set of polynomials $\psi_1, \ldots, \psi_l \in \mathbf{R}[s]$ not all zero and $l > 1$, their *greatest common divisor* is unique and it is denoted by $\gcd\{\psi_1, \ldots, \psi_l\}$. If $\gcd\{\psi_1, \ldots, \psi_l\} = 1$, then we say $(\psi_1, \ldots, \psi_l)$ is *coprime*. The derivative of ψ is denoted by ψ'. The set $\mathcal{H}$ of Hurwitz stable polynomials are

$$\mathcal{H} = \{\psi(s) \in \mathbf{R}[s] : \psi(s) = 0 \implies s \in \mathbf{C}_-\}. \quad (1)$$

The *signature $\sigma(\psi)$ of a polynomial* $\psi \in \mathbf{R}[s]$ is the difference between the number of its $\mathbf{C}_-$ roots and $\mathbf{C}_+$ roots. Given $\psi \in \mathbf{R}[s]$, *the even-odd components* (a, b) *of* $\psi(s)$ are the unique polynomials $a, b \in \mathbf{R}[u]$ such that

$$\psi(s) = a\left(s^2\right) + sb\left(s^2\right). \quad (2)$$

It is possible to state a necessary and sufficient condition for the Hurwitz stability of $\psi(s)$ in terms of its even-odd components (a, b). Stability is characterized by the interlacing property of the real, negative, and distinct roots of the even and odd parts. This result is known as the Hermite-Biehler theorem. Below is a generalization of the Hermite-Biehler theorem applicable to not necessarily Hurwitz stable polynomials. Let us define the *signum function* $\mathcal{S} : \mathbf{R} \to \{-1, 0, 1\}$ by

$$\mathcal{S}r = \begin{cases} -1, & \text{if } r < 0, \\ 0, & \text{if } r = 0, \\ 1, & \text{if } r > 0. \end{cases} \tag{3}$$

Lemma 1 (see [8]). *Let a nonzero polynomial $\psi \in \mathbf{R}[s]$ have the even-odd components (a, b). Suppose $b \not\equiv 0$ and (a, b) is coprime. Then, $\sigma(\psi) = r$ if and only if at the real negative roots of odd multiplicities $v_1 > v_2 > \cdots > v_l$ of b the following holds:*

$$r = \begin{cases} \mathcal{S}b(0)\big[\mathcal{S}a(0) - 2\mathcal{S}a(v_1) + 2\mathcal{S}a(v_2) \\ \qquad + \cdots + (-1)^l 2\mathcal{S}a(v_l)\big] \deg \psi \text{ odd}, \\ \mathcal{S}b(0)\big[\mathcal{S}a(0) - 2\mathcal{S}a(v_1) + 2\mathcal{S}a(v_2) \\ \qquad + \cdots + (-1)^{l+1}\mathcal{S}a(-\infty)\big] \deg \psi \text{ even}. \end{cases} \tag{4}$$

The following result determines the number of real negative roots of a real polynomial.

Lemma 2 (see [6]). *A nonzero polynomial $\psi \in \mathbf{R}[s]$, such that $\psi(0) \neq 0$, has r real negative roots without counting the multiplicities if and only if the signature of the polynomial $\psi(s^2) + s\psi'(s^2)$ is $2r$. All roots of ψ are real, negative, and distinct if and only if $\psi(s^2) + s\psi'(s^2) \in \mathcal{H}$.*

We now describe a slight extension of the constant stabilizing gain algorithm of [8]. Given a plant

$$g(s) = \frac{p(s)}{q(s)}, \tag{5}$$

where $p, q \in \mathbf{R}[s]$ are coprime with $m = \deg p$ less than or equal to $n = \deg q$, the set

$$A_r(p, q) := \{k \in \mathbf{R} : \sigma[\phi(s, k)] = \sigma[q(s) + kp(s)] = r\} \tag{6}$$

is the set of all real k such that $\phi(s, k)$ has signature equal to r.

Let (h, g) and (f, e) be the even-odd components of $q(s)$ and $p(s)$, respectively, so that

$$q(s) = h(s^2) + sg(s^2), \tag{7}$$
$$p(s) = f(s^2) + se(s^2).$$

Let (H, G) be the even-odd components of $q(s)p(-s)$. Also let $F(s^2) := p(s)p(-s)$. By a simple computation, it follows that (we replace s^2 by u):

$$H(u) = h(u)f(u) - ug(u)e(u),$$
$$G(u) = g(u)f(u) - h(u)e(u), \tag{8}$$
$$F(u) = f^2(u) - ue^2(u).$$

With this setting, we have

$$[q(s) + kp(s)]p(-s) = \big[H(s^2) + kF(s^2)\big] + sG(s^2). \tag{9}$$

If $G \not\equiv 0$ and if they exist, let the *real negative roots with odd multiplicities* of $G(u)$ be $\{v_1, \ldots, v_l\}$ with the ordering $v_1 > v_2 > \cdots > v_l$, with $v_0 := 0$ and $v_{l+1} := -\infty$ for notational convenience.

The following algorithm determines whether $A_r(p, q)$ is empty or not and outputs its elements when it is not empty [6].

Algorithm 3. (1) Consider all the sequences of signums

$$\mathcal{I} = \begin{cases} \{i_0, i_1, \ldots, i_l\}, & \text{for odd } r - m, \\ \{i_0, i_1, \ldots, i_{l+1}\}, & \text{for even } r - m, \end{cases} \tag{10}$$

where $i_j \in \{-1, 1\}$ for $j = 0, 1, \ldots, l + 1$.

(2) Choose all the sequences that satisfy

$$r - \sigma(p) = \begin{cases} i_0 - 2i_1 + 2i_2 - 2i_3 + \cdots + 2(-1)^l i_l \\ \qquad \text{for odd } r - m, \\ i_0 - 2i_1 + 2i_2 - 2i_3 + \cdots + (-1)^{l+1}i_{l+1} \\ \qquad \text{for even } r - m. \end{cases} \tag{11}$$

(3) For each sequence of signums $\mathcal{I} = \{i_j\}$ that satisfy step (2), let

$$\alpha_{\max} = \max\left\{-\frac{H}{F}(v_j)\right\}$$

$\forall v_j$ for which $F(v_j) \neq 0$ and $i_j \mathcal{S}F(v_j) = 1$,

$$\alpha_{\min} = \min\left\{-\frac{H}{F}(v_j)\right\} \tag{12}$$

$\forall v_j$ for which $F(v_j) \neq 0$ and $i_j \mathcal{S}F(v_j) = -1$.

The set $A_r(p, q)$ is nonempty if and only if for at least one signum sequence $\mathcal{I}$ satisfying step (2), $\alpha_{\max} < \alpha_{\min}$ holds.

(4) $A_r(p, q)$ is equal to the union of intervals $(\alpha_{\max}, \alpha_{\min})$ for each sequence of signums $\mathcal{I}$ that satisfy step (3).

The algorithm above is easily specialized to determine all stabilizing proportional controllers $c(s) = k$ for the plant $g(s)$. This is achieved by replacing r in step (3) of the algorithm by n, the degree of $\phi(s, k)$.

Remark 4. By step (2) of Algorithm 3, a necessary condition for the existence of a $k \in A_r(p, q)$ is that the odd part of

$$[q(s) + kp(s)]p(-s) \tag{13}$$

has at least $\bar{r} = \max\{0, \lfloor(|r - \sigma(p)| - 1)/2\rfloor\}$ real negative roots with odd multiplicities. When solving a constant stabilization problem, this lower bound is $\bar{r} = \max\{0, \lfloor(n - \sigma(p) - 1)/2\rfloor\}$.

3. Stabilizing with PI Controllers

In this section, we consider PI controllers

$$c(s) = \frac{k_p s + k_i}{s}, \tag{14}$$

applied to a plant transfer function

$$g(s) = \frac{1}{\tau s + a} e^{-\theta s}, \tag{15}$$

where $\tau > 0$ is the time constant, a is a constant whose sign and value determines the open-loop stability and steady-state gain, respectively, and θ represents the time delay. Our aim is to find all values of (k_p, k_i) such that the closed-loop system is stable. We replace the time delay by a second-order Padé approximation

$$e^{-\theta s} \approx \frac{\theta^2 s^2 - 6\theta s + 12}{\theta^2 s^2 + 6\theta s + 12}. \tag{16}$$

In what follows, we show how to find stabilizing values of (k_p, k_i). The closed-loop characteristic polynomial is given by

$$\phi(s, k_p, k_i) = q(s) + (k_p s + k_i) p(s), \tag{17}$$

where

$$q(s) = s(\tau s + a)(\theta^2 s^2 + 6\theta s + 12),$$
$$p(s) = \theta^2 s^2 - 6\theta s + 12. \tag{18}$$

Multiplying $\phi(s, k_p, k_i)$ by $p(-s)$, we obtain

$$\begin{aligned} \psi(s, k_p, k_i) &= \phi(s, k_p, k_i) p(-s) \\ &= H(s^2) + k_i F(s^2) \\ &\quad + s\left[G(s^2) + k_p F(s^2)\right]. \end{aligned} \tag{19}$$

Note here that $p(-s) \in \mathcal{H}$, therefore the odd part $G(u) + k_p F(u)$ of $\psi(s)$ must have all its roots real and negative. The idea behind this method which determines the set of stabilizing parameters (k_p, k_i) is to divide the problem into two subproblems: first we find all values of k_p for which the polynomial $G(u) + k_p F(u)$ has all its roots real and negative. Next, by sweeping over all values of k_p found in the first step and using Algorithm 3, we determine stabilizing values of k_i. In the first step, we construct a new polynomial

$$\psi_1(s, k_p) = \left[G(s^2) + sG'(s^2)\right] + k_p\left[F(s^2) + sF'(s^2)\right], \tag{20}$$

using Lemma 2, finding values of k_p such that $G(u) + k_p F(u)$ has all its roots real and negative is equivalent to stabilizing the new constructed polynomial $\psi_1(s, k_p)$. Hence our method consists of applying Algorithm 3 to two specially constructed plants:

$$\begin{aligned} g_1(s) &= \frac{p_1(s)}{q_1(s)} \\ &= \frac{F(s^2) + sF'(s^2)}{G(s^2) + sG'(s^2)}, \\ g_2(s) &= \frac{p_2(s)}{q_2(s)} \\ &= \frac{p(s)}{q(s) + k_p s p(s)}. \end{aligned} \tag{21}$$

Remark 5. For a fixed value of k_p, the set of stabilizing values of k_i, if they exist, are a single interval. In fact, using Padé approximation $p(-s)$ is always Hurwitz stable. By the interlacing property of the roots of the even and odd parts of a Hurwitz stable polynomial, it follows that only one sequence of signums satisfies step (2) in Algorithm 3 and therefore there is only one interval as a solution. This conclusion still holds if we use higher-order Padé approximations.

4. Robust Stabilizing PI Controllers

We now consider a second-order uncertain plant with an interval type uncertainty for the coefficients

$$g(s) = \frac{[a_2^-, a_2^+] s^2 + [a_1^-, a_1^+] s + [a_0^-, a_0^+]}{[b_2^-, b_2^+] s^2 + [b_1^-, b_1^+] s + [b_0^-, b_0^+]} e^{-\theta s}. \tag{22}$$

Our aim is to find robust stabilizing PI controllers. Replacing the time delay by a second-order Padé approximation, we get the following rational transfer function

$$\begin{aligned} g(s) &= \frac{p(s)}{q(s)} \\ &= \left([c_4^-, c_4^+] s^4 + [c_3^-, c_3^+] s^3 + [c_2^-, c_2^+] s^2 \right. \\ &\quad \left. + [c_1^-, c_1^+] s + [c_0^-, c_0^+]\right) \\ &\quad \times \left([d_4^-, d_4^+] s^4 + [d_3^-, d_3^+] s^3 + [d_2^-, d_2^+] s^2 \right. \\ &\quad \left. + [d_1^-, d_1^+] s + [d_0^-, d_0^+]\right)^{-1}. \end{aligned} \tag{23}$$

Let $p_i(s)$ and $q_j(s)$, $i, j = 1, 2, 3, 4$ be the four Kharitonov polynomials corresponding to $p(s)$ and $q(s)$, respectively. Let $p_i^\lambda(s)$, $i = 1, 2, 3, 4$ be the four Kharitonov segments of $p(s)$, that is,

$$\begin{aligned} p_1^\lambda(s) &= (1 - \lambda) p_1(s) + \lambda p_2(s), \\ p_2^\lambda(s) &= (1 - \lambda) p_1(s) + \lambda p_3(s), \\ p_3^\lambda(s) &= (1 - \lambda) p_2(s) + \lambda p_4(s), \\ p_4^\lambda(s) &= (1 - \lambda) p_3(s) + \lambda p_4(s), \end{aligned} \tag{24}$$

where $\lambda \in [0, 1]$. The four Kharitonov segments $q_j^\lambda(s)$, $j = 1, 2, 3, 4$ of $q(s)$ can be defined similarly. Let $g_{\mathrm{seg}}(s)$ denote the family of 32 segment plants

$$g_{\mathrm{seg}}(s) = \left\{ g_{ij}(s, \lambda) \mid g_{ij}(s, \lambda) = \frac{p_i^\lambda(s)}{q_j(s)} \text{ or} \right.$$

$$g_{ij}(s, \lambda) = \frac{p_i(s)}{q_j^\lambda(s)}, \ i, j = 1, 2, 3, 4, \qquad (25)$$

$$\left. \lambda \in [0, 1] \right\}.$$

It is well known [9] that the family $g(s)$ is stabilized by a particular controller, if and only if the 32 segment plants g_{seg} are stabilized by the same controller. Let $\tilde{g}_{\mathrm{seg}}(s)$ denote the family of 16 segment plants

$$\tilde{g}_{\mathrm{seg}}(s) = \left\{ g_{ij}(s, \lambda) \mid g_{ij}(s, \lambda) = \frac{p_i^\lambda(s)}{q^j(s)}, \right.$$

$$\left. i, j = 1, 2, 3, 4, \lambda \in [0, 1] \right\}. \qquad (26)$$

It is shown in [10] that "the entire family $g(s)$ is stabilized by a particular PID controller, if and only if each segment plant $g_{ij}(s) \in \tilde{g}_{\mathrm{seg}}(s)$ is stabilized by that same PID controller." In reaching this result the structure of the PID controller was used to reduce the 32 segment plants to only 16. Since we are working with PI controllers, the numerator and denominator of the controller are convex directions [9]. Therefore, stabilizing the interval plant $g(s)$ by a PI controller is equivalent to stabilizing 16 vertex plants, namely,

$$g_v(s) = \left\{ g_{ij}(s) \mid g_{ij}(s) = \frac{p_i(s)}{q_j(s)}, \ i, j = 1, 2, 3, 4 \right\}. \qquad (27)$$

The stabilizing controller, if any, can be determined by first calculating k_p which is the intersection of k_p's found for the 16 plants mentioned above. We can then sweep over the values of k_p and apply Algorithm 3 to the 16 vertex plants to find k_i.

5. Application to the PT-326 Thermal Process

An application example for the theoretical approaches presented in the precedent sections can be given by the temperature control of an air stream heater (process trainer PT-326), Figure 1. This type of process is found in many industrial systems such as furnaces, air conditioning, and so forth. The PT-326 Thermal Process Control models the industrial situation commonly found in such equipments as air conditioning plants where temperature control is achieved through a combination of more than one means. The process contained within the PT-326 comprises an air duct through which air may be circulated using an electrically driven variable speed fan. An electrically heated process block is

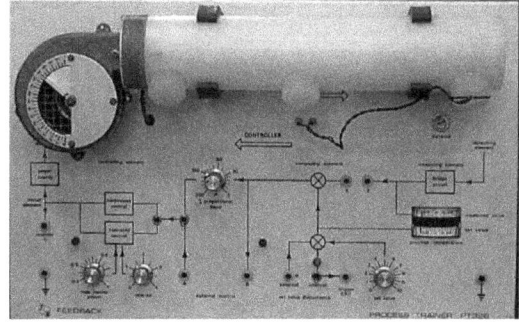

FIGURE 1: Front panel of the PT-326 apparatus.

mounted in the air flow path such that it attains thermal equilibrium by balancing the heat gained through the energy supplied to it via the heater coil and the heat lost through convection and conduction. Temperature control is achieved either by

(1) varying the heat energy input to the system by regulating electrical power to the heater coil or,

(2) varying the heat transfer rate by regulating the air flow rate either by

   (i) controlling the speed of the circulating fan or,

   (ii) restricting the actual flow channel itself using a controlled Vane mounted in the flow path.

Two platinum resistance thermometers (T_1 and T_2) monitor the actual temperature of the block, being in the direct thermal contact with the block and being mounted on an insulation spacer to introduce thermal inertia and additional time constant effects into the control loop. Figure 1 below shows the front panel of the apparatus.

The physical principle which governs the behavior of the thermal process in the PT-326 apparatus is the balance of heat energy. The rate at which heat accumulates in a fixed volume V enclosing the heater is

$$q_a = q + q_i - q_o - q_t, \qquad (28)$$

where q is the rate at which heat is supplied by the heater, q_i is the rate at which heat is carried into the volume V by the coming air, q_o is the rate at which heat is carried out of the volume V by the outgoing air, and q_t is the heat lost from the volume V to the surroundings by radiation and conduction. Figure 2 below depicts the volume V. The behavior of the PT-326 thermal process is governed by the balance of heat energy. When the air temperature inside the tube is supposed to be uniform, a linear delay system model can be obtained. Thus, the transfer function between the heater input voltage and the sensor output voltage can be obtained as $V_o(s)/V_i(s) = ke^{-\theta s}/(\tau s + 1)$ as shown in Figure 3.

A typical control objective in thermal systems is to maintain the temperature of some component at a user specified value called the set point. Figure 4 below depicts a closed-loop system designed to maintain the output temperature of the PT-326 apparatus at a desired set point. For the experiment,

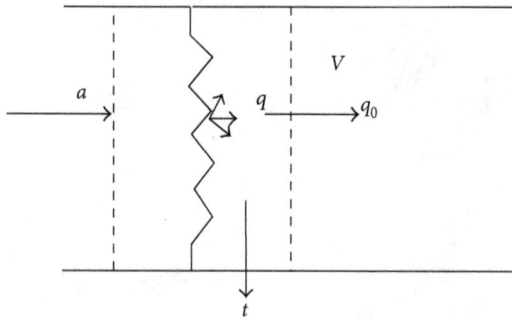

FIGURE 2: Heat transfer from the volume V.

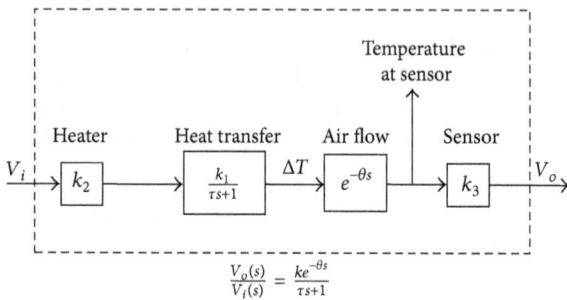

$$\frac{V_o(s)}{V_i(s)} = \frac{ke^{-\theta s}}{\tau s+1}$$

FIGURE 3: Block diagram of the PT-326 process.

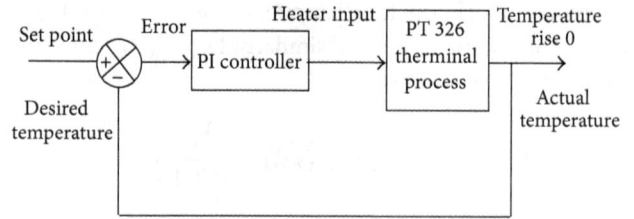

FIGURE 4: Closed-loop control of a thermal system.

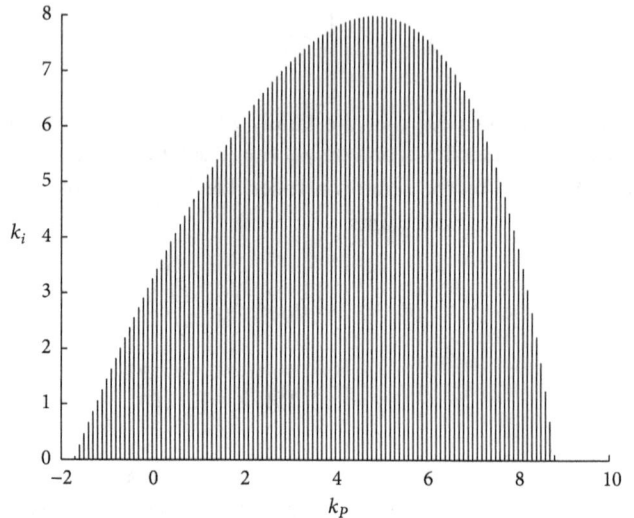

FIGURE 5: The stabilizing set of (k_p, k_i) values.

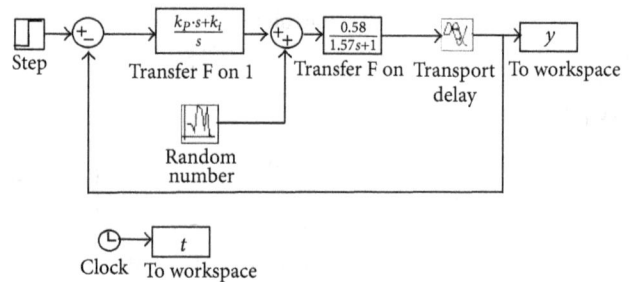

FIGURE 6: Simulink diagram of the system.

the damper position is set to 30, and the temperature sensor is placed in the third position. The transfer function is described by the following expression:

$$g(s) = \frac{V_o(s)}{V_i(s)} = \frac{0.58e^{-0.56s}}{1.57s + 1}. \tag{29}$$

5.1. Regions of All Stabilizing Values of (k_p, k_i). Approximating the time delay by a second-order Padé approximation we get

$$g(s) = \frac{0.58s^2 - 6.214s + 22.19}{1.57s^3 + 17.82s^2 + 70.79s + 38.27}. \tag{30}$$

The closed-loop characteristic polynomial is given by

$$\psi(s, k_p, k_i) = s\left(1.57s^3 + 17.82s^2 + 70.79s + 38.27\right)$$
$$+ \left(k_p s + k_i\right)\left(0.58s^2 - 6.214s + 22.19\right). \tag{31}$$

Using Lemma 2, we solve subproblem 1: find all values of k_p such that $\psi(s, k_p) = 20.0916s^4 + 80.3664s^3 + 857.5115s^2 + 1715s + 849.2113 + kp(0.3364s^4 + 1.3456s^3 - 12.8734s^2 - 25.7468s + 492.3961)$ is Hurwitz stable. Applying Algorithm 3, we get $k_p \in (-1.7247, 9.855)$. The complete solution is given in Figure 5.

5.2. Simulation Results. In this section, we simulated the process PT-326 using MATLAB Simulink, we tested the evolution of output y of the studied system in the presence of a disturbance. For the simulation, we took a step unit. We also took the block (Random Number) as a disturbance where its mean equal to 0 and variance equal to 0.01 and a simple time equal to 0.5. Figure 6 shows the MATLAB Simulink schematic of our application. Figure 7 shows the evolution of the output y as a function of time. Notice that from this figure a choice of couple (k_p, K_i) belonging the domain of stabilizing parameters can stabilize the studied system.

6. Conclusions

In this paper, we determine the set of all stabilizing PI controllers for first-order systems with time delay. By using a second-order Padé approximation for the time delay, we show that the set of stabilizing k_i values is a single interval when k_p is fixed. Robust stabilizing PI controllers

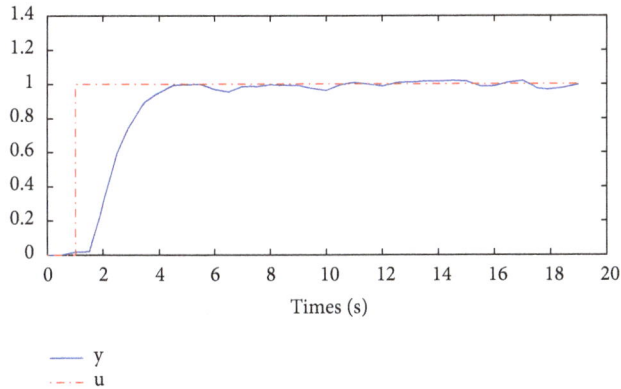

FIGURE 7: Simulation result for $k_p = 1.8$ and $k_i = 1.1$.

are determined for plants with uncertain parameters. Note that higher-order Padé approximations can be used without changing the analysis given in this paper, as the Hermite-Biehler framework is applicable to plants irrelevant of the order.

References

[1] A. Datta, M. T. Ho, and S. P. Bhattacharyya, *Structure and Synthesis of PID Controllers*, Springer, New York, NY, USA, 2000.

[2] J. P. Richard, "Time-delay systems: an overview of some recent advances and open problems," *Automatica*, vol. 39, no. 10, pp. 1667–1694, 2003.

[3] G. J. Silva, A. Datta, and S. P. Bhattacharyya, "PI stabilization of first-order systems with time delay," *Automatica*, vol. 37, no. 12, pp. 2025–2031, 2001.

[4] G. J. Suva, A. Datta, and S. P. Bhattacharyya, "New results on the synthesis of PID controllers," *IEEE Transactions on Automatic Control*, vol. 47, no. 2, pp. 241–252, 2002.

[5] A. Roy and K. Iqbal, "PID controller design for first-order-plus-deadtime model via Hermite-Biehler theorem," in *Proceedings of the American Control Conference*, pp. 5286–5291, Denver, Colo, USA, June 2003.

[6] K. Saadaoui and A. B. Özgüler, "A new method for the computation of all stabilizing controllers of a given order," *International Journal of Control*, vol. 78, no. 1, pp. 14–28, 2005.

[7] K. Saadaoui and A. B. Özgüler, "On the set of all stabilizing first-order controllers," in *Proceedings of the American Control Conference*, Denver, Colo, USA, June 2003.

[8] A. B. Özgüler, A. A. Koçan et al., "An analytic determination of stabilizing feedback gains," Report 321, Institut für Dynamische Systeme Universität Bremen, 1994.

[9] B. R. Barmish, *New Tools for Robustness of Linear Systems*, Macmillan Publishing Company, 1994.

[10] N. Munro and M. T. Söylemez, "Fast calculation of stabilizing PID controllers of uncertain parameter systems," in *Proceedings of ROCOND*, Prague, Czech Republic, June 2000.

Permissions

The contributors of this book come from diverse backgrounds, making this book a truly international effort. This book will bring forth new frontiers with its revolutionizing research information and detailed analysis of the nascent developments around the world.

We would like to thank all the contributing authors for lending their expertise to make the book truly unique. They have played a crucial role in the development of this book. Without their invaluable contributions this book wouldn't have been possible. They have made vital efforts to compile up to date information on the varied aspects of this subject to make this book a valuable addition to the collection of many professionals and students.

This book was conceptualized with the vision of imparting up-to-date information and advanced data in this field. To ensure the same, a matchless editorial board was set up. Every individual on the board went through rigorous rounds of assessment to prove their worth. After which they invested a large part of their time researching and compiling the most relevant data for our readers. Conferences and sessions were held from time to time between the editorial board and the contributing authors to present the data in the most comprehensible form. The editorial team has worked tirelessly to provide valuable and valid information to help people across the globe.

Every chapter published in this book has been scrutinized by our experts. Their significance has been extensively debated. The topics covered herein carry significant findings which will fuel the growth of the discipline. They may even be implemented as practical applications or may be referred to as a beginning point for another development. Chapters in this book were first published by Hindawi Publishing Corporation; hereby published with permission under the Creative Commons Attribution License or equivalent.

The editorial board has been involved in producing this book since its inception. They have spent rigorous hours researching and exploring the diverse topics which have resulted in the successful publishing of this book. They have passed on their knowledge of decades through this book. To expedite this challenging task, the publisher supported the team at every step. A small team of assistant editors was also appointed to further simplify the editing procedure and attain best results for the readers.

Our editorial team has been hand-picked from every corner of the world. Their multi-ethnicity adds dynamic inputs to the discussions which result in innovative outcomes. These outcomes are then further discussed with the researchers and contributors who give their valuable feedback and opinion regarding the same. The feedback is then collaborated with the researches and they are edited in a comprehensive manner to aid the understanding of the subject.

Apart from the editorial board, the designing team has also invested a significant amount of their time in understanding the subject and creating the most relevant covers. They scrutinized every image to scout for the most suitable representation of the subject and create an appropriate cover for the book.

The publishing team has been involved in this book since its early stages. They were actively engaged in every process, be it collecting the data, connecting with the contributors or procuring relevant information. The team has been an ardent support to the editorial, designing and production team. Their endless efforts to recruit the best for this project, has resulted in the accomplishment of this book. They are a veteran in the field of academics and their pool of knowledge is as vast as their experience in printing. Their expertise and guidance has proved useful at every step. Their uncompromising quality standards have made this book an exceptional effort. Their encouragement from time to time has been an inspiration for everyone.

The publisher and the editorial board hope that this book will prove to be a valuable piece of knowledge for researchers, students, practitioners and scholars across the globe.

List of Contributors

Chantal Baril
Department of Industrial Engineering, Université du Québec à Trois-Rivières, 3351 Boulevard des Forges, Trois-Rivières, QC, Canada

Soumaya Yacout and Bernard Clément
Department of Mathematical and Industrial Engineering, École Polytechnique de Montréal, 2900 Boulevard Édouard-Montpetit, Campus de l Université de Montréal, Montréal, QC, Canada

Prasad Karande
Mechanical Engineering Department, Government Polytechnic Mumbai, Maharashtra 400 051, India

Shankar Chakraborty
Department of Production Engineering, Jadavpur University, Kolkata, West Bengal 700 032, India

Mehmet Savsar
Kuwait University, College of Engineering and Petroleum, P.O. Box 5969, 13060 Safat, Kuwait

Zhicong Zhang, Kaishun Hu, Shuai Li, Huiyu Huang and Shaoyong Zhao
Department of Industrial Engineering, School of Mechanical Engineering, Dongguan University of Technology, Songshan Lake District, Dongguan, Guangdong 523808, China

T. J. Roosen and D. J. Pons
Department of Mechanical Engineering, University of Canterbury, Private Bag 4800, Christchurch 8140, New Zealand

Vahid Zharfi and Abolfazl Mirzazadeh
Industrial Engineering Department, Faculty of Engineering, Kharazmi University, Tehran 31979-37551, Iran

Lifang Yang, Tianjiao Zhao and Fanyu Meng
Department of Industrial Design, Harbin Institute of Technology, Harbin 150001, China

Rajarshi Mukherjee, Debkalpa Goswami and Shankar Chakraborty
Department of Production Engineering, Jadavpur University, Kolkata, West Bengal 700 032, India

Kris Lawry
Department of Chemical and Process Engineering, University of Canterbury, Private Bag 4800, Christchurch 8020, New Zealand

Dirk John Pons
Department of Mechanical Engineering, University of Canterbury, Private Bag 4800, Christchurch 8020, New Zealand

Karan Forghani, Abolfazl Mirzazadeh and Mehdi Rafie
Department of Industrial Engineering, University of Kharazmi, Tehran, Iran

Hubert Gattringer, Roland Riepl and Matthias Neubauer
Institute for Robotics, Johannes Kepler University, Altenbergerstraße 69, 4040 Linz, Austria

Biswajit Sarkar
Department of Applied Mathematics with Oceanology and Computer Programming, Vidyasagar University, Midnapore 721-102, India

Shib Sankar Sana
Department of Mathematics, Bhangar Mahavidyalaya, University of Calcutta, Kolkata 743-502, India

Kripasindhu Chaudhuri
Department of Mathematics, Jadavpur University, Kolkata 700-032, India

Mohammad Abdolshah
Department of Engineering Faculty, Islamic Azad University, Semnan Branch, P.O. Box 35136-93688, Semnan, Iran

Mohsen Moradi
Department of Industrial Engineering, Semnan University, Semnan, Iran

Maryam Ghoreishi, Alireza Arshsadi khamseh and Abolfazl Mirzazadeh
Industrial Engineering Department, Kharazmi University, University Square, Dr. Beheshti Street, Karaj 31979-37551, Tehran, Iran

Abolfazl Doostparast Torshizi, Jamshid Parvizian and Farshad Tooyserkani
Department of Industrial Engineering, Isfahan University of Technology, Isfahan 84156, Iran

Yong He and Hongfu Huang
Institute of Systems Engineering, School of Economics and Management, Southeast University, Nanjing 210096, China

Nadia Jamil, Rosli Besar and H. K. Sim
Faculty of Engineering and Technology, Multimedia University, Jalan Ayer Keroh Lama, 75450 Bukit Beruang, Melaka, Malaysia

Sami Elmadssia, Karim Saadaoui and Mohamed Benrejeb
Unité de Recherche LARA-Automatique, Ecole Nationale d Ingénieurs de Tunis, BP 37, le Belvédère, 1002 Tunis, Tunisia